PLANT ECOLOGY

By

T.W. Woodhead

SONALI PUBLICATIONS

NEW DELHI-110002

Edition : 2019

ISBN 81-88836-04-4

Published by

SONALI PUBLICATIONS
4228/1, Ansari Road,
New Delhi - 110002
Phone: 91-11-23266109
Fax: 91-11-23283267
email.: sonalipub@yahoo.co.in

Printed at: Tarun Offset Printers, (Delhi)

Preface

The course of work followed in this book is directed, in the main, to the establishment of the fundamental principles of Plant Physiology. Plant Morphology receives a less extended treatment; but this aspect on the subject is freely introduced in the discussion of Plant Ecology, i.e. the relation of the structure and functions of plants to their habitat. More space has been devoted to Ecology than is usual in an elementary text-book, but the Author believes that this respect of plant life gives to field work a more definite aim, and broadens the outlook of the student by linking up Botany with the study of climate, geology, and topography. Similarly, to avoid the weariness of lessons dealing merely with the comparison of forms, the Author has throughout treated the forms of roots, stems, and leaves in relation to their functions and to the habitat of the plant.

The plants selected for study are common species: nearly all of them can be obtained from the fields, hedgerows, and gardens, and it is expected that specimens will be in the hands of students using the book. The experiments suggested are usually so simple and require such inexpensive apparatus that every pupil in a class ought to be able to do them. Details of structure occasionally require the compound microscope; where this instrument is not available, a general idea can be obtained by the aid of a pocket lens. It is hoped that the photo-micrographs and drawings in the book will clear up any difficulties. They have been made specially for this book, and are designed not as substitutes for actual specimens, but as aids to the practical observation of plants.

Technical words have been introduced when necessary for accurate description, but they have been avoided whenever simpler terms were adequate. The common as well as the systematic names of plants employed in the book are to be found in *The Botanist's Pocket Book*, by W. Hayward (thirteenth edition, revised by G. Claridge Druce. Bell).

Although no particular syllabus has been followed, the subject-matter covers the work necessary for Matriculation, Senior Local Examinations, and the Elementary Teacher's Certificate Examination; with suitable omissions, the book can also be used for Preliminary and Junior Local Examinations and for Scholarship Examinations for entrance into Secondary Schools.

The Author records his obligations to Miss M.M. Brierley for the great care and interest which she has taken in the preparation of the illustrations, and for help in many ways; to Miss H. Rigby for some of the drawings; and to Mr. A.W. Sykes, Mr. W.H. Sikes, Miss H.M. Sikes, and the late Mr. H.G. Brierley, for many of the photographs. To Miss D. Ventham, M.A., Miss E.M. Poulton, M.Sc., and especially to the Rev. T.A. Jefferies, F.L.S., he is indebted for many helpful criticisms.

CONTENTS

For permission to reprint the examination papers I have to acknowledge the courtesy of the Delegates of Oxford Local Examinations, the Syndics of Cambridge Local Examinations, the University of London, the Controller of His Majesty's Stationery Office, the Central Welsh Board, the Universities of Manchester, Liverpool, Leeds, and Sheffield Joint Matriculation Board, the Oxford and Cambridge Schools Examination Board, the College of Preceptors, the Universities of Birmingham, Wales, Durham, Bristol, and the Assistant Commissioners of Intermediate Education, Ireland.

PART I

THE VEGETATIVE ORGANS

CHAPTER I

THE GARDEN STOCK

WE will begin the study of plants by the examination of a familiar example, in order (1) to discover its parts; (2) to compare the parts growing in the air with those in the soil; and (3) to notice how they are related one to the other, and their uses to the plant. It is necessary to describe the structures methodically and in suitable terms, and to illustrate by means of sketches all the features we have observed.

The several forms of **Stock** which are commonly cultivated in our gardens are derived from plants growing wild in the south of England, western Europe, the Mediterranean region, and elsewhere.[1] Any of these will answer our purpose, plants with single flowers being the best. One of these is shown in Fig. 1.

Vegetative organs.—Obtain a plant, examine it carefully, and draw the parts you see. Two regions are at once

[1] The plants from which the chief garden Stocks have been derived by cultivation and selection are known to botanists as *Matthiola incana*, *M. annua*, and *M. sinuata*. Plants which have come from one parent or one kind are said to belong to one *species*. Thus the Queen Stock belongs to the species *incana*; the Ten-weeks' Stock belongs to the species *annua*. Both these, however, agree in many important details of structure, and are therefore grouped together under one *genus*, *Matthiola*. Each plant-name thus consists of two words, a substantive or generic name, and an adjective or specific name. The generic name, *Matthiola*, was given to the Stock by Robert Brown, in honour of the Italian botanist, Mattioli (1501–72).

recognized: (A) a part which grows downwards into the soil, and (B) a part which grows upwards into the air and is green. The former we may call the **descending axis** or **root**. This consists of a main tapering root called the tap-root (often distorted in cultivated plants) which gives off numerous branches growing obliquely downwards. These in turn produce branches which are white, and on the younger parts minute root-hairs are developed. The root-hairs are almost too small to be seen with the naked eye, and are generally broken off in digging up the plant. At the end of each root-branch is the growing-point, covered by a protective root-cap. The older root-branches are brown, being covered with a layer of cork, and are unable to absorb water.

The part of the plant above ground is the **ascending axis**, and consists of organs of two kinds: (1) a **stem** which is erect, cylindrical, strong, and somewhat woody below; more tender, slightly ridged, and green above, with a grey covering of branched hairs. A tender green stem is said to be *herbaceous*. From this arise (2) thin, flat **leaves** (Fig. 1, *f*). In the angle or axil between the leaf and the stem, buds may be found which develop into leafy shoots or **axillary branches** (*b*). At the end of the stem and of each branch is the growing-point, protected, not as in the root by a cap, but by the overlapping young leaves of the bud. In older plants the leaves have fallen from the lower part of the stem, leaving scars (*s*) on the surface. The base or region of attachment of a leaf is somewhat enlarged, and this passes imperceptibly into the blade without a definite leaf-stalk or **petiole**. The shape of the blade is oblong-lanceolate with an even or entire margin; and the apex is bluntly pointed or sometimes rounded. Like the stem, the leaf is covered with branched hairs. Running through the leaf from base to apex is the **midrib**, which gives off a branching network of

FIG. 1. THE STOCK PLANT.—A, the root system; B, the shoot system; *b*, axillary branch; *f*, foliage-leaf; *fl*, flower; *fr*, young fruit; *g*, ground level; *p*, pedicel; *s*, leaf-scar.

veins; but the leaf is so fleshy that the veins are not easily seen. Notice the arrangement of the leaves on the stem and the way in which they are related one to another.

Perform the following experiment: Tie one end of a piece of thread round the base of a leaf; then wind the thread round the stem from right to left in such a way that it touches the base of each leaf in turn as it ascends. Eventually you will meet with a leaf standing vertically above the one with which you began. Count the leaves passed by the thread, omitting the first, and determine the number of times the thread has passed round the stem.

FIG. 2. DIAGRAM OF LEAF ARRANGEMENT.

Commonly you will find that the spiral goes twice round the stem and touches five leaves; thus we see that the leaves are arranged spirally on the stem, and each is separated from the one above or below it by two-fifths of the circumference (Fig. 2). Exceptions to this arrangement are not uncommon in the Stock. The same test might be applied to other plants, e.g.: Groundsel, Oak, Deadnettle, Elder; and Hazel. The arrangement of leaves on a stem is called **phyllotaxy** (Gr. *phyllon* = leaf, *tasso* = arrange), and is usually such as to place the leaves in a favourable position with regard to sunlight.

The three structures—root, stem, and leaf—are concerned with the growth of the plant, and are hence known as **vegetative organs**.

Reproductive organs.—Eventually other organs, viz. **flowers** (Fig. 1, *fl*), appear on the upper part of the plant, and these are concerned with reproduction.

They are produced in considerable numbers both on the main stem and on the axillary branches. These flowering shoots form the **inflorescence**. Notice that in either case the flowers at the bottom are the oldest, those next above them are younger and so on until at the top the youngest flowers are still in bud. Each flower arises independently of a leaf and is attached to the stem by a short stalk or **pedicel** (Fig. 1, *p*). Such an inflorescence is termed a **raceme**.

Examine the parts of the flower, commencing at the outside. Remove the parts one by one and lay them out

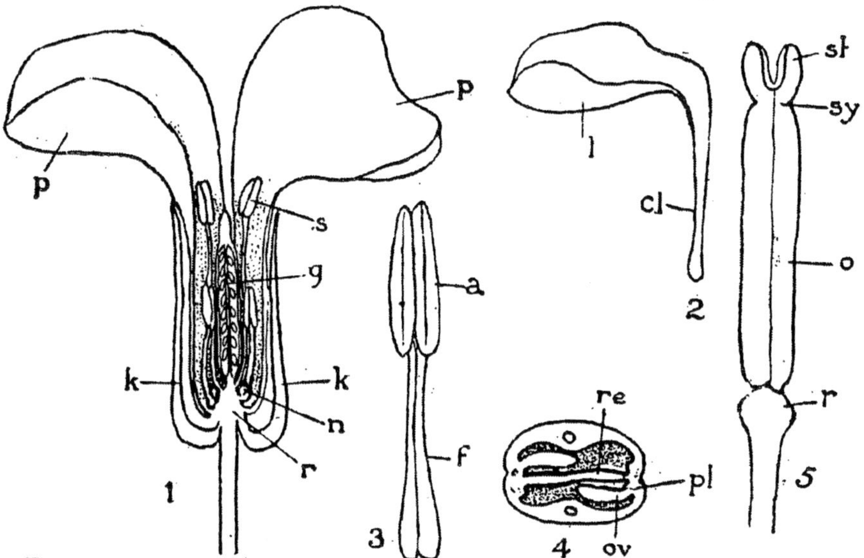

FIG. 3. DISSECTION OF A STOCK FLOWER.—1, Vertical section of flower; 2, petal removed; 3, stamen removed; 4, transverse section of ovary; 5, pistil; *a*, anther; *cl*, claw; *f*, filament; *g*, pistil; *k*, calyx; *l*, limb; *n*, nectary; *o*, ovary; *ov*, ovule; *p*, petal; *pl*, placenta; *r*, receptacle; *re*, replum; *s*, stamen; *st*, stigma; *sy*, style.

before you so as to show their interrelations. The portion of the stalk on which these parts are borne is called the **receptacle** (Fig. 3, *r*). On the outside are four erect distinct green leaves; these are called **sepals**, and together form the **calyx** (*k*). They serve to protect the inner parts when the flower is in bud. When the sepals of a calyx are distinct or free one from another, the calyx is said to be **polysepalous** (Gr. *polys* = many). Note that two of the sepals are bulged or saccate at the base. Now press the flower backwards against the stem and determine whether these two sepals stand right and left, i. e. are lateral with reference to the stem; or are anterior and posterior. These sepals, though apparently lower, are fixed a little higher on the receptacle than the other two. Now notice the next four inner leaves, the **petals**. These, like the sepals, are free from one another and form the **corolla**, which is therefore **polypetalous**. The petals alternate with the sepals and are said to be placed diagonally in the flower. Each petal (Fig. 3, 2) consists of a long stalk, the **claw** (*cl*), which reaches to the top of the narrow tube formed by the sepals, then spreads out at right angles as a broad, white or highly coloured, thin blade, known as the **limb** (*l*). Further inwards, and higher on the receptacle, are the **stamens** (*s*), six stalked bodies together forming the **androecium** (Gr. *aner*, *andros* = man, *oikos* = house). Each stamen consists of a stalk or **filament** (*f*) bearing a two-lobed yellowish body termed the **anther** (*a*), and each lobe contains two parallel chambers called **pollen-sacs** filled with minute bodies known as **pollen-grains**. The stamens are not all alike: two are short and lateral in position, fixed a little lower on the receptacle than the remaining four longer ones, which are in two pairs, anterior and posterior. Examine the base of each short stamen and you will find on the inner side two swellings, known as **nectaries** (*n*), which secrete honey. The bulged sepals provide accommodation for these. When sepals, petals, or stamens are free from and arise below the pistil they are said to be **hypogynous** (Gr. *hypo* = under, *gyné* = female).

When the stamens are removed we find in the centre of the flower an elongated green body, the **pistil** or gynoecium (*g*). All the other parts—sepals, petals, and stamens—stand below, i. e. are inferior to this. In other words, the pistil stands highest on the receptacle and is therefore superior to the other parts of the flower. If the pistil be cut across (Fig. 3, 4) and examined by the aid of a pocket lens, it will be seen to consist of two chambers, each containing two rows of minute bodies called **ovules** (*ov*). The structure from which the ovules arise is called the **placenta** (*pl*); and the ovule-bearing part of the pistil is known as the **ovary** (5 *o*). On the top of the ovary is a very short neck—the **style** (*sy*), terminating in a two-lobed structure—the **stigma** (*st*). The pistil may thus be seen to consist of two bodies fused together; these are called **carpels.** When two or more carpels are united, the pistil is said tc be **syncarpous** (Gr. *syn* = together, *karpos* = fruit). These organs are called floral leaves; the sepals and petals are leaf-like, but the stamens and carpels bear little resemblance to leaves. Fig. 4 is a plan showing the relative positions of the parts. Such a plan is known as a 'floral diagram'.

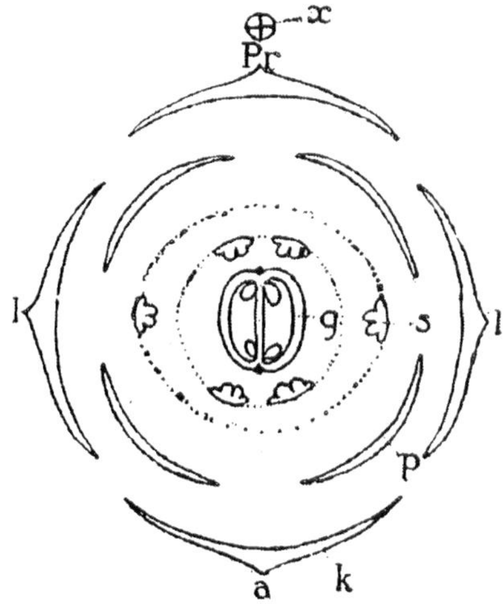

FIG. 4. FLORAL DIAGRAM.—*a*, anterior; *g*, pistil; *k*, sepal; *l*, lateral; *p*, petal; *Pr*, posterior; *s*, stamen; *x*, axis.

In the lower, older part of the inflorescence it will be noticed that all the parts of the flower have fallen off, with the exception of the pistil, and that this has grown enormously to form a long narrow **fruit** (Fig. 1, *fr*). Select a mature fruit and dissect it (Fig. 5). Remove the two side lobes, which separate easily from the base upwards. It will then be found that a frame is left, called the **replum** (*re*), with a thin membrane stretching across it. Attached to the frame by slender stalks are the flattened **seeds** (*sd*), each surrounded by a thin wing. The wall of the ovary, which forms the coat of the fruit, is called the **pericarp** (Gr. *peri* = around).

The flowers of Stocks are often sweetly scented, especially at night. This attracts night-flying moths, which visit the flowers and search for the pollen and honey. In the process they become dusted with pollen, and, carrying it to other Stock flowers, may deposit grains on the stigma and thus

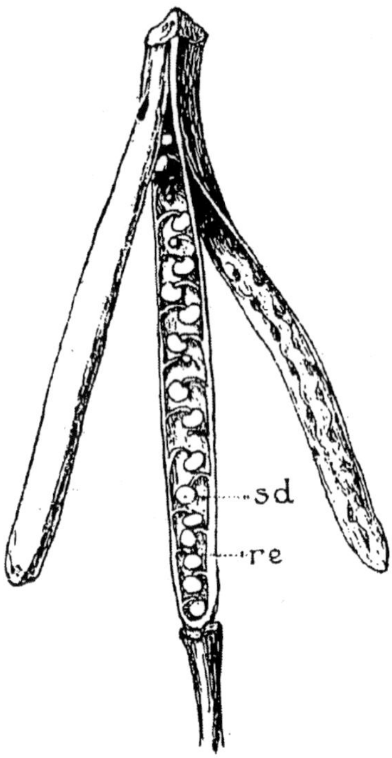

FIG. 5. DEHISCENT FRUIT.
re, replum ; *sd*, seed.

secure fertilization of the ovules and the formation of seeds, from which a new generation of plants arises.

From our study of the Stock we learn that the organs of a plant are of two distinct kinds, (1) vegetative organs—roots, stems, and leaves—which are concerned with obtaining food and building up the main body of the plant ; and (2) reproductive organs—the flowers—whose function is to produce seed, from which arises a new generation of plants.

CHAPTER II

STRUCTURE AND GERMINATION OF SEEDS

(*a*) Dicotyledons

OUR analysis of the growing Stock showed that the plant consisted of a number of organs. We shall now show that these organs come into existence successively, and how

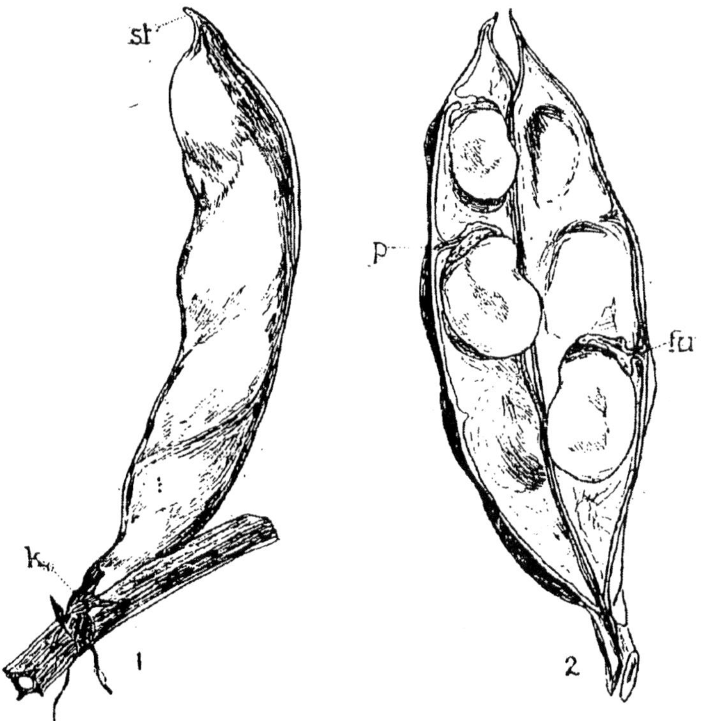

FIG. 6. THE BEAN POD.—1, side view; 2, pod opened along the inner or ventral suture; *fu*, funicle; *k*, calyx; *p*, placenta; *st*, stigma.

they perform their twofold function, (1) in helping to sustain the life of the plant, and (2) in providing materials for the growth of succeeding organs. The requisite data are easily derived from observation and experiment.

Pod and seeds of the Bean.—The **Broad Bean** provides

us with excellent material, and we will begin with the Bean-pod (Fig. 6, 1). This is a fruit derived from the pistil of the Bean flower, but, unlike the Stock fruit, it consists of only one carpel. The parts of the pistil and some remnants of the flower may be found. At the base is the calyx (*k*), and often parts of the stamens are to be seen. The ovary has

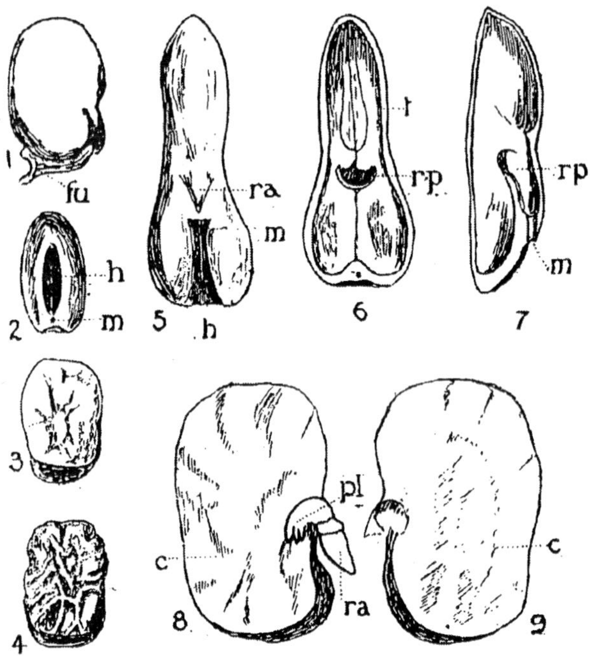

FIG. 7. DISSECTION OF THE BEAN SEED.—1, seed with funicle attached; 2, end view of seed; 3, dry seed; 4, partly soaked seed; 5, concave edge of seed; 6, testa showing entrance to radicle pocket; 7, radicle pocket in side view; 8 and 9, cotyledons separated; *c*, cotyledon; *fu*, funicle; *h*, hilum; *m*, micropyle; *pl*, plumule; *ra*, radicle; *r.p*, radicle pocket; *t*, testa.

enlarged greatly, and now contains the seeds, or beans as we call them, while at the tip are the remains of the style and stigma (*st*). If we cut the pod along its upper edge and open it (2), we find it is to this edge that the seeds are attached. The seed-stalk or **funicle** (*fu*) is curious; it grows from the edge or **placenta** (*p*) of the carpel, and it enlarges into a much-thickened body clasping the seed.

Remove a seed (Fig. 7, 1). Notice that it is flattened and oval; one edge is convex and the other slightly concave; it is covered on the outside by a thick, tough, light-brown skin—the **testa** (6 *t*); and when the seed-stalk or funicle (*fu*) is removed, a scar or **hilum** (2 *h*) is left, indicating the point of attachment. Compare this seed with a dry one (3) as supplied by the seedsman, and note the darker wrinkled skin of the latter and the prominent dark-brown scar. Evidently such a seed has lost much water. If one of these seeds is soaked in water, a marked change occurs. In six or seven hours the skin becomes more wrinkled (4), then the seed swells so much that the skin is tightly stretched, and if it be squeezed laterally, a little drop of water will be seen to ooze out from a small hole –the **micropyle** (Gr. *mikros* = small, *pyle* = gate)—at one end of the scar (2 and 5 *m*). Wipe off the water and repeat the experiment. As we shall see later, this hole represents the micropyle of the ovule, through which the pollen-tube entered when fertilization took place.

Remove the skin (6 *t*) from the seed. Note its thickness; examine the inner surface of the coat covering the concave edge; find the little pocket and determine its use (6 and 7 *r.p*). The structure enclosed by the skin consists of two large fleshy lobes—the seed-leaves or **cotyledons** —(8 and 9 *c*), and between these is a bluntly-pointed structure—the **radicle** (8 *ra*), the tip being directed towards the micropyle. The pocket in which it rested is called the **radicle pocket.**

Separate the two cotyledons and look for the young, curved shoot, bearing tiny leaves at its tip. This is the **plumule** (8 *pl*). Between the plumule and the radicle thick stalks are given off to the cotyledons. These structures—the two cotyledons, radicle, and plumule—form a young, dormant plantlet called the **embryo.**

Compare the pod and seeds of the Garden Pea or the

Sweet-Pea with those of the Bean. What are the most important points of agreement or difference?

Food stored in the cotyledons.—Seeds like the Garden Pea and the Bean are common articles of food. In what does the nutriment consist, and where is it contained? An instructive, though only partial, answer is easy to find.

Place a little powdered laundry starch in a test tube, add water, and boil for a few minutes. Allow it to cool, add a drop of iodine solution,[1] and note the dark violet

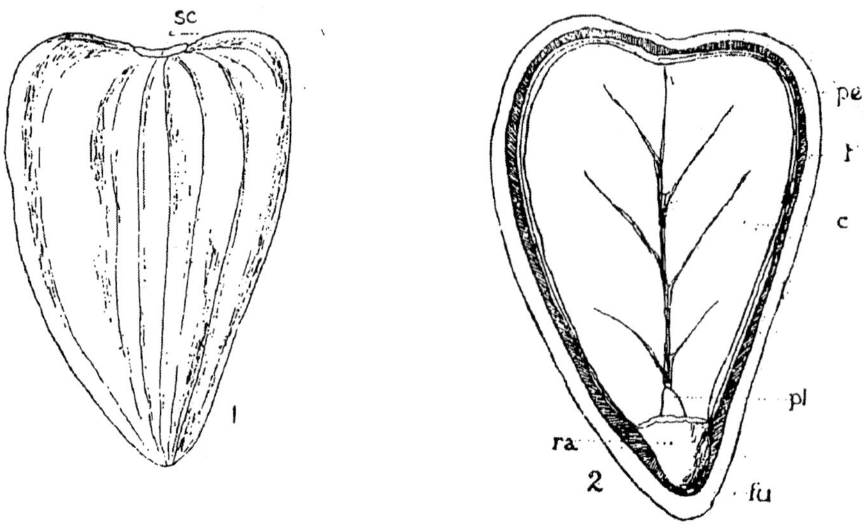

FIG. 8. FRUIT OF SUNFLOWER.—1, side view; 2, fruit opened and one cotyledon removed; *c*, cotyledon; *fu*, funicle; *pe*, pericarp; *pl*, plumule; *ra*, radicle; *sc*, scar; *t*, testa.

colour produced. Repeat the experiment with flour, and a slice of potato; the same violet coloration is seen. This coloration indicates the presence of starch; in other words, the addition of a solution of iodine to a starch-containing substance produces a violet coloration. Now take a cotyledon of the Bean and place a drop of iodine solution on its uninjured surface. Scratch the surface of another cotyledon, add iodine solution to the scratched

[1] A solution of iodine and potassium iodide in water.

portion, and compare the two. Repeat this experiment with the Pea. We thus see that the cotyledons of the Bean and Pea contain much starch, and it is chiefly this which gives them their value as food.

The seeds of the Bean and Pea agree closely in their general structure, e.g. they consist merely of a skin and an embryo. Many seeds, however, are more complex, while some so-called 'seeds' are really fruits, e.g. those of the **Sunflower** (Fig. 8, 1). Here we find a triangular fruit with a narrower end, which was attached to the receptacle, and also a broader end on which is a scar (*sc*), left when the corolla and style dropped off. The outer ribbed fruit-coat (2 *pe*) is hard and brittle, and on removing it a single seed will be found covered by a thin testa (2 *t*). Such a dry, hard, one-seeded fruit is called an **achene** or **nutlet.** Look for the short stalk (*fu*) which attaches the seed to the pointed end of the fruit. Remove the seed-coat and examine the embryo, noting the radicle (*ra*) at the pointed end and the two flat cotyledons (*c*), between which is the small plumule (*pl*). Test the cotyledons with iodine solution: do they contain starch? If a thin section, treated with iodine solution, is examined under the microscope, the cells will be seen to contain a number of small yellowish granules. These are protein or nitrogenous bodies called **aleurone grains** (similar granules may also be found in the Bean, Pea, and Potato); there will also be seen many bright globules which do not stain with the iodine solution. If, however, a section is placed for a while in ether the globules dissolve; also, if a drop of 1 per cent. solution of osmic acid is placed on another section, the globules stain a blackish brown. These tests prove that the cotyledons contain protein or aleurone grains and much fatty oil. Oil is a common storage material in seeds, and often replaces starch.

Endosperm. Food stored outside the embryo.—The Ash

fruit (Fig. 9, 1) is a different type. Examine its curious, slightly-twisted, and winged fruit-case (1), which is swollen at one end and contains a single seed. Cut open the fruit-coat (2 *pe*) and notice the mode of attachment of the seed (2 *fu*); then remove the seed-coat and examine the contents. Split open the seed, and between the two flat lobes you will find the embryo, consisting of a radicle (2 *ra*), above which are the two cotyledons (2 *c*), having

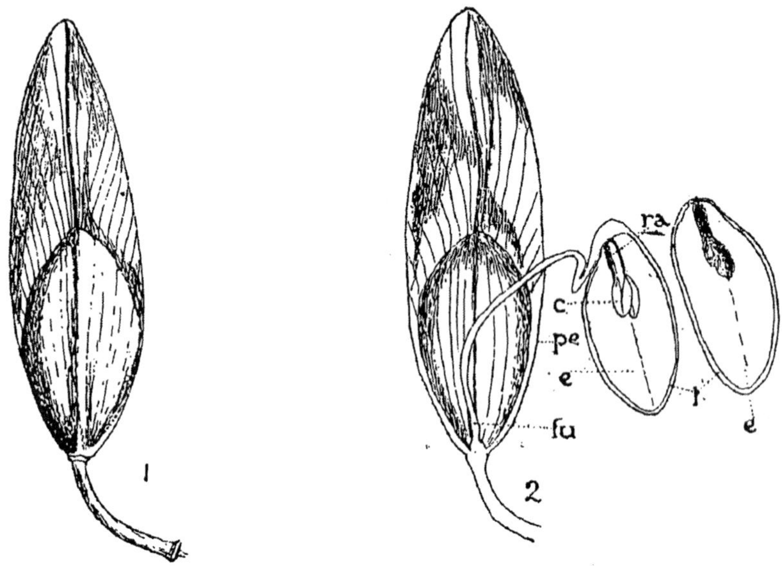

FIG. 9. FRUIT OF ASH.—1, side view; 2, fruit opened and seed dissected; *c*, cotyledon; *e*, endosperm; *fu*, funicle; *pe*, pericarp; *ra*, radicle; *t*, testa.

between them a very small plumule. This seed contains not only an embryo, but, in addition, two large lobes (*e*) stored with food-materials. Such food-reserve stored outside the embryo is called **endosperm**, and the seed is said to be **endospermous** (Gr. *endon* = within, *sperma* = seed).

Plants similar to the above, which contain two cotyledons in the embryo, are placed together in a large class called **Dicotyledons**.

Germination. Growth of root and shoot.—Let us now

determine the uses of the structures found in a Bean seed. Soak a number of dry seeds in water for a day, and then sow several in a pot of damp coco-nut fibre, sand, or sawdust. In a few days remove some of them carefully, allowing the rest to continue their growth. Note what has happened. Which structure is the first to emerge from the seed—radicle or plumule ? As the radicle emerges you will see that it bursts through the skin near the micropyle. Later on, look for the plumule as it pushes its way through the soil. Of what form is it ? Do you see any advantage in this mode of emergence ?

At the end of the crook-like plumule note the tender young leaves, observing that they are carried upwards as the stem grows, with little risk of injury, and that when well above the surface the plumule grows more quickly on the under side than on the upper, and so straightens itself. The leaves expand, new ones appear in succession, and in time buds form in their axils, which grow into leafy shoots similar to that from which they spring. In cases where the plumule has been injured, note that buds arise in the axils of the cotyledons.

Although a seedling is able to obtain little or no food from the coco-nut fibre, sawdust, or sand, for some time it continues to grow vigorously, increasing its roots and enlarging its stem and leaves. Upon what is the plant feeding ? Where does the material come from ? If we examine an older Bean seedling, we find that the cotyledons remain below ground and do not emerge from the seed-coat, but gradually lose their contents and shrivel up as the plant grows : i.e. the cotyledons, at first swollen with starchy and other food, provide the materials upon which the seedling feeds and grows in the earlier period of its life.

In the Kidney-Bean or Scarlet-Runner the cotyledons, though fleshy, come above ground and turn green (Fig. 10).

Hypogeal and epigeal germination.—Sow seeds of Mustard

and Cress on damp blotting-paper or flannel, covering them with a jar to keep the air around them moist, and note the mode of emergence of radicle and plumule; observe also

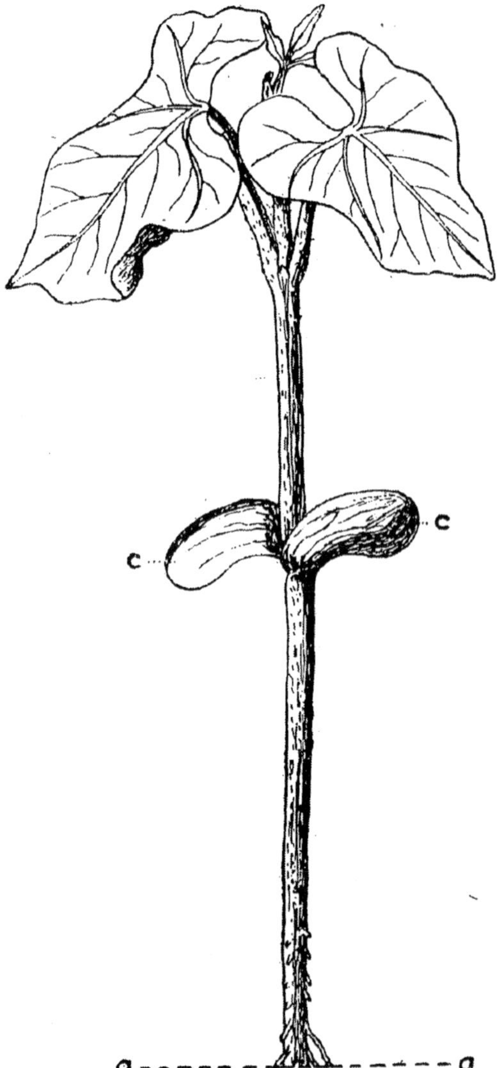

FIG. 10. SEEDLING OF KIDNEY-BEAN.
g, ground level; *c*, cotyledon.

how the cotyledons are folded, and that they come above ground, turn green, and, while simple in the Mustard, (Fig. 11, 1–5), they are three-lobed in the Cress, a rather unusual occurrence (6). As the roots of these seedlings grow they become covered with a rich crop of root-hairs (*r.h*). Notice carefully how they are distributed, and from what part of the root they are absent.

Seeds of the Edible Pea, Sweet-Pea, Vegetable Marrow,

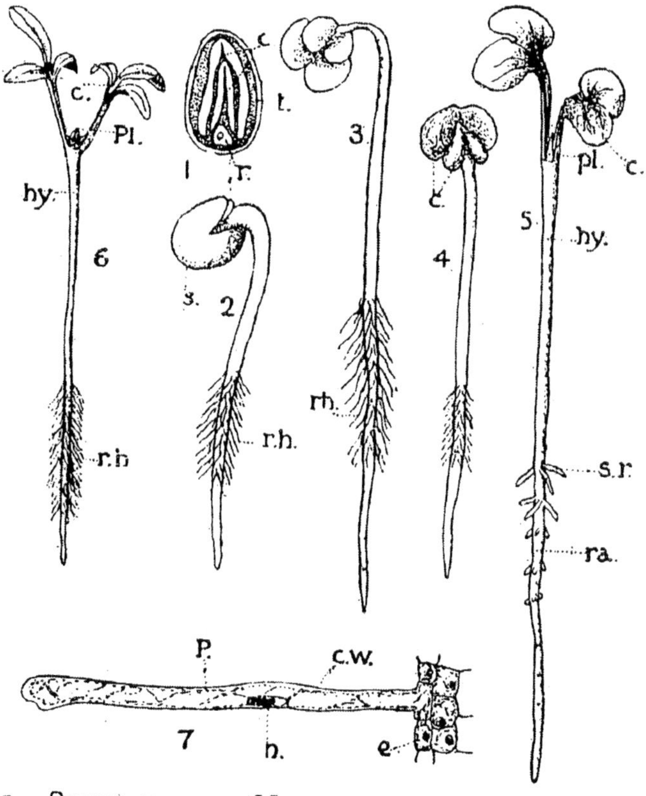

FIG. 11. SEEDLINGS OF MUSTARD AND CRESS.—1, section of Mustard-seed showing the folded cotyledons; 2, 3, 4, and 5, different stages in germination of Mustard-seed; 6, Cress seedling; 7, root-hair, magnified; *c*, cotyledon; *c.w*, cell-wall; *e*, epidermal cell; *hy*, hypocotyl; *n*, nucleus; *P*, protoplasm; *pl*, plumule; *r*, section of radicle; *ra*, radicle; *r.h*, root-hair; *s*, seed; *s.r*, secondary root; *t*, testa.

fruits of the Sunflower, Common Ash, Sycamore, and Oak, should be germinated in the same way and their modes of growth compared. In the case of the Sunflower (see Fig. 22), note that when the radicle emerges, the fruit-coat splits into two halves along the edge and, unlike the Bean, the stalk below the cotyledon (**hypocotyl**, Fig. 11, *hy*) grows in such a way as to bend in the form of a crook and, continuing to elongate, carries the cotyledons and the split fruit-coat above ground. The cotyledons now separate, throw off the fruit- and seed-coats, and turn green.

Seeds of the Vegetable Marrow should be sown with their flat faces horizontal; in a few cases cut away a third of one side of the seed-coat at the micropylar end and sow these with the cut surface downwards, and determine the mode of emergence of the plumule in each case, and note how it separates from the seed-coat. As the radicle emerges and turns downwards from the uninjured seeds, a peg is formed on the under side, at the base of the hypocotyl, which fixes the lower edge of the split seed-coat and holds it down. The hypocotyl above elongates, the upper half of the seed-coat splits off, and the cotyledons are withdrawn and carried above ground, where they expand as two flat, oval, green leaves. In those cases where part of the seed-coat has been removed, the peg is unable to act as holder, and the cotyledons, unable to extricate themselves, carry the seed-coat upwards, and it is only thrown off when the cotyledons expand. The use of the peg, therefore, is to enable the cotyledons and plumule to free themselves more readily from the seed-coat. The position of the peg is always on the under side, and, like the downgrowth of the radicle, is determined by gravity.

Two modes of germination are seen in these types. In the Bean, Edible Pea, and Sweet-Pea, the seed remains below ground and only the plumule grows into the air; this mode of germination is known as **hypogeal** (Gr. *hypo* = beneath, *gē* = earth). In the Kidney-Bean, Mustard, Cress, Sunflower, Vegetable Marrow, Common Ash, and Sycamore, the part of the stalk below the cotyledon (the hypocotyl) grows and carries the cotyledons and plumule together into the air. In these cases the cotyledons turn green and act as the first green leaves. This mode of germination is called **epigeal** (Gr. *epi* = upon).

Allow the seedlings to continue their growth, compare the successive leaves as they appear, and note that the first-formed foliage-leaves are often simpler than the later ones. In this respect the Sweet-Pea is an interesting example to study.

Development of the Sweet-Pea.—Sow a few seeds in soil and allow them to grow for several weeks, supporting the tender stems with thin sticks. Note carefully each leaf as it appears until the fully-matured ones are formed. Compare the first or 'juvenile' foliage-leaves with the later adult leaves and try to determine the structure and uses of the parts.

As the arched shoot comes above ground (Fig. 131, 5, p. 189), very small leaves appear; the first is seen to consist of two pieces or lobes, with a narrow pointed lobe between them (Fig. 131, 6, *t*); the second leaf is a little larger, and the lobes are better developed and toothed (Fig. 131, 7); the third leaf still shows the two lobes close to the stem; and above these is a short leaf-stalk, then a pair of oval lobes, between which is a slender green thread (Fig. 131, 8). The two lower lobes, or **stipules** (*st*), are outgrowths of the leaf-base; they cover and protect the rest of the leaf in the bud. The leaf-stalk and blade are represented at first only by the narrow terminal lobe; in the later leaves the blade develops two large opposite lobes, and the terminal thread becomes longer. Finally, the terminal part of the blade divides to form paired structures agreeing in position with the lobes of the blade, of which they are special modifications. Note their behaviour on coming into contact with a stick, and you will see at once the use of this curious modification as a clinging organ. In the older plant (Fig. 131, 1) these organs are well developed, and by twining round a support enable the stems to grow up above overshadowing plants. These clinging organs are known as **tendrils**, and are of great use to plants with stems too long and too slender to support themselves in an erect position. In some plants, however, tendrils are formed from organs other than leaflets (see pp. 144–7).

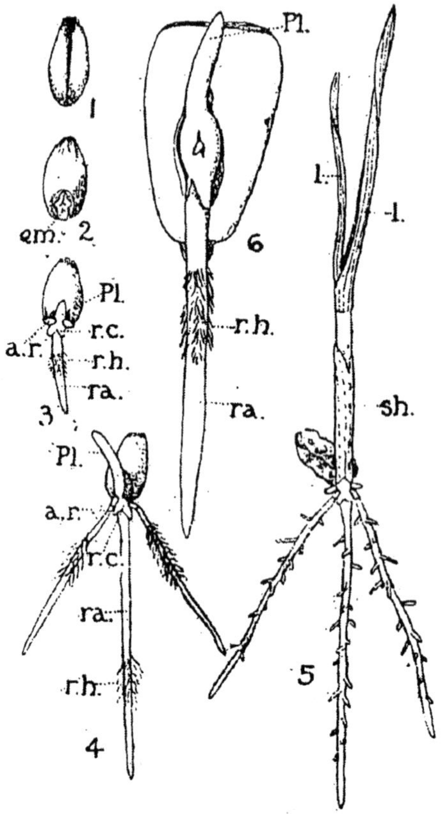

FIG. 12. GRAINS OF WHEAT AND MAIZE.—1, grooved surface; 2, convex surface of Wheat grain; 3, 4, and 5, stages in germination of Wheat grains; 6, germinating grain of Maize; *a.r*, adventitious roots; *em*, embryo; *l*, foliage-leaves; *Pl*, plumule; *ra*, radicle; *r.c*, root-collar; *r.h*, root-hair; *sh*, colourless sheath.

(*b*) Monocotyledons

The Wheat grain. Embryo with one cotyledon; the endosperm.—Soak a few grains of **Wheat** in water for a day; then place some of them in a jar lined with wet blotting-paper, and others in coco-nut fibre. Compare dry and soaked grains as to size and shape. A grain of wheat (Fig. 12) is the product of the **pistil**, and is a fruit, not a seed. The fruit-coat and the seed-coat, however, adhere firmly, and cannot be easily distinguished. The outer fruit-coat, or **pericarp**, is smooth, grooved on one side (Fig. 12, 1) and convex on the other (2). Note the tuft of

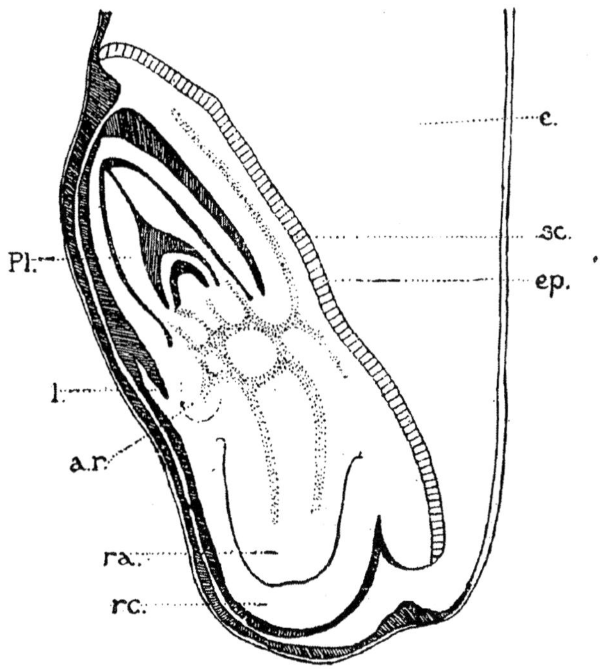

FIG. 13. VERTICAL SECTION OF WHEAT EMBRYO.—*a.r*, adventitious root; *e*, endosperm; *ep*, epithelium of the scutellum; *l*, ligule; *Pl*, plumule; *ra*, radicle; *r.c*, root-collar; *sc*, scutellum.

hairs at one end, and at the other an oval area in which is a small wrinkled body (*em*). Cut the grain in two along the groove and apply a drop of iodine solution to the cut surface. Do all parts stain equally? The little structure at the end is unstained. This is the **embryo**. The parts of which it is composed can be well made out on examining soaked grains with the aid of a pocket lens.

A section of the embryo is shown in Fig. 13. Below is the short **radicle** (*ra*); above is the straight **plumule** (*pl*). On one side is a shield-shaped structure, the cotyledon or **scutellum** (*sc*), with its convex face applied to the starchy food-reserve; and opposite the point of attachment of the shield is a tiny scale, the **ligule** (*l*).

The Wheat grain thus consists of a fruit-coat and seed-coat fused together, an embryo consisting of a radicle, a plumule, a single cotyledon (the scutellum), and a ligule; and, occupying the larger part of the grain, is the food-reserve, which is outside the embryo and therefore called endosperm (*e*). (Compare this with the fruit of the Ash, Fig. 9.)

Plants which contain only one cotyledon in the embryo are placed in a large class called **Monocotyledons**.

Germination of Wheat and Maize.—Examine germinating grains in different stages of growth, and note the behaviour of the parts of the embryo (Fig. 12, 3, 4, and 5). As the radicle (*ra*) elongates, it bursts through the thick tissue which surrounds it, and now encircles its base as a root-collar (*r.c*). Continuing to grow, it becomes clothed with delicate root-hairs (*r.h*) everywhere except at the tip. Look for the root-branches (*a.r*) and determine their point of origin. Note that some spring from the base of the stem. Roots arising from the stem are called **adventitious roots** (Fig. 12, 3 and 4 *a.r*). These also bear root-hairs. Note how the particles of fibre adhere to them and are not easily washed off. In Wheat the radicle does not become the main root: the fibrous roots of the mature plant all arise from the nodes of the stem.

Compare the mode of emergence of the plumule (*pl*) with that of the Bean and other seedlings examined, and note the differences. In the Wheat the leaves are rolled up and enclosed in a smooth, colourless, tubular leaf, the whole forming a compact structure well adapted for boring its way upwards and uninjured through the soil. Examine the grain of an older seedling (Fig. 12, 5) and note the changes that have occurred. It is no longer a firm, solid mass, but has become wrinkled; and when squeezed, readily collapses, ejecting a milky-looking fluid.

As the green foliage-leaves grow they push their way out of the colourless sheath (Fig. 12, *sh*) and unroll. Soon a drop of water appears on the tip of each leaf. Where has this come from? Considering the conditions under which the seedlings have been grown, do you think they are drops of dew? As the drops dry off, note that a whitish deposit is left. Is such a deposit left when dew-drops evaporate? Note that such drops occur freely when the seedlings are well supplied with water. The root-hairs and young roots absorb more than the plant is able to use, and the excess exudes from pores at the leaf-tips. This water contains mineral salts in solution, and when the exuded drops evaporate, the salts remain as a powder on the leaf-tips.

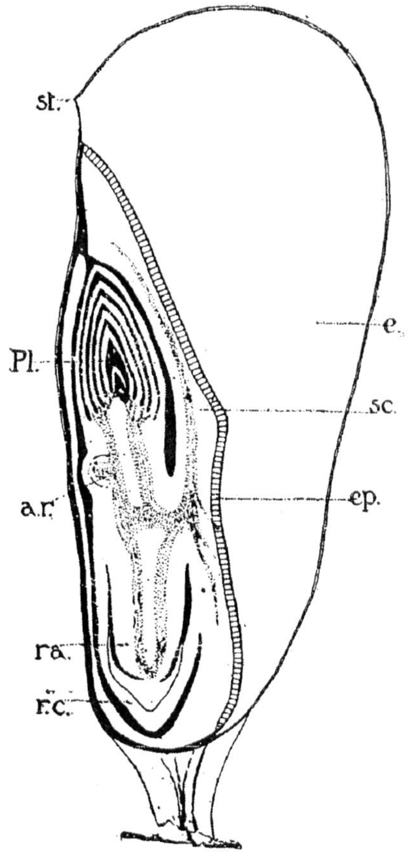

FIG. 14. VERTICAL SECTION OF MAIZE GRAIN.—*a.r*, adventitious root; *e*, endosperm; *ep*, epithelium of the scutellum; *Pl*, plumule; *ra*, radicle; *r.c*, root-collar; *sc*, scutellum; *st*, stigma.

Grains of **Maize** (Fig. 12, 6 and Fig. 14) and **Oat** should be treated in the same way as the Wheat, and their parts compared. In the Maize, note the smooth pericarp, and the remains of the stigma as a little point above the embryo (Fig. 14, *st*), also the hard endosperm (*e*). A germinating grain is shown in Fig. 12, 6. The Oat differs from the Wheat and Maize in that the grain is covered with chaffy scales. Other interesting examples belonging to the same class are the Onion and the Wild Hyacinth. In these the tip of the cotyledon remains in the endosperm and acts as a sucking or absorbing organ (see Fig. 87, A, C, D).

After seeds have ripened in the fruit they commonly require to pass through a dormant period, which varies in length in different species, before they are able to recommence growth. So long as a seed is kept dry, or at a very low temperature, growth does not take place, and some seeds may lie dormant while retaining their vitality for very many years; but under the conditions we have provided —viz. moisture, air, and warmth—germination begins.

CHAPTER III

STRUCTURE OF ROOTS

Tissues of a mature dicotyledonous root.—We have seen how roots arise, also their form and mode of growth; let us next consider how roots are constructed. Fig. 15, 1 is a photo-micrograph of a cross-section of an old root of a Dicotyledon showing the different tissues of which it is composed. The outer surface is composed of a layer of cells called the **epidermis** (*e*) (Gr. *epi* = upon, *derma* = skin). In the young root some of these cells grow out to form root-hairs (see Fig. 11, 7). Beneath the epidermis is a wide ring of cellular tissue—the **cortex** (Fig. 15, *co*) with small air-spaces between the cells. The innermost layer of

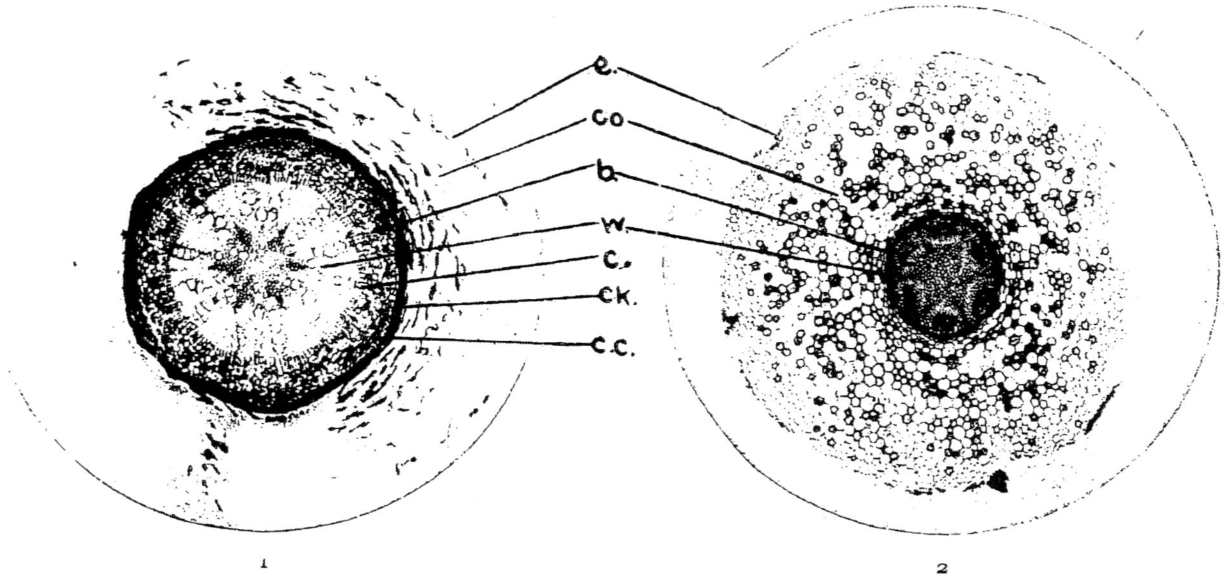

FIG. 15. TRANSVERSE SECTIONS OF (1) AN OLD ROOT AND (2) A YOUNG ROOT OF THE LIME TREE, the young root showing alternating groups of wood and bast.—*b*, bast; *c*, cambium; *c.c*, cork cambium; *ck*, cork; *co*, cortex; *e*, epidermis; *w*, wood.

the cortex consists of closely-fitting cells with thickenings on their radial and outer walls. This layer is the **endodermis** (Gr. *endon* = within). The inner tissues form the **central cylinder** or **stele**, and consist of a layer of delicate cells next to the endodermis, called the **pericycle** (Gr. *peri* = around, *kyklos* = circle); within this are the **veins** or **vascular bundles**, each being composed of three kinds of tissue—an outer tissue, the **bast** or **phloem** (Fig. 15, *b*) (Gr. *phloios* = bark); an inner tissue, the **wood** or **xylem** (*w*) (Gr. *xylon* = wood); and between the two a very delicate, actively growing tissue, the **cambium** (*c*).

The water, which has been absorbed by the root-hairs, is transmitted through the cortex to the wood, and through this conducted upwards to the stem, to which the veins go in the form of a complicated network. Excellent skeletons consisting of the veins of a Radish can often be obtained from a garden rubbish-heap.

Tissues of a young root.—The structure of a very young root, however, is different (Fig. 15, 2). The bast is not placed on the outside of the wood, i.e. collaterally, but the two tissues are arranged in alternating groups, i.e. radially. The groups of first-formed or primary wood (Fig. 15, 2, *w*, and Fig. 16, 1, 2, 3, *p.w*) develop from outside inwards (the ends of the rays being the oldest parts), and by further growth they form a solid mass of wood in the centre. In such a root there is no pith. It is not until a cambium ring is formed, winding to the outside of the wood and to the inside of the bast, that the arrangement found in an old root is developed. Fig. 16, 1, 2, 3, makes this point clear.

Secondary growth.—From the cambium and by the division of its cells to form new tissues, arises new wood to the inside of it (Fig. 16, 2, 3, *s.w*), and new bast to the outside (Fig. 16, 2, 3, *s.b*); and in this way a ring of vascular bundles is formed, each bundle consisting of the three tissues—bast, cambium, and wood. At first the cambium is a wavy line (Fig. 16. 3, *c*), but as growth proceeds it becomes a uniform ring (Fig. 15, 1, *c*), to the inside of which is the wood and to the outside the bast. The root has now lost its radial structure and grows in thickness by additions from the cambium in the same manner as in the stem. Finally, a layer of cork is formed from a ring of cambium, the cork cambium (Fig. 15, 1, *c.c*), which arises in the tissues immediately to the outside of the bast. The cork layer thus produced cuts off communication between the vascular

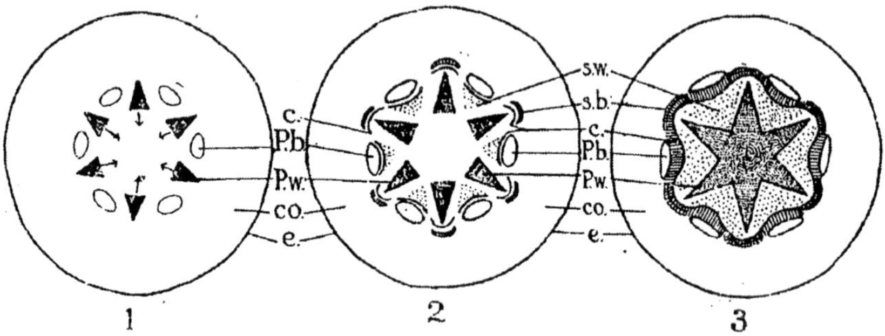

FIG. 16. DIAGRAMS ILLUSTRATING THE DEVELOPMENT OF A ROOT. —1, very young root, showing alternating groups of primary wood (*P.w*) and primary bast (*P.b*); 2, older root, showing cambium (*c*), arising on the outside of the wood and on the inside of the bast; 3, old root, showing a complete cambium (*c*); *co*, cortex; *e*, epidermis; *s.w*, secondary wood, formed on the inside of the cambium; *s.b*, secondary bast, formed on the outside of the cambium.

bundles and the cortex, and the latter dies and crumbles away, as shown in Fig. 15, 1, *co*.

The loss of the cortex greatly reduces, for a time, the diameter of the root, but growth goes on steadily and the root continues to increase in thickness throughout life.

Structure of a monocotyledonous root.—The root of a Monocotyledon differs from the above in an important respect: no cambium is formed between the wood and the bast, and when these tissues are once developed, no further increase in thickness can take place. The bundles of alternating wood and bast are also more numerous (often ten or more of each); the smallest and first-formed tissues of the wood are to the outside, the later ones are well developed, large, and in some species they meet and fill the centre of the root with wood; in others the middle is occupied by a pith.

The tissues of the root are of several kinds and modified to serve special purposes; the root-hairs absorb from the soil water which is passed through the cortex to the wood, and the latter conducts it upwards to the shoots. The organic food, which, as we shall see later, is formed in the leaves, is conducted by the bast and neighbouring delicate tissues to the growing organs and storage-tissues.

CHAPTER IV

WORK OF THE ROOT

In our study of germinating seeds we found that the root was the first organ to be formed, and that its appearance was followed by the emergence of the young shoot. What is the future of such a root? How does it grow? Of what special use is it to the plant? How does it do its work? A few experiments and observations will help us to answer these questions.

Direction of growth of root and shoot.—In a pot of fibre or soil, sow three soaked Bean seeds, one with the radicle pointing downwards, another horizontally, and a third upwards. After a few days examine them. In what directions have the radicle and plumule grown in each? (Fig. 17, 1, 2, 3.) We observe that the radicle endeavours to grow downwards into the dark, moist soil, independently of the position in which the seed was placed in the ground, and that the plumule just as persistently grows upwards into the air and sunlight.

The stimulus of gravity. Geotropism.—What force is at work which determines these directions of growth? If a growing seedling is placed on its side and attached to a rod which is caused to revolve horizontally by means of clockwork (Fig. 18), the radicle and plumule will continue to grow in that direction. An instrument constructed for the purpose of rotating plants in various positions is called a

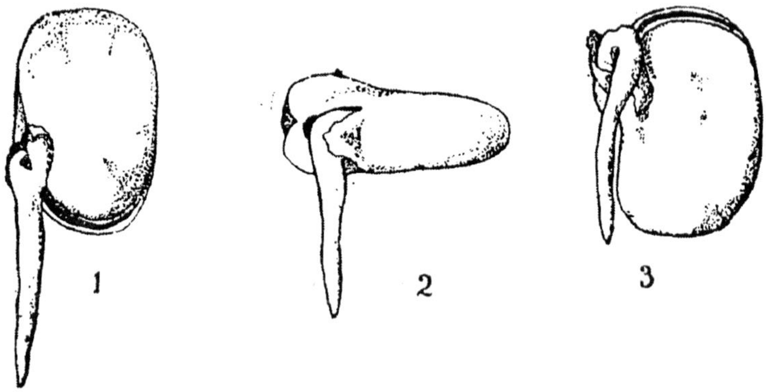

FIG. 17. SEEDS OF BROAD BEAN SOWN IN DIFFERENT POSITIONS.
In each case the radicle grows downwards.

klinostat.[1] Further, if seedlings of the Pea are pinned on a vertical disk with their roots pointing towards the centre, and the disk is then revolved rapidly for four or five hours, the growing roots turn outwards away from the centre of rotation, and the plumules turn inwards towards the centre. In this experiment the seedlings have grown under the influence of a force stronger than the attraction of gravity.

The influence of centrifugal force affects the radicle and

[1] A simple form of klinostat may be made from a small clock by removing the hands and fixing a short tube to the axle of the minute-hand. Fit up a light bottle as a moist chamber and bore a hole in the cork, into which the tube may be firmly fitted. Seedlings pinned to the cork may be rotated as in the experiment described.

plumule in opposite ways. The radicle grows in the same direction as that of the force, and the plumule grows in the contrary direction. Such experiments establish the fact that growing organs are sensitive to a physical force in nature, and the force acting on plants in this manner is gravity, the radicle growing along the line of action of gravity towards the earth while the plumule grows in the opposite direction. The response of growing organs to the attraction of

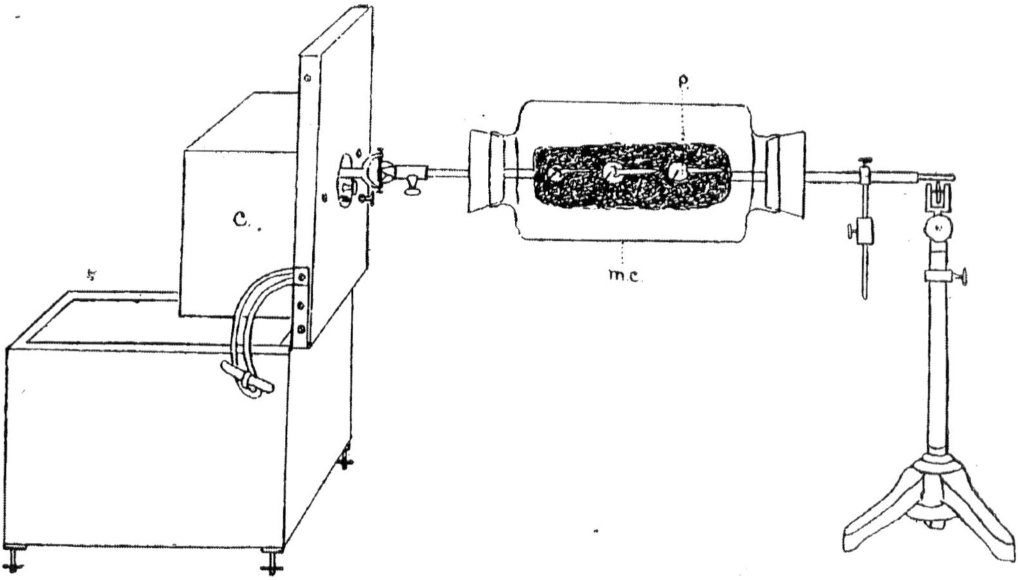

FIG. 18. KLINOSTAT.—*c*, case containing a clock, by means of which the glass cylinder (*m.c*) is slowly revolved; in this are seedlings (*p*) growing on moist turf.

gravity is known as **geotropism** (Gr. *gē* = earth, *tropos* = turning), the radicle being positively geotropic and the plumule negatively geotropic. Any influence which acts upon the living organs of a plant, and induces in them a change of behaviour, is called a **stimulus**. In addition to gravity, plant organs respond to many other stimuli, e. g. light, heat, contact, electrical currents; also to water and other chemical substances. These stimuli are important factors in the environment of a plant, and as they vary frequently, it is necessary for the plant to respond and adjust itself to the changing conditions. The power of response and adjustment is the most characteristic feature of life, and it is important that we should pursue the subject a little further.

Contact stimulus.—In an ordinary soil, it will commonly happen that roots will meet with obstructions, such as stones. Under these circumstances, how will they behave? Take a wide-mouthed bottle, half-filled with stones or fragments of broken plant pots, moistened with a little water. Then attach two or three seedlings to the cork, and suspend them with the radicles directed downwards into the bottle. Notice what happens as they come into contact with the hard fragments. The roots turn away, escaping the injury which would result if the tip were forced against a solid object.

The sensory region of the root.—This shows that some part of the root must be sensitive, and the following experiments determine the sensory region of the root.

Take four seedlings of the Broad Bean (*a* to *d*) with radicles about 1½ inches long, and treat them as follows:

1. Place seedling *a* horizontally on moist coco-nut fibre.

2. Take a razor and cut off one-sixteenth of an inch from the tip of the radicle of seedling *b* and place it alongside *a*.

3. Place seedling *c*, uninjured, on its side for an hour, then cut off the tip as in *b* and lay it horizontally on the moist fibre.

4. Place seedling *d* on its side for a day until its tip has curved downwards, then cut off the tip as with *b* and *c*, but place the seedling with the root pointing downwards.

Allow the seedlings to grow and carefully compare the results.

a turns downwards; *b* grows, but does not bend; *c* bends as in *a*; *d* does not turn downwards, but continues to grow horizontally.

If the seedlings are allowed to grow under favourable

conditions a new tip is formed and grows downwards as in an uninjured root. We thus see that, though able to grow in length, seedlings *b*, *c*, and *d* had lost their sensitiveness, and that the last one-sixteenth of an inch includes the region which is able to receive and respond to a stimulus. This is known as the **sensory region** of the root.

The stimulus of water. Hydrotropism.—To pursue the subject of root sensitiveness further, perform the following experiment. Obtain a shallow box, remove the bottom, and replace it by wire gauze (Fig. 19). Fill the box with wet coco-nut fibre and sow in it a number of peas. Tilt

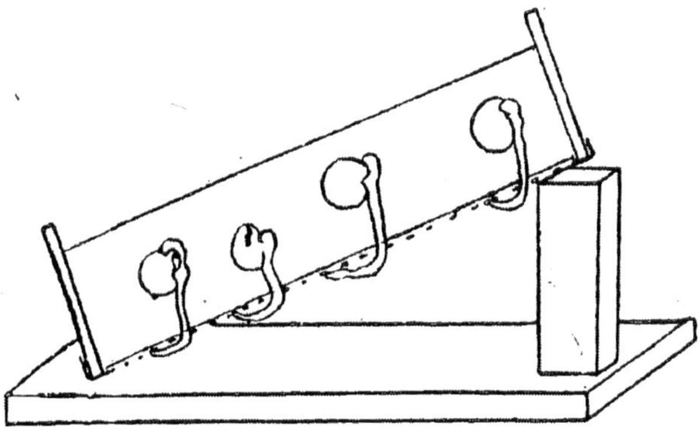

FIG. 19. EXPERIMENT TO SHOW THAT ROOTS ARE SENSITIVE TO THE STIMULUS OF WATER.

the box at an angle of 45° and protect the bottom from strong light. As the seeds germinate, the roots, owing to the shallowness of the box, soon grow through the gauze into the air. Note the behaviour of these roots. We see that they cling to the surface and may even bend back and grow upwards into the wet fibre. From previous experiments we should expect the roots to grow vertically downwards in response to the stimulus of gravity. What other force is now operating to draw the roots away from the vertical direction? We see that the attraction of water is, under these circumstances, more powerful than that of gravity. This tendency of roots to turn towards or be attracted by water is called **hydrotropism** (Gr. *hydor* = water). Roots are attracted to, and grow best in, the moist layers of the soil; and the roots of trees and shrubs may be drawn considerable distances in the direction of a suitable water-supply. A dry soil retards root growth.

In this connexion it is interesting to note how the distribution of roots below ground is related to the shoot system above ground. The leaves of some plants are so placed on the stem that drops of rain falling on them are directed towards the stem; the drainage from the leaves, therefore, is concentrated on that part of the ground immediately round the stem. Such plants have commonly a long tap-root but no wide-spreading root-branches. In other cases the leaves are so arranged as to throw the water from their tips and the water falls over a larger circle to the outside of the plant. The watershed of such a plant is therefore a large one, and on examining its root-system we find that the young absorbing branches spread extensively, and collect water from a corresponding area.

The stimulus of light. Heliotropism.—If we observe the roots of the Ivy we shall find that they very generally turn towards the surface along which the plant is growing. This will commonly be the moister surface, and hence we have a case of hydrotropism. But it is equally true that they are growing towards the shady side and away from the light. That roots can be tempted to grow from the usually exposed surface of the Ivy stem can be shown by the simple experiment of keeping a portion covered with a wet cloth; the roots will soon be seen to grow under the cloth.

Suppose, however, that seedlings are grown on gauze over a jar of water so that the roots hang downwards into the water. If we now exclude light from the jar at every point except a narrow slit on one side, through which a strong light can be admitted, the growing roots will turn away from the light towards the shaded side. Thus Ivy roots are probably not merely growing towards the moist surface, but also away from the sunlight. The sensitiveness of growing organs to sunlight is termed **heliotropism** (Gr. *hēlios* = sun), but as roots usually turn away from, and not towards, the light they are said to be negatively heliotropic.

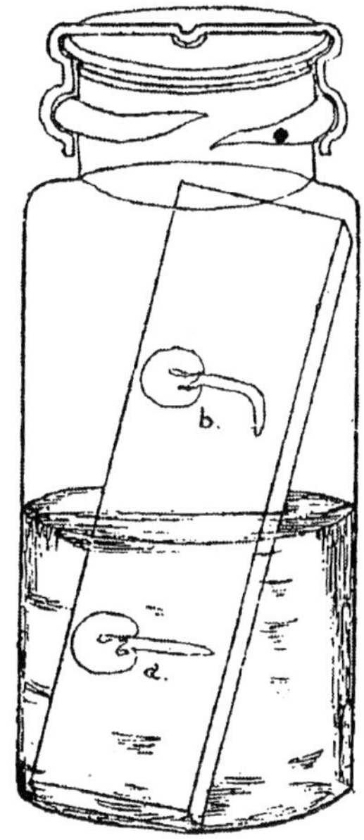

Fig. 20. Experiment to show that the Root is not sensitive in the Absence of Oxygen. — *a*, seedling in boiled water; *b*, seedling in air.

Our experiments have shown that roots are sensitive to a number of different stimuli, namely, gravity, contact, water, and light, and that in response to these stimuli roots grow in a definite manner. We will now determine the important fact that roots are sensitive only in the presence of oxygen.

Necessity for oxygen.—Take a wide-mouthed bottle fitted with a good cork or stopper, and fill half of it with water that has been boiled so as to expel the dissolved air. Select two seedlings of the Pea with radicles about an inch long. Take a board (which has been previously boiled to destroy mould-spores) and pin the seedlings to it in such a way that the radicles are directed horizontally as in Fig. 20. Place them in the bottle, submerging one in the water (*a*), and placing the other well above the water in the air (*b*); then close the bottle with a stopper. Ensure that both seedlings have their radicles directed downwards and place them horizontally only when putting them into the bottle. Leave them for a few days and note what happens. The one in the air has turned downwards, as we should expect from our previous experiment, but the one in airless water has continued to grow in the direction in which it was placed in the water. In other words, the root only possesses this power of turning when supplied with air containing oxygen. This suggests a further question. Do plants utilize the oxygen of the air in the ordinary process of growth?

Respiration.—A simple experiment will enable us to understand the important rôle played by plants in changing the composition of the air. Soak a number of peas in water for a day, and then place them in two jars (A and B) lined with wet blotting-paper. Put on the stoppers and keep them on for a day or two. A similar jar may be prepared, but without peas, for comparison. Then test the air in the jars as follows: (*a*) insert a lighted taper, and note whether the air in the jar containing the peas will now support combustion; (*b*) pour in a little lime-water and note the result. From these tests we see that the germinating peas have removed from the air the gas which supports combustion, viz. oxygen, and have given up to the air a gas which turns lime-water milky, viz. carbon dioxide. If we breathe on to lime-water we observe a similar effect. In other words, Peas, during their growth, are using up oxygen and giving off carbon dioxide, just as we are when breathing. This process, which is called **respiration**, is necessary to the existence of plants; they would soon die if kept wet in a closed bottle and without air.

The oxygen taken into the tissues of the plant during respiration acts chemically upon the complex organic substances which constitute the plant, with the result that they are converted into simpler compounds such as carbon dioxide and water. During these changes energy is set free, and thus the work necessary to the life of the plant is performed. Some of this energy appears as heat which helps to maintain a suitable temperature within the living tissues of the plant.

In describing the Stock we spoke of the growing-points of root and stem. The root in pushing its way through the soil meets with much resistance, but is protected at the tip by a root-cap, the end of which is constantly dying and wearing away, to be replaced by new tissue. On the other hand, the stem-tip has no cap, but is protected by the overlapping leaves which form the end bud.

Growing-region of the root.—It will be interesting to compare the manner of growth of root and stem and to determine experimentally the mode of elongation in each case.

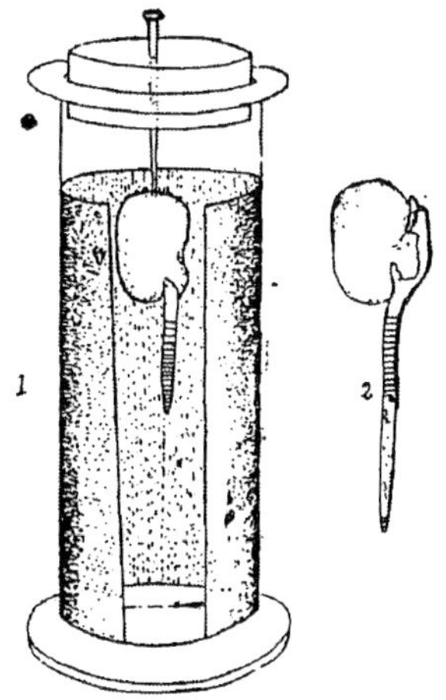

Fig. 21. Bean Seedling marked to determine the Region of Elongation in the Root.—1, at the commencement of the experiment; 2, at the end of $30\frac{1}{2}$ hours.

Select a germinating Bean with a radicle about an inch long, wipe off adhering particles, taking care not to injure the Bean in any way, and mark about a dozen lines across it in Indian ink. This can be done as follows: Take a piece of cotton thread and hold it by both ends; bend it, and dip the middle of the thread into the ink. Then lay it across the root so as to make a clear transverse mark. Begin quite at the tip and make a series of marks about $\frac{1}{16}$ inch apart backwards from the tip. Obtain a large, wide-mouthed bottle or jar fitted with a cork (Fig. 21). Line it with blotting paper so that the lower edge dips into about half an inch of water; this provides a moist chamber in which to grow the seedling. Push a large pin through the cork and fix the Bean seed firmly to the point; then replace the cork in such a way that the marked radicle hangs downwards within the bottle so as to avoid touching the sides. Make a sketch of the radicle at the commencement of the experiment, showing the exact number of lines and their distance apart, and at the end of two or three days make another sketch and note what has happened (Fig. 21, 2). Has the root grown? Count the lines and compare them with the original sketch. What is their position now? Which lines are most widely separated? What changes have occurred in some of the lines? Has elongation occurred at the extreme tip, or is the position of the first line still unchanged? We observe that the extreme tip (the end of the root-cap) has not grown, but that elongation has been most active in the region immediately behind this and included within the next two or three lines, which, as the root has elongated, have been drawn out and now appear as a number of dots.

Region of curvature.—Mark the radicle of another Bean seedling and place it in a moist chamber as before, but fix it with the radicle horizontal. Allow it to grow, and note where curvature takes place. Observe that the regions of curvature and elongation coincide and are included within the last quarter of an inch, but the sensory region (p. 40) is confined to the last sixteenth of an inch of the root-tip.

If the experiments have been carried out successfully, the seedlings may be used for observing the manner in which the branch roots emerge. Replace the seedlings and allow them to grow a few days longer. Note that they come out in four, sometimes five, vertical rows; that they are not in any way related to leaves, and grow out obliquely from the radicle. In cases where the radicle has been injured and its growth stopped, look for roots emerging from the base of the stem. In a still older plant, note what a large part of the soil is drained by the later branches which grow in various directions.

Growing-region of the stem.—On seedlings still growing in the pots carry out a similar experiment with the stem, selecting plants in which the plumule has emerged an inch

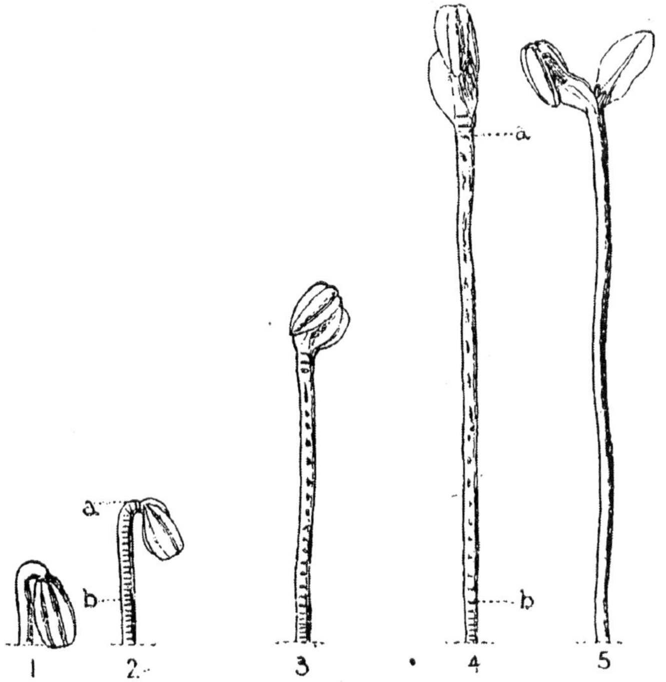

FIG. 22. MODE OF EMERGENCE OF THE COTYLEDONS OF SUNFLOWER SEEDLINGS.—2, 3, and 4 show the region of elongation (*a–b*) in the hypocotyl.

or more above the soil. Beginning at the tip, make a series of lines backwards as far as convenient. Make a sketch true to scale and show in it the exact number of lines (Fig. 22, 2). Seedlings of Sunflower and Kidney-Bean (Fig. 10) serve well for this experiment. Allow them to grow for a few days and note the result (Fig. 22, 3 and 4). We see that the region of growth in a stem (*a–b*) is not confined to such a small area as it is in a root.

Absorption by Roots

Root-hairs the organs of absorption.—If young roots, bearing root-hairs, are examined by means of a pocket lens or under a microscope, it will be seen that each root-hair consists of a long, tubular outgrowth of an epidermal cell (see Fig. 11, 7, *e*). The wall (*c.w*) consists of a thin membrane of cellulose, the substance of which the fibres of cotton are composed. The outer exposed surface is somewhat slimy, and to this the soil-particles adhere firmly. Within the tube is the living substance known as *protoplasm* (*p*), together with a little rounded body—the *nucleus* (*n*). The protoplasm forms a thin living lining to the tube; and from the lining, strands of protoplasm stretch across the cavity. The centre of the tube is occupied by sap. Farther back, in the older parts of the root, the surface tissues are corky and no root-hairs are found. Root-hairs are also absent from the region protected by the root-cap. That part of the young root which bears root-hairs is called the root-hair region.

The different regions of a root are strikingly shown in a plant which a gardener would describe as 'pot-bound', i. e. a plant which has grown so long in a pot that its roots have spread themselves out in a tangled mat between the soil and the pot. On turning out such a plant numerous tender, white roots, crowded with root-hairs, are found covering the surface; while the roots buried in the soil are tough, wiry, and covered with a firm brown layer of cork.

What are the uses of these different parts? The root-cap doubtless serves to protect the young growing root as it pushes its way through the soil. The older parts, surrounded and protected by cork, are no longer able to absorb water, but they fix the plant in the soil and bear at the ends of their branches the tender roots clothed with root-hairs. We will determine the important function of the root-hair region with the help of a few experiments.

Acids excreted by roots.—Obtain a piece of polished marble, place it in the bottom of a shallow box, and cover it two to three inches deep with fine wet sand. In this, place a few seeds of Sunflower, allow them to germinate, and in ten to twelve days examine the seedlings, noting how the roots, unable to descend vertically, have spread over the marble slab. Now examine the surface of the slab and see the change that has taken place wherever the roots of the seedlings have touched it. The tracks of the roots are clearly marked by the corrosion of the polished surface. Now take a little dilute hydrochloric acid, dip into it a small camel-hair brush, and write the date of the experiment on the polished surface. When this is dry it will be seen that the course taken by the brush is etched on the surface, just as was the course taken by the roots. A small piece of cotton thread dipped in the acid and laid on the slab will similarly leave its trail. Can it be that the young roots and their root-hairs during their active growth excrete a substance capable of etching marble? They do, for we have already seen (p. 44) that germinating seeds give off much carbon dioxide, and this gas in the presence of water is able to dissolve marble. It is probable that young roots are able to excrete other acids capable of dissolving mineral substances in the soil. Thus they may bring into solution substances otherwise difficult to dissolve, and these may be absorbed as food by the plant.

Wilting.—Allow a plant growing in a flower-pot to remain unwatered for a few days, and note what happens. Now water the soil and notice what changes take place. Obtain a young, leafy shoot of Laburnum and notice how soon the leaves droop. Place the cut end in water, cut off a small piece (under water), and note how quickly the shoot revives. We see that, as the soil loses water, the shoots above ground droop or 'wilt', but on renewing the water-supply they become turgid, i.e. the tissues absorb water. The leaves expand and once more become firm. Water is essential to enable a plant to maintain its firmness.

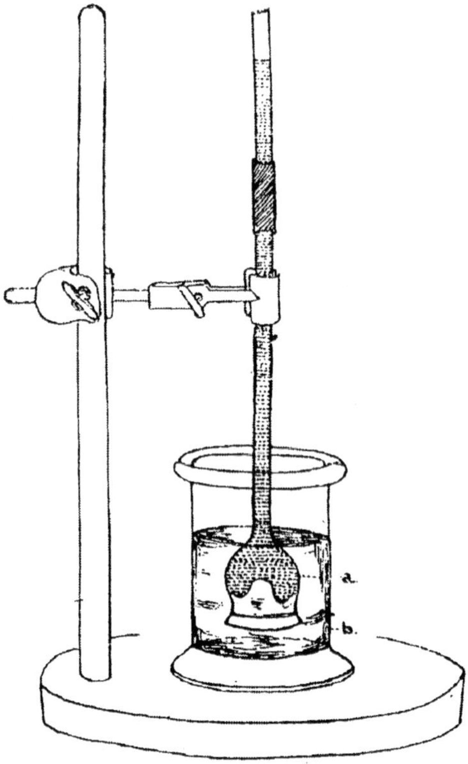

FIG. 23. OSMOSIS EXPERIMENT.
a, thistle funnel containing sugar solution; *b*, jar of water.

Osmosis

Osmosis is a natural process which has an important bearing on the work of roots. We shall best understand it by means of the following experiment. Obtain a thistle funnel and with a piece of moistened parchment paper close the mouth of it as in Fig. 23. Tie the paper firmly round the rim and seal the junction carefully with vaseline, paraffin, or plasticine. Fill the bulb of the funnel with a strong sugar solution and suspend it in a jar of water.

At definite periods of time mark the height of the rising column by means of narrow strips of gummed paper. If the solution is sufficiently strong, the liquid will rise considerably, and it will be interesting to increase the length of the tube and determine how high a column can be raised. This can easily be done by connecting a long piece of glass tubing to the funnel by means of a piece of rubber tubing.

Consider the conditions of the experiment. In the bulb of the funnel is a dense solution separated from a weak solution by a permeable membrane. Under these conditions an exchange takes place, but, as we see from the experiment, a very unequal one. On the one hand, the water passes quickly through the membrane, diluting the sugar solution, whilst, on the other, the sugar solution passes slowly into the water. If this exchange goes on, the two solutions will in time become equal in density. The passage of liquids of different densities through a permeable membrane which originally separated them is called **osmosis.**

Let us now compare the conditions of the roots in the soil with those of their counterpart in this experiment. The root-hairs may be regarded as corresponding to the closed funnels, their contents a dense solution, and their walls permeable membranes. The soil-water will now be the weak solution separated from the dense solution within the root-hair by (1) the outer slimy or mucilaginous layer which readily absorbs the soil-water, (2) the permeable cellulose wall, and (3) a very thin lining layer of the living protoplasm known as the plasmatic membrane, which, unlike the parchment membrane, can determine what substances shall enter or leave the cell. From such a comparison we are able to obtain some idea as to how water is taken up by the root-hairs and passed on to the inner tissues of the plant. The process, however, is extremely complicated, partly owing to the complex nature of the cell-contents and the selective power of the plasmatic membranes.

Each of the fine particles of which the soil is composed is coated with a thin film of water containing mineral salts in solution. The root-hairs press against and become moulded to the particles, and water (together with a selection of the dissolved mineral salts) is drawn into the root-hair. This continues as long as the osmotic attraction of the cell-contents can overcome the surface tension of the film of water surrounding the soil-particle. If water is renewed from adjacent particles absorption will continue: if not, the plant will be unable to obtain a sufficient supply, and the effect on the plant will be seen by the drooping of the leaves.

Turgidity.—When a plant is dug up and transplanted it wilts, but after a time recovers. Why does it wilt? What changes take place that enable it to recover? We have seen how very small and tender the root-hairs are and how easily injured, and we have noticed also their great importance as absorbing organs. The movement of the soil in digging up and transplanting will obviously break off a large number of these hairs and so reduce the absorbing power of the root, hence wilting will occur. New root-hairs, however, are gradually formed, which absorb water and soon bring the plant back again to its fresh condition. The condition of living cells, whereby their elastic walls are stretched by the pressure of internal fluids, is called *turgidity.* The freshness of a shoot depends upon its turgidity; loss of turgidity is the cause of wilting.

Potato osmometer.—Another experiment will enable us to realize how the water taken up by the root-hairs may travel through the outer to the inner tissues of the root. Obtain two long and similar potatoes with uninjured skins. Boil one of these for fifteen minutes and allow it to cool. Then prepare both the boiled and the unboiled potatoes as follows (Fig. 24): Cut a slice from one end so that each potato will stand upright, and pare from this end a ring of skin to the height of three-quarters of an inch. Cut a slice from the other end of each potato, and bore a hole an inch in diameter through the middle nearly to the lower end. Fill half of this hole with sugar and add a little water to moisten it. Now stand each potato upright in a dish containing sufficient water to cover the peeled surface. Place the two dishes side by side and allow them to stand for two or three hours; then compare them. In the living (unboiled) potato the liquid rises steadily in the cavity and eventually runs over the margin. The dense sugar solution has withdrawn water from the cells lining the cavity, the sap of these cells thereby becoming concentrated. These cells now withdraw water from those farther outwards; and this is repeated until the cells of the pared surface outside, which draw water from the dish, are reached. Very different is the behaviour of the cells of the other potato, which have been killed by boiling. Compare the amount of water absorbed in the two cases and the difference in level of the liquid in the two cavities. In the living potato the liquid rises in the tube and eventually overflows, in the other exchange is very slow indeed. This will help us to realize the activity of living tissues in taking up and transmitting water, as compared with the action of a dead tissue.

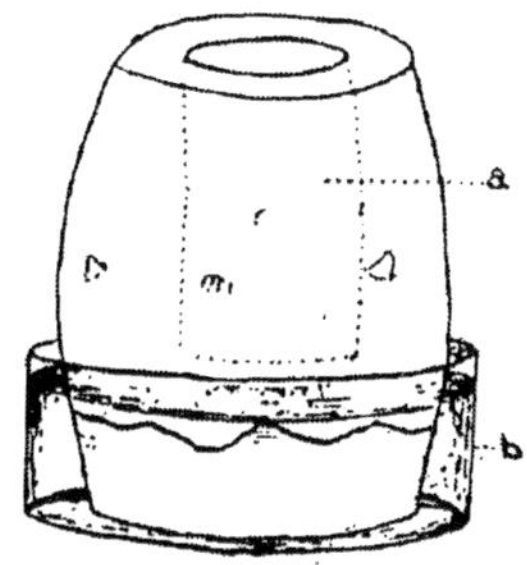

FIG. 24. POTATO OSMOMETER.—*a*, cavity in potato containing sugar; *b*, dish of water.

We are now in a position to understand the importance of root-hairs to a plant. The root-hair region, together with the younger part of the root which has not yet formed root-hairs, is the special organ for the absorption of water.

The root-hairs are often of considerable length, and their form and number very greatly affect the absorbing power of a plant, e. g. the root-hairs of Maize increase the surface of the root five and a half times; while those of Barley increase the surface twelve times. In some plants with very small leaves (e. g. Heaths) few or no root-hairs are formed; and in many water-plants they are absent or nearly so, for these plants are able to absorb water through the general surface of the epidermis, not only of the root but often of the shoot as well; but most plants depend for their supply of water and mineral food on the absorbing activity of the root-hair region.

The Food absorbed by a Root

Soil-organisms and their work.—Ordinarily the soil in which the roots of plants grow is a very complex mixture of substances—solid, liquid, and gaseous; inorganic and organic; living and dead; animal and vegetable. Myriads of tiny organisms find a home there, and also bigger ones such as earthworms and the larvae of many insects. These feed upon the living and dead materials in the soil, reducing them into simpler compounds, breaking the soil itself into a finely divided state, so that eventually a number of substances are brought into solution which are essential to the food of a green plant. The organisms, however, differ much in their usefulness in this respect; earthworms are great ploughers and pulverizers of the soil, and minute organisms, like bacteria, are valuable or even indispensable; but others prey upon them and, by reducing their numbers and therefore their usefulness, retard the formation of soluble food-materials, thus rendering more and more difficult the sustenance of plants. But these foes in turn are checked by others, and so this complex society of inter-dependent and ever-changing members is actively at work reducing the complex materials and preparing from them suitable food for succeeding generations. It is important, however, that a suitable balance should be maintained between the organisms in the soil, if higher plants are to thrive in it. From what we have seen, the food which an ordinary green plant can take up from the soil must be a weak solution of inorganic substances, e.g. compounds which form the mineral matter of the soil as distinct from organic substances which are carbon compounds built up by living organisms, e.g. cellulose, sugar, starch, and proteins.

Water-cultures.—If soil-water or ordinary tap-water be placed in a shallow vessel and covered by a sheet of paper so as to prevent access of dust or other matter, and allowed to evaporate, a sediment will be left at the bottom of the vessel, consisting of substances which were held in solution. These, when analysed, are found to consist of a number of mineral salts which, in suitable proportions, are able to sustain a green plant grown under the usual conditions of air and light. Such a sediment, however, may possibly contain substances not necessary to the plant.

Experiments have been made to determine which of these compounds are essential. The following solution contains the inorganic substances commonly present in a natural soil :—

Distilled water (H_2O)	1,000 c.c.
Potassium nitrate (KNO_3) . .	1·0 grm.
Sodium chloride ($NaCl$)	0·5 ,,
Calcium sulphate ($CaSO_4$) . . .	0·5 ,,
Magnesium sulphate ($MgSO_4$) . . .	0·5 ,,
Calcium phosphate ($Ca_3(PO_4)_2$) . . .	0·5 ,,
Ferric chloride ($FeCl_3$)	a trace.

Such a solution is known as a normal water-culture solution, and in it plants may be grown up to the flowering and fruiting stages (Fig. 25, *b*).

FIG. 25. WATER-CULTURES OF BUCKWHEAT (Pfeffer).
a, without potassium; *b*, normal solution; *c*, without iron; *d*, cover split to receive the plant stem; *g*, jar containing culture solution.

In order to show the importance of the various constituents, plants should be grown in the following incomplete culture solutions and the results compared: (1) Culture solution without potassium nitrate (Fig. 25, *a*), (2) without magnesium sulphate, (3) without calcium phosphate, and (4) without ferric chloride (Fig 25, *c*).

It will be interesting to carry out such experiments, but as the results are often very variable and contradictory, it is well to try several of each and express the average results by means of curves. The jars containing the food solutions should be covered with opaque paper in order to exclude light, and the solutions should be renewed at least once a fortnight and the vessels thoroughly cleaned and sterilized with boiling water, as Algae and other organisms are liable to develop in them. The plant may be supported by a split cork (*d*), as shown in the figure. Keep the cork and the part of the stem passing through it dry, otherwise it may be attacked by Fungi and decay.

Analyses of plants show that a number of substances are commonly present which water-culture experiments prove to be non-essential. Silica, which is present in large quantities in Grasses, Horsetails, &c., is one of these. Chlorine also is necessary to only a very few species. On the other hand, if potassium compounds or nitrates are omitted, the plant suffers. If iron is omitted, chlorophyll is not developed, and the leaves are sickly yellow in colour (Fig. 25, *c*). By means of water-culture experiments we learn that the food of a green plant must contain the following elements: oxygen, hydrogen, nitrogen, calcium, magnesium, potassium, phosphorus, sulphur, and iron; all of which a plant obtains in solution from the soil. One important element, carbon, forms about half the dry weight of a plant, yet is not present in a culture solution, nor is the plant able to obtain it from the soil. The question therefore arises—How does a plant obtain the carbon which forms so large a proportion of its substance? But this is a question we cannot answer until we come to the work of leaves.

From the above observations we learn that in many plants the radicle gives rise to a root which grows downwards as the tap-root, e.g. Bean, Pea, Oak. In others, e.g. Wheat and most Grasses, the radicle soon dies and is replaced by adventitious roots from the stem. Roots fix the plant in the soil and absorb weak solutions of mineral salts; the absorbing area being increased by branching and to a greater extent by root-hairs. Usually, roots contain no chlorophyll and bear no leaves, therefore the branches are not axillary. In a young root, the groups of wood and bast alternate with each other, and the first-formed wood develops towards the centre (i.e. centripetally). From cells of the pericycle and opposite the groups of primary wood, branch roots arise in vertical rows. The tip of each root is protected by a root-cap and possesses a sensory region which is able to perceive a stimulus and transmit an impulse to the neighbouring tissue, where growth occurs. The root is sensitive only in the presence of oxygen. The direction of growth is determined by the nature of the stimulus, i.e. towards the soil, water, and food; and away from obstacles and light. Some roots, swollen with a large cellular tissue, store much starch, sugar, inulin, and other reserves of food. Some of these we will now consider.

CHAPTER V

FORMS OF ROOTS

WE have seen that in Dicotyledons (e.g. Bean, Pea, and Stock) the radicle of the embryo grows downwards and becomes the primary root of the plant. From the tap-root, branches or secondary roots arise which in turn give off numerous branching fibres. In Wheat, Oat, Maize, roots are developed not as branches of the radicle, but from the stem. The radicle is usually short-lived, and the roots are all similar and slender, and known as *fibrous roots* (Fig. 26, 4). As they are developed from some part of the plant

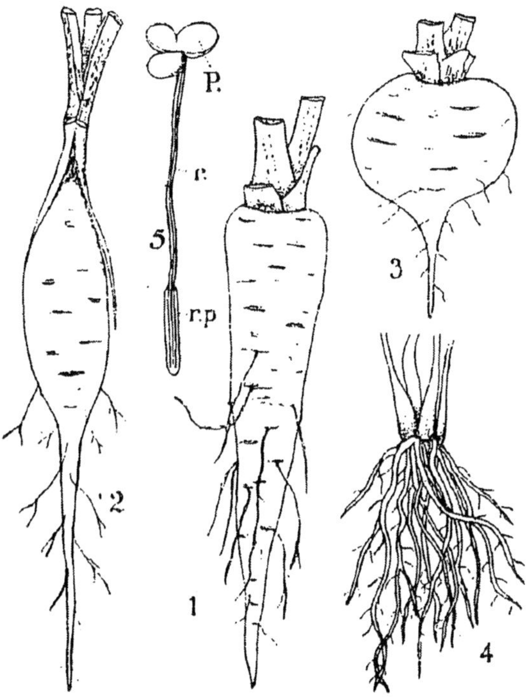

FIG. 26. ABNORMAL FORMS OF ROOTS.—1, conical root of Carrot; 2, spindle-shaped root of Radish; 3, globular root of Turnip; 4, fibrous roots of Grass; 5, plant of Duckweed; *P*, flattened leaf-like stem; *r*, root; *r.p*, root-pocket.

other than the radicle and its branches, they are known as *adventitious roots*, and in Monocotyledons they are the most common kind.

Storage, climbing, and aquatic roots.—In exceptional cases (Fig. 26, 1–3) roots thicken considerably. The tap-root becomes *conical* in the Carrot and Beet, and *spindle-shaped* and sometimes *globular* in the Radish. The Lesser

Celandine (Fig. 28) and Dahlia have swollen or *tuberous* roots. Such swollen roots serve as important food-stores for the plant, and some of them, if tested with iodine solution, will be found to contain much starch. Others, like Beet, contain cane sugar, and the Dahlia contains an allied substance called inulin. Frequently plants produce roots of more than one kind and serving different functions, (1) some being fibrous, absorbing roots; (2) others swollen and stored with food (e.g. Lesser Celandine and Dahlia). Some have roots which after greatly elongating, contract and pull the stem down deeper into the ground, e.g. Dandelion, Crocus (Fig. 84), Bluebell (Fig. 87). The adventitious roots of the Ivy, which arise in clusters on the aerial stems, serve rather as holdfasts and climbing organs than for the purpose of absorption. The roots of some water-plants like the Duckweed (*Lemna*) (Fig. 26, 5) and Frog-bit (*Hydrocharis*) dangle in the water, from which they absorb nutriment, and do not enter the soil. They are truly aquatic. Those of some tropical aquatic plants contain large air-spaces and serve as floats.

Aerial roots.—Roots, though rarely green, do sometimes develop the green colour characteristic of leaves, as in the roots of a few water-plants such as the Duckweed, and in the aerial roots of Orchids. Many tropical Orchids, growing perched on the trunks of trees, produce roots of three kinds: (1) holdfasts, which fix the plant like a bracket to the tree; (2) long aerial roots which hang down in, and absorb moisture from, the air; and (3) nutritive roots, which grow among, and absorb substances from, the humus that collects on the bracket of leaves.

Adventitious shoots: suckers.—One of the most constant characteristics of roots is that they give rise to members similar to themselves, viz. root-branches. Thus they differ from stems, which produce members unlike themselves, viz. leaves, which are usually green. It often happens, however, that roots give rise to leafy shoots. Familiar examples are Dock and Dandelion. If, in attempting to eradicate these from a lawn, we cut the plants so as to leave

FIG. 27. PLANT OF RASPBERRY, bearing a sucker (*s*) on its root (*r*).

part of the tap-root in the ground, leafy shoots eventually spring from the buried portion of the root. In the case of the Dandelion, often five or six shoots, each bearing a rosette of leaves, will appear in the place of the part we have removed. Such shoots are called *adventitious shoots*. Some shrubs and trees often produce adventitious shoots from horizontal root-branches, e.g. Raspberry (Fig. 27), Rose, Bramble, Hawthorn, Poplar, and Hazel. Shoots arising in this manner from roots are called *suckers*.

Roots vary greatly in their duration: they may be annual, a fresh crop being produced each season; biennial, living two years only; or perennial, living for many years. Even among perennials, some roots, such as those of bulbs and corms, often live only one season; the bulb of one season dying away and leaving an offshoot or bud to continue growth which forms roots of its own.

Tuberous roots of Lesser Celandine.—The Lesser Celandine has roots which show some specially interesting features. The plant flowers in the early spring. Sometimes it grows in open, sunny places and receives frequent visits from insects, but often in wet, shady hollows in woods, where insects are scarce and fewer seeds are set. The plants growing under such conditions should be carefully studied. We have already seen (p. 60) that this plant produces two kinds of roots (Fig. 28). From what part of the plant do the tubers spring? How do they grow? How is the compact cluster of tubers formed? Of what uses are the tubers to the plant?

Take a few seedlings (Fig. 28, 2, 3, 4) and note the coloured scale (*sc*) at the base and the small green foliage-leaf. In the axil of the scale a tuber (*t*) is formed, which bursts through it and grows parallel to the root (*r*). Later, a young shoot elongates, uses up the food-reserves in the tuber, and forms one or more leaves (*l*). In the axils of these leaves new tubers develop, and may often be found to elongate and turn sharply over the edge of the sheathing-base on their way to the soil (1, *t*). Examine them closely, and note that they are clothed with root-hairs, especially when growing in damp air. If a transverse section is examined, you will find the cortex to be very large, the cells crowded with starch grains, and the central cylinder, a small strand in the middle, to contain four groups of wood alternating with four groups of bast.

If a shoot, bearing axillary tubers, be placed in water for

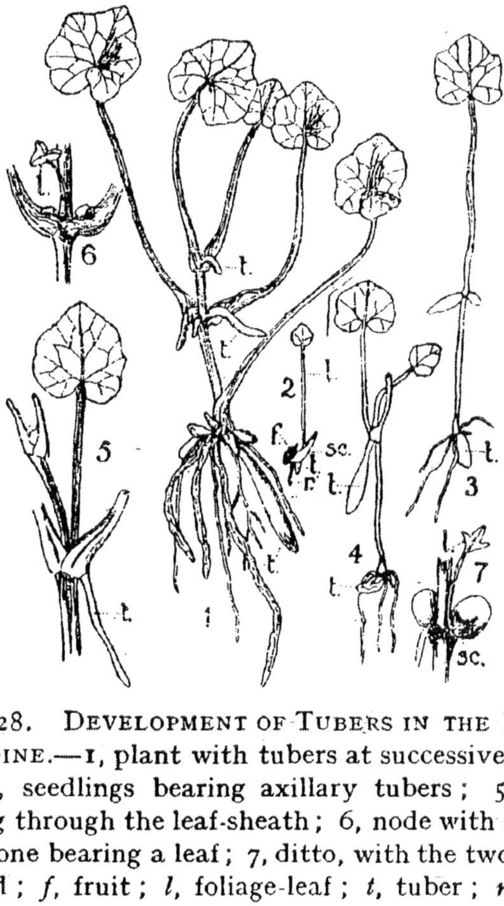

FIG. 28. DEVELOPMENT OF TUBERS IN THE LESSER CELANDINE.—1, plant with tubers at successive nodes; 2, 3, 4, seedlings bearing axillary tubers; 5, tuber bursting through the leaf-sheath; 6, node with axillary tubers, one bearing a leaf; 7, ditto, with the two leaves removed; *f*, fruit; *l*, foliage-leaf; *t*, tuber; *r*, root; *sc*, scale-leaf.

a few days, leaves may be seen to arise from the axils of scale-leaves at the bases of the tubers (Fig. 28, 6 and 7). Tubers, therefore, are compound structures: the lower leaf-bearing portion is a stem; and the outer or distal part is a root with a double function: (*a*) it absorbs by means of root-hairs, and (*b*) stores up starch in the bulky cortex.

Such axillary tubers occur commonly on the older plants; and, as the other parts decay, the tubers fall to the ground, and in time produce new plants from the buds which arise at their base. Two or three tubers may arise at a node, and when several leaves are produced close together a large cluster of tubers results.

Fig. 28, 1, shows tubers springing from three successive nodes. By the elongation of the tubers they may enter the soil and become independently rooted. The decay of the internodes at the end of the season will result in several independent plants. This mode of origin of roots at the nodes, and the formation of new plants by vegetative means, is of common occurrence.

CHAPTER VI

STRUCTURE OF THE SHOOT

Environment of the root and shoot.—The environment of the shoot is totally different from that of the root. In the soil the root is surrounded by a moist medium, and is in the dark. It is less exposed than the shoot to drying winds or heavy rain, to biting cold or the bright rays of the sun, to the heat of the day or the chills of the night. The conditions of life below ground are, on the whole, more uniform, and the parts are not exposed to such sudden and often extreme changes as are those parts growing above ground. For healthy existence, their form and structure must be adapted, not only to withstand, but to make the best use of, these conditions. We have, therefore, to regard the shoot of a plant from these different points of view.

As we have seen in the Stock, the stem is directly continuous with the root, and is the means by which leaves are spread out to the best advantage as regards light and air. It is obvious, too, that the stem is the means of communication between root and leaf. In order to bear the weight of leaves and branches, and to withstand the strain of heavy winds, it needs to be strong; and to resist the attacks of numerous enemies, its outer surface must be tough or otherwise resistant. To prevent the escape of sap, which passes along the stem, it must be impervious. Structures thus exposed to many and varied conditions, and having to serve so many purposes, are likely to show a wide range in duration and modification of form and structure.

Though the leaf and stem differ in many details from the root, they are built up of the same general tissues—an epidermis on the outside enclosing a cortex, and, within, a ring of vascular bundles surrounding the central pith. In a leaf, however, the blade of which is usually in the form of a thin plate, the veins spread out in the form of a flat meshwork.

Structure of the Box leaf.—A simple dissection will show the relationship of these different tissues of which the leaf is composed. Boil a few Box leaves in a solution of caustic potash for fifteen to twenty minutes, then wash gently in water, and place them on a glass slip. Dissect off carefully first the lower skin, then the upper skin, and mount them on separate slips (Fig. 29, 1 and 2). There now remains the skeleton, the meshes of which are covered by and filled in with a soft green tissue. With a camel-hair brush carefully remove this tissue, and so prepare a clean skeleton (Fig. 30). Examine all these parts carefully with a pocket lens. How do these skins differ? Is one more readily removed than the other? and if so, which? What structures do you find on the under skin which are absent from the upper one? Examine these with a microscope; and also the green tissue you have removed from the meshes of the veins. Each of the dots seen with a lens on the under skin consists of two sausage-shaped cells joined end to end, leaving a pore or mouth between them (Fig. 29, 1 and 4 *s*). These openings are called **stomata** (sing., stoma, Gr. *stoma* = the mouth), and their function is to communicate between the interior of the leaf and the air outside. On the upper skin (Fig. 29, 2), however, these are almost or entirely absent.

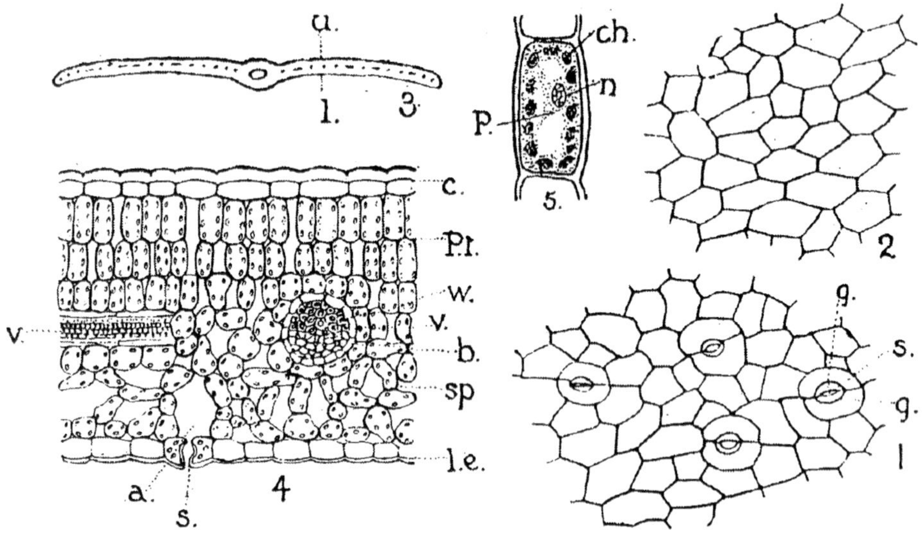

FIG. 29. STRUCTURE OF A BOX LEAF.—1, part of lower epidermis; 2, part of upper epidermis; 3, cross-section of Box leaf; 4, part of 3 at the point *u*, *l*, highly magnified; 5, a cell of the mesophyll; *a*, air-chamber; *b*, bast; *c*, cuticle of epidermis; *ch*, chloroplast; *g*, guard cells; *l*, lower surface; *l.e*, lower epidermis; *n*, nucleus; *P*, protoplasm; *P.t*, palisade tissue; *s*, stoma; *sp*, spongy tissue; *u*, upper surface; *v*, vein; *w*, wood.

A thin transverse section of the leaf should be examined with a lens or microscope and the details shown in Fig. 29, 3 and 4, identified. The green tissue between the two skins is seen to be arranged in two distinct layers; the upper one, called the **palisade tissue**, consists of perpendicular elongated cells, closely packed together and attached above to the upper skin (Fig. 29, 4 *p.t*). The cells below, known as the **spongy tissue**, are loosely arranged, leaving large spaces between them filled with air (Fig. 29, 4 *sp*). These cells contain the small rounded green bodies to which the colour of the leaf is due. The bodies are known as **chlorophyll corpuscles** or **chloroplasts**, and the green colouring matter as **chlorophyll** (Gr. *chloros* = green, *phyllon* = a leaf). It will be noticed that each stoma on the under surface always opens into one of these air-chambers (Fig. 29, *a*). Between the two layers runs the meshwork of veins (Fig. 30). We can understand from such a section why the lower skin should be more easily removed than the upper.

Structure of the stem.—Turning now to the tissues of a stem like the **Buttercup** (Fig. 31), **Dead-nettle** (Fig. 34), or **Bean**, we find these similar to and continuous in structure with those of the root and leaf; and they are also similar in function. A layer of cells—the epidermis (Fig. 31, *e*)—covers the outer surface, the exposed walls of which are thickened with a protective layer — the cuticle. Stomata occur here and there as in the epidermis of a leaf. Beneath the epidermis is the cortex (*co*), followed by a ring of veins, or vascular bundles (*v.b*). In the centre is the pith (*p*), which in older plants breaks down, making the stem hollow. Broad rays of tissue pass between the bundles from pith to cortex. These are the medullary rays. These tissues—epidermis, cortex, vascular bundles, rays, and pith—occur generally in plant stems, but become greatly modified according to the requirements of the plant.

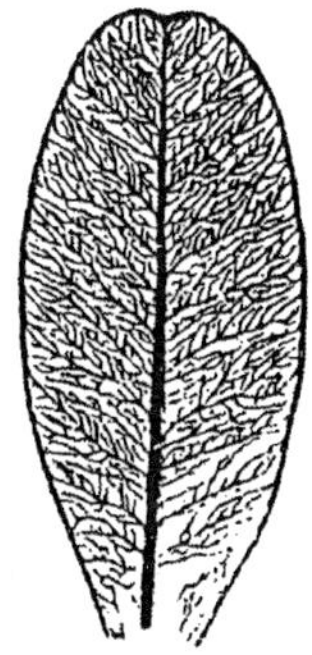

FIG. 30.
SKELETON OF
BOX LEAF.

The epidermis may develop a very thick cuticle in evergreens, and is usually waxy and more or less impervious. In water-plants the cuticle is often very thin or absent. In some plants, hairs are so abundant as to produce a woolly covering. Not uncommonly some hairs pour out a sticky secretion, e. g. Campion and Catchfly; while in other plants they secrete a poisonous acid and the hairs become formidable stings, e. g. Nettle.

The cortex and mechanical supporting tissues.—Beneath the epidermis is the cortex, and the cells of the outer part, especially in herbaceous stems, may be strongly thickened so as to form a firm supporting tube (Fig. 36 *sc*[1]). In the Deadnettle, the cortical cells of the angles of the stem are strongly thickened at their corners (Fig. 34), and these, together with the wood of the large bundles, are effective in protecting the stem against the stresses of tension and compression; in fact the whole stem is built on the principle of a box girder. A tissue consisting of cells which retain their living contents, and whose walls are thickened at the corners, is called **collenchyma** (Gr. *collen* = gelatinous matter, *chyma* = an infusion). A ring of collenchyma occurs in the outer cortex of the Sunflower stem (Fig. 37), and it is frequently found in leaf-stalks. Bands of cells in the cortex of the Bracken and Birthwort are uniformly thickened and devoid of living contents. Those of the Birthwort (Fig. 36) form a strengthening ring immediately beneath the epidermis (*sc*[1]), and a second one (*sc*[2]) as a strengthening cylinder midway between the epidermis and the bast. Such a mechanical supporting tissue is called **sclerenchyma** (Gr. *skleros* = hard).

The ring of wood in the Elder (Fig. 38, *w*) forms a supporting mechanism on the principle of a hollow pillar, while in older woody stems the arrangement is that of a solid pillar. Woody tissues thus perform a double function —conduction of crude sap and mechanical support. Many ingenious devices for mechanical support may be found in the stems of plants, and several common species should be compared in this respect.

Structure and arrangement of vascular bundles.—A comparison of Figs. 31 to 34 and 36 to 39, which are transverse

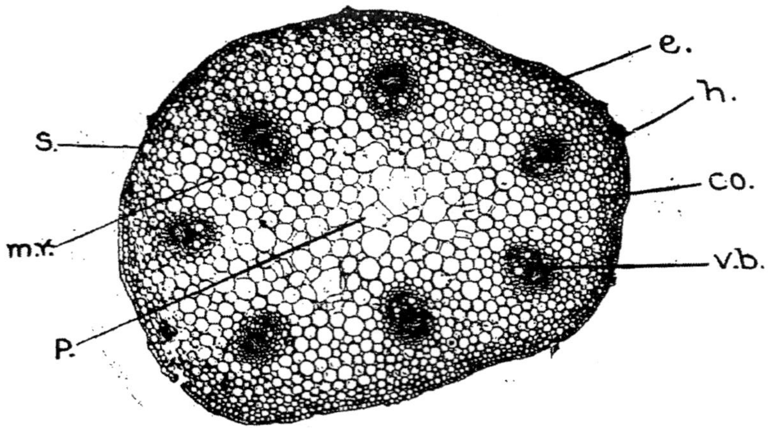

FIG. 31. TRANSVERSE SECTION OF THE RUNNER OF A BUTTERCUP.—*co*, cortex; *e*, epidermis; *h*, hair; *m.r*, medullary ray; P, pith; *s*, stoma; *v.b*, vascular bundle.

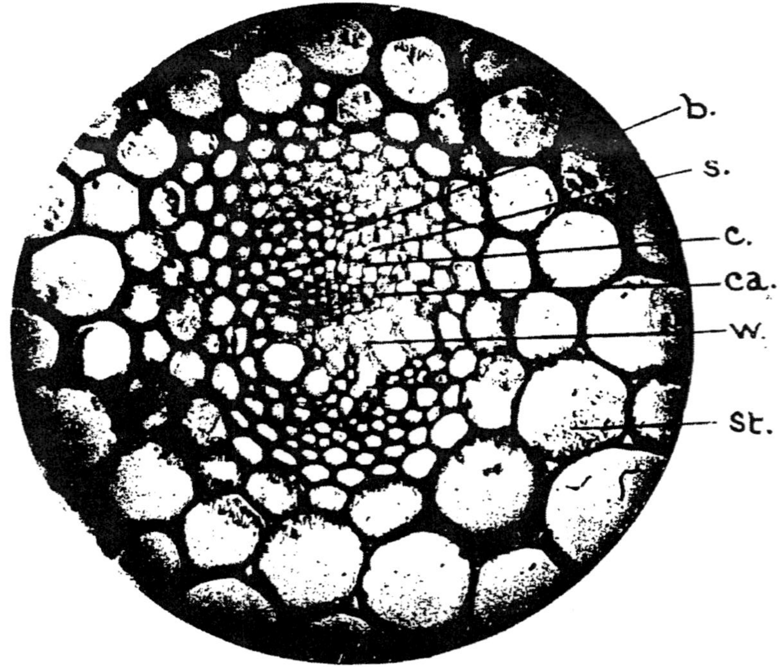

FIG. 32. A VASCULAR BUNDLE OF THE BUTTERCUP, HIGHLY MAGNIFIED.—*b*, bast; *c*, companion cell; *ca*, cambium; *s*, sieve tube; *st*, starch grains in cell of cortex; *w*, wood.

sections through the internodes of several dicotyledonous stems, shows that the vascular bundles are arranged in the form of a ring. Fig. 32 shows one of the vascular bundles of the Buttercup, highly magnified. Note the three distinct groups of tissue of which it is composed. On the outside is a group of delicate cells, the bast or **phloem** (*b*); the larger elements—**sieve tubes** (*s*)—are accompanied by very small ones, called **companion cells** (*c*); both are filled with organic materials. On the inner side of the bast is a band of narrow, flattened cells—the **cambium** (*ca*). In that portion adjoining the pith is the wood or **xylem** (*w*), composed of wide, tubular, thick-walled vessels, among which are narrower, thick-walled woody fibres. The veins, as in the root, are arranged in the form of a network and do not continue the parallel course which a section through the internode might suggest. Numerous examples of this may be obtained from waste heaps where shoots such as old cabbage-stalks are undergoing decay.

With a little trouble, the above details may be made out by carefully dissecting the stem of a Deadnettle which has previously been boiled in water for about twenty minutes. Tougher stems may be boiled in water to which a little caustic potash has been added. This softens the cortical tissues so that they may be brushed away from the veins, as was done in preparing the skeleton of the Box leaf. If a piece of stem including two or three pairs of leaves be selected, it will be seen that the veins, passing from the leaves down the leaf-stalks, enter the stem, branch at the nodes, and join on to neighbouring veins. This will be clear from a study of Fig. 33. By means of these veins direct communication is set up between roots, stem, and leaves.

Fig. 34 shows the arrangement of these tissues in a transverse section of the Deadnettle stem. At the corners are the large fused bundles from the leaves (*v.b*), while at the sides are the small bundles. In this stem the innermost layer of the cortex stands out clearly as a layer of larger cells—the **endodermis** (*en*). Examine the mature stem of the Bean and compare it with that of the Deadnettle. Although these stems are square, the general arrangement of the tissues is the same as that which is found in the Buttercup.

Scattered bundles of Monocotyledons.—The stems of Monocotyledons differ in several important respects from those just described; the vascular bundles are scattered in the ground tissue (Fig. 35), and there is no cambium between the wood and bast, so that when once these bundles are formed, no further increase in thickness is possible in them. They are therefore called 'closed' bundles to distinguish them from the 'open' bundles of Dicotyledons which possess a cambium.

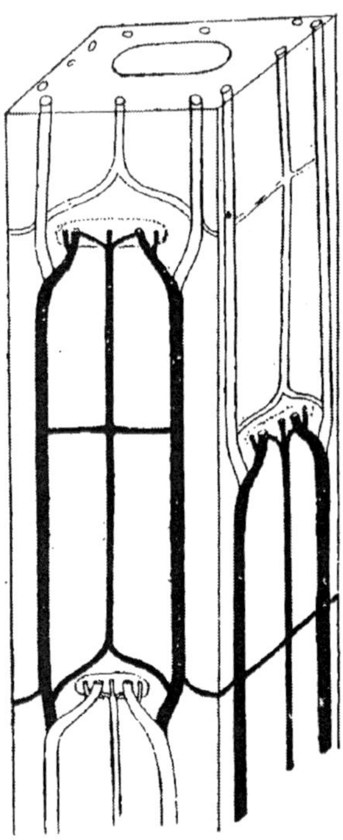

FIG. 33. DIAGRAM SHOWING THE ARRANGEMENT OF VEINS IN A DEADNETTLE STEM (after Farmer).

Woody stems. Secondary growth.—The cambium of Dicotyledons, though a small and inconspicuous tissue, is a very important one, inasmuch as its cells are able to divide repeatedly and form new tissue. That formed on its outer side becomes part of the bast, while that formed on its inner side adds to, and increases the thickness of, the wood. In a woody stem which lives for a number of years, cambium is formed across each medullary ray, so as to form a **cambium ring** (Fig. 36, *i.f.c*). This later-formed cambium, called interfascicular cambium, produces small new bundles between the original large ones, as shown in Fig. 37. Growth

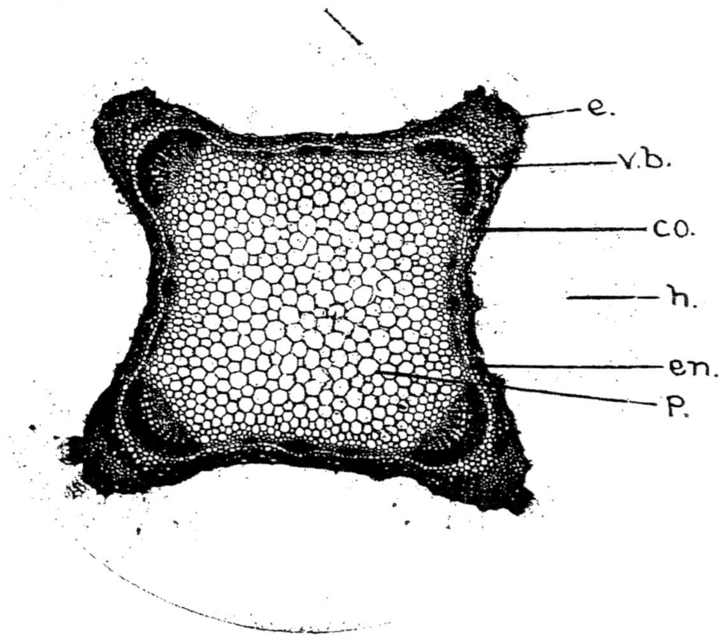

FIG. 34. TRANSVERSE SECTION THROUGH THE INTERNODE OF THE DEADNETTLE STEM.—*co*, cortex ; *e*, epidermis ; *en*, endodermis ; *h*, hair ; P, pith ; *v.b*, vascular bundle.

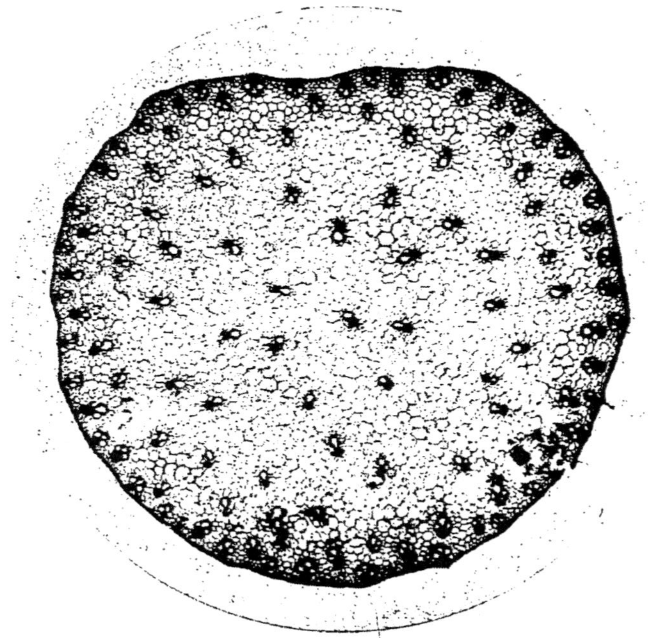

FIG. 35. TRANSVERSE SECTION OF STEM OF MAIZE, showing numerous scattered Vascular Bundles.

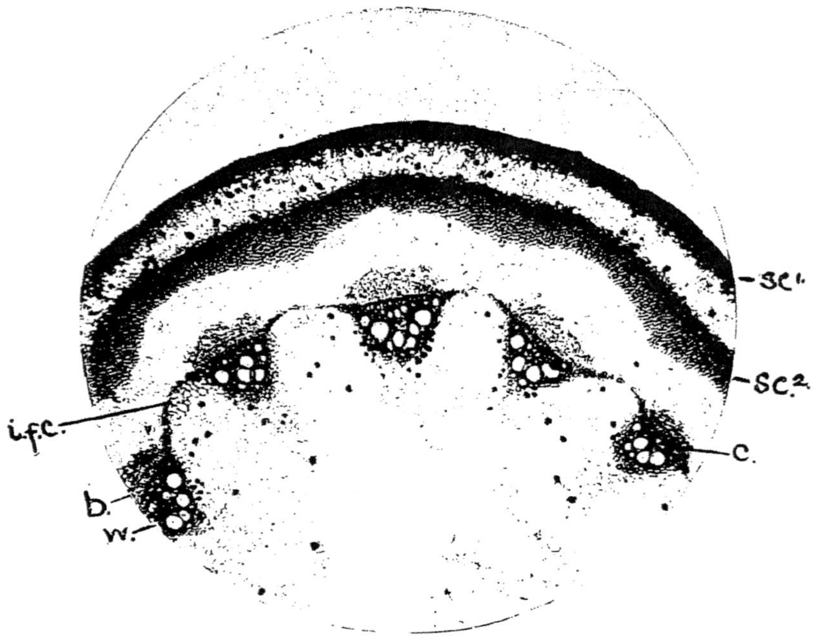

Fig. 36. Part of Transverse Section of Birthwort.—*b*, bast; *c*, bundle cambium; *i.f.c*, cambium between the bundles; sc^1, outer ring; sc^2, inner ring of mechanical tissue; *w*, wood.

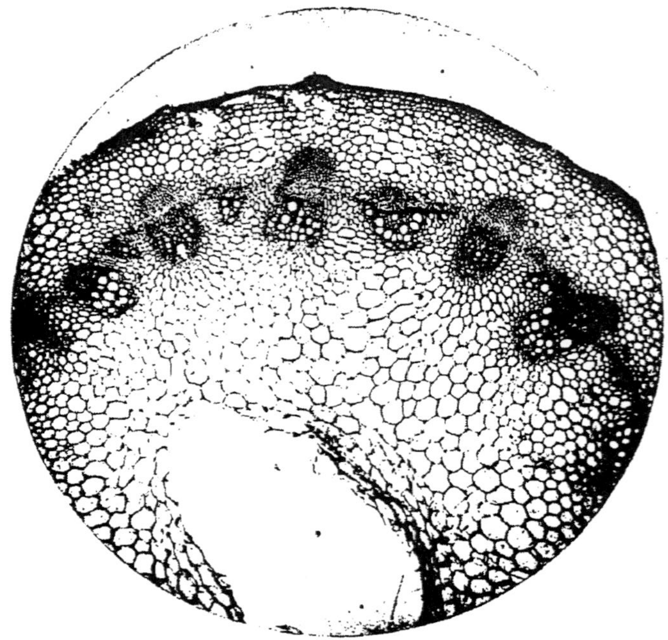

Fig. 37. Part of Transverse Section of Stem of Sunflower, showing small secondary bundles formed by the cambium between the large primary bundles.

continues, bundles are introduced, and ultimately a compact ring of wood is formed with thin medullary rays between the bundles (Fig. 38).

The formation of new tissue and especially of woody tissue goes on actively in the spring and summer ; less is formed in the autumn ; and little or none in the winter. On the return of spring, the process is repeated ; and as the wood formed in spring consists of elements with much larger cavities than those formed in the late summer and autumn, the successive zones stand in strong contrast with each other, and may be clearly seen in a transverse section. These **annual rings** (Fig. 39, *a*) are often irregular, and sometimes more than one ring may be formed in a season, but their number enables us to obtain a fairly accurate idea of the age of a tree. By such increase some stems may grow to a great age and size, and, unlike animals, they may add to their body-substance year by year.

These different tissues may be determined by dissecting a piece of **Elder** stem. Outside is the thin dying epidermis with a layer of cork below ; next the green cortex, followed by the slimy tissues of bast and cambium. The firm wood is easily discovered surrounding the central pith. In some stems, e. g. the Laburnum, the wood in the centre becomes dense and dark-coloured and is known as 'heart-wood'; while the newer, outer wood is soft and light in colour, and is known as 'sap-wood'. The heart-wood usually serves chiefly for the storage of water, the main ascending current passing along the sap-wood, whence its name.

Cork

Cork and lenticels.—During the early stages in the development of the woody tissue of Dicotyledons, the outer cells of the cortex and epidermis keep pace with it, but eventually they lose their power of increasing and tend to give way under the internal strain. Meanwhile, provision is being made for a new protective covering. There arises in the cortex another kind of cambium, known as the cork-cambium, because its cells by repeated division form, not wood and bast, but cork. Fig. 38 shows this cork-cambium (*c.c*) arising just below the epidermis (*e*) in the stem of the Elder, and its cells have divided in such a way as to form somewhat regular rows of cells. These enlarge, lose their living contents, and their walls become transformed into cork (*ck*). The epidermis gives way under this extra strain, producing the cracks which may be easily seen on the surface of a twig. Thus the stem becomes covered by a layer of cork, which is a dead impervious layer, well adapted as a protective coat. As the stem thickens from year to year, the outer cork layers split, new layers are formed beneath and the bark thickens, and in time takes on the ruggedness characteristic of the species. Often an irregular group of cork cells is formed beneath a stoma, the cells being so arranged as to permit air to enter or leave the stem. These structures, which take the place of stomata, are called **lenticels** (Fig. 42, *l*) and are formed on most trees, their shape being peculiar to the species.

Other stems should be examined and compared. In the Laburnum the cork-cambium is formed in the middle of the cortex, while in the Black Currant it arises deeper still, near the ring of vascular bundles (Fig. 40 *c.c.*). As the cork develops, the cortex to the outside of it dies and is eventually thrown off.

Callus and separation-layer.—The formation of cork is useful to plants in many ways, and especially as a means of healing wounds. Examine the trees in a wood, look for examples of pruning, and note the change taking place around the cut surface of a branch. You will be able to find all stages of cork-formation from a narrow ring just outside the wood to others broader and broader, gradually encroaching and growing over the surface of the wound

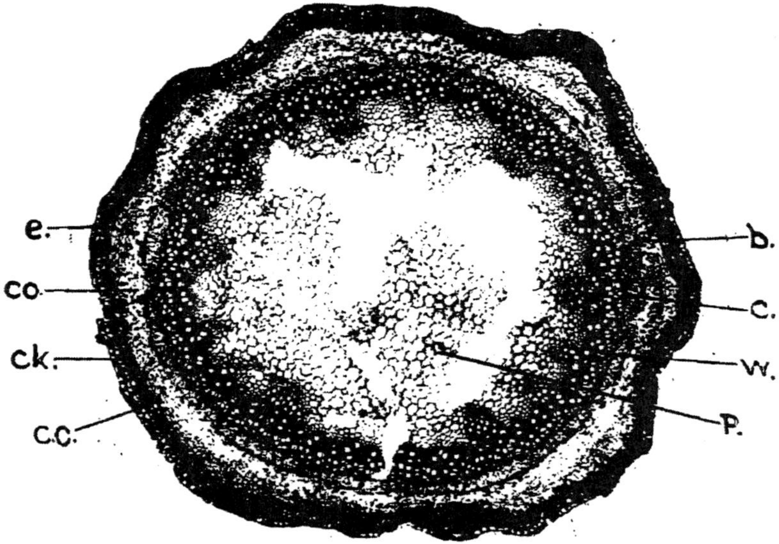

Fig. 38. Transverse Section of Stem of Elder.—*b*. bast; *c*, cambium; *c. c*, cork-cambium; *ck*, cork; *co*, cortex; *e*, epidermis; p, pith; *w*, wood.

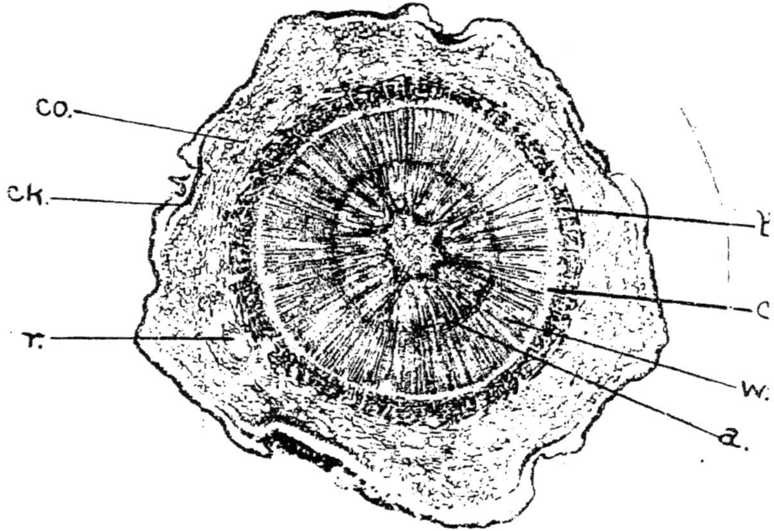

Fig. 39. Transverse Section of Stem of Pine.—*a*, annual ring, *b*, bast; *c*, cambium; *ck*, cork; *co*, cortex; *r*, resin canal; *w*, wood.

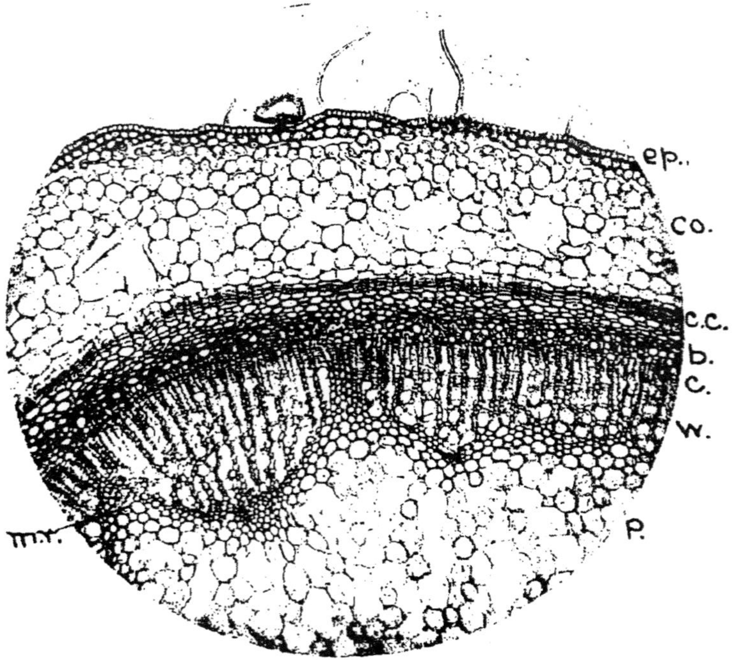

FIG. 40. PART OF TRANSVERSE SECTION OF STEM OF BLACK CURRANT.—*b*, bast; *c*, cambium; *c.c*, cork-cambium; *co*, cortex; *ep*, epidermis; *m.r*, medullary ray; P, pith; *w*, wood.

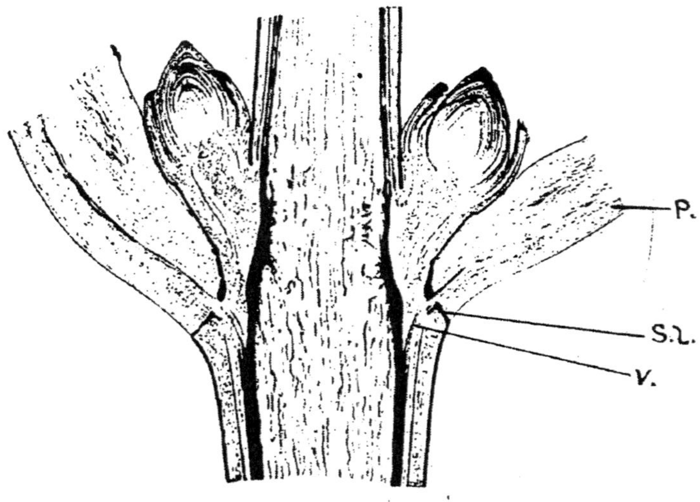

FIG. 41. LONGITUDINAL SECTION THROUGH NODE OF SYCAMORE TWIG.—P, leaf-stalk; *s.l*, separation-layer; *v*, vein from leaf to stem.

and eventually closing it up. Such a healing tissue of cork is called **callus.**

An interesting case occurs in leaves. Examine shoots of Privet, Ash, or other common shrubs or trees in the summer, and look carefully at the leaf-bases; bend the leaf backwards, and note where it tends to break. Here a distinct ring is clearly seen (Fig. 42, *s. l*), and as the leaves grow older, they break off along this line very easily. Observe what happens later in the season and examine shoots just before the leaves begin to fall. By means of a lens a healing scar of cork may easily be seen stretching like a plate across the leaf-base. Note how easily the leaf breaks off here, and also that the leaf has lost its freshness, and is often torn and withered, having clearly served its purpose for the plant which bore it. In compound leaves like the Common Ash and Horse-Chestnut a separation-layer is formed, not only across the leaf-base, but across the bases of the leaflets as well.

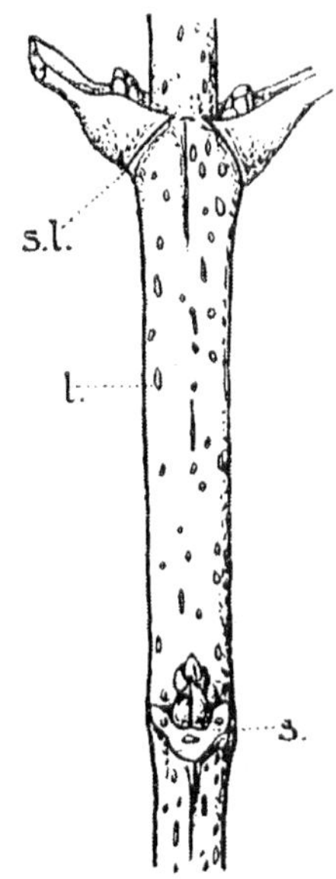

FIG. 42. PART OF SYCAMORE TWIG.—*l*, lenticel; *s*, leaf-scar; *s.l*, separation-layer.

Fig. 41 is a longitudinal section of a **Sycamore** twig through a node, and at *s.l* it is seen that the separation-layer (or absciss layer) is already formed before the leaf falls. Notice that cork has not formed across the vein (*v*). This is kept open to the last, for along it much of the nutrient material is passed backwards into the stem before the leaf is finally snapped off by the wind.

The tissues we have considered not only form the structure of the shoot, but it is by their means that the work of the shoot is carried on. It will be of interest to determine something of this work. How and under what conditions is it performed ? Do changes in any of the conditions affect the behaviour and growth of the shoot ? Are the differences we find in the form and structure of plants growing in different habitats correlated in any way with the differences in their environment ? These are problems we must now endeavour to solve.

CHAPTER VII

WORK OF THE SHOOT

We have seen in the experiments with germinating seeds that the main shoot grows upwards towards the light, in the direction opposite to that of the main root, the stimuli producing this result being gravity and centrifugal force. We must now extend our knowledge by making one or two further observations with older plants.

Perception of and response to stimuli.—Lay on its side a plant of **Geranium** (*Pelargonium*) or **Balsam** in a horizontal position, as in Fig. 43. Note at intervals the behaviour of both stem and leaves (*a–e*) ; the stem-tip turns upwards and continues to grow vertically. Take another actively growing plant, lay it on its side for an hour, then place it in a normal upright position and observe its mode of growth. Does the tip show signs of turning while lying horizontally ? Does any subsequent bending take place ? We see that the stem did not bend during the short time it lay in a horizontal position, but bending does occur later, even after the plant is placed upright. This experiment shows that a stimulus was received, though response was not immediate ; and it is clear that the stimulus persists for a time, as bending occurs even in the altered position of the plant.

FIG. 43. SHOOT OF BALSAM PLACED HORIZONTALLY, SHOWING SUCCESSIVE PHASES OF CURVATURE (*a* to *e*) (Pfeffer).

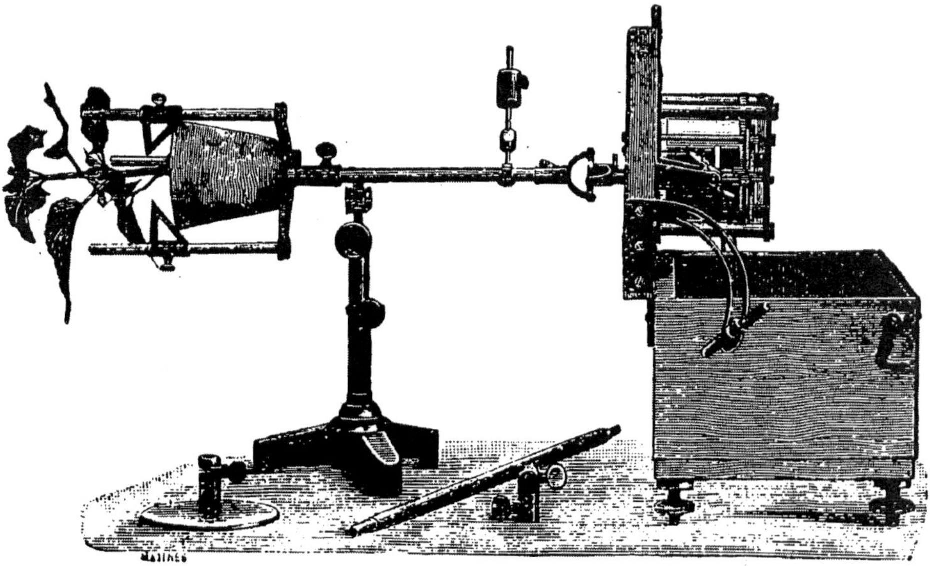

FIG. 44. PLANT PLACED IN A HORIZONTAL POSITION AND REVOLVED SLOWLY BY MEANS OF A KLINOSTAT (Jost).

A few trials will enable you to determine the time required for the stem to perceive the stimulus, and also that required for curving. If a plant is fixed in a klinostat (Fig. 44), and caused to revolve in a horizontal position, the stem, like the root, does not curve, since it receives the stimulus equally on all sides.

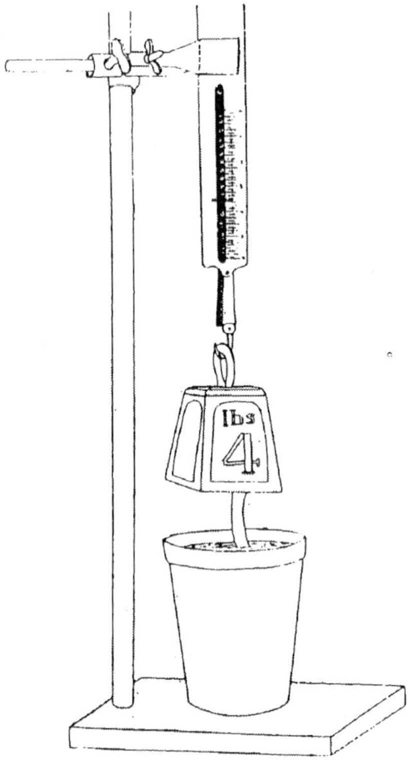

Fig. 45. Experiment to determine the Lifting Power of a Shoot.

Force exerted by a growing stem.—Just as roots exert much force in their downward growth in response to the stimuli of gravity and centrifugal force, so in the reverse direction do shoots, and their lifting power is considerable. Obtain a spring-balance, attach a weight, and arrange it as in Fig. 45, over the stem of a Bean seedling so that the weight is raised as the stem elongates. Fix the weight firmly to the hook so that it is not readily tilted, and determine the lifting power of the shoot. In the experiment illustrated, the shoot in three days supported a weight of seven ounces.

Stimulus of light. Heliotropism.—Consider next the heliotropism of the shoot.

Place a plant in a window so that light falls on it on one side only, and note the behaviour of both stem and leaves (Fig. 46). In what direction has the stem turned? Which is the longer side of the stem, the one facing the light or the one which has grown in the shade? Has the light hastened

FIG. 46. SEEDLINGS TURNING TOWARDS THE LIGHT.

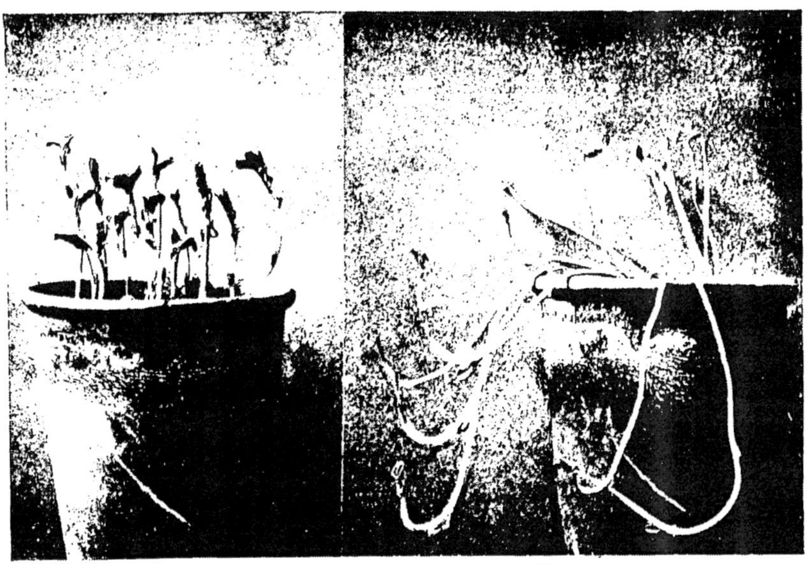

1 2

FIG. 47. BEAN SEEDLINGS.– 1, grown in the light; 2, grown in the dark.

or retarded growth ? These experiments show that shoots are sensitive to light, and that their growth is retarded by it. Thus shoots turn to the light because they grow more quickly on the shaded than on the illuminated side. Such movements of plants due to the stimulus of light are called heliotropic movements. And since shoots turn towards the light they are positively heliotropic. Note that the leaves also turn their blades towards the light. The leaves of some plants, however, grow erect and expose their tips and edges to the light, e.g. the Iris, Daffodil, and Hyacinth ; this habit is common in plants growing in very sunny situations.

Rates of growth in light and in darkness.—Take two pots of seedling Beans, allow one to grow for two or three weeks under ordinary conditions of light, and the other for a similar time in the dark (Fig. 47, 1 and 2). Compare them as to colour, length of internodes, and size of leaf. Measure them day by day, compare their behaviour and plot the results in curves on squared paper as in Fig. 48.

Etiolation. Conditions necessary for the formation of chlorophyll.—From these observations we learn that shoots which develop in the dark are yellowish-white in colour, greatly elongated and tender, and that the leaves are much smaller than normal. These changes brought about by growth in darkness are known as **etiolation** (Fr. *étioler* = to blanch). Shoots grown under ordinary conditions of light develop the green pigment chlorophyll, and have tougher tissues and larger leaves. Normally, chlorophyll is not developed in plants from which light is excluded, but cases of the contrary are not rare, e.g. Pine seedlings develop chlorophyll in the dark, and the embryo of the Sycamore is green while still enclosed within the thick and opaque fruit-coat.

Not only is light usually necessary for the formation of chlorophyll, but we learned from water-culture experiments that iron is essential. The shoots grown in solutions from which iron is excluded are a sickly yellow colour (Fig. 25, *c*). Further, a supply of oxygen and a suitable temperature are also requisite. If you examine the young shoots of plants as they emerge from the soil in the cold early spring, you will find they are often very pale, and contain little chlorophyll.

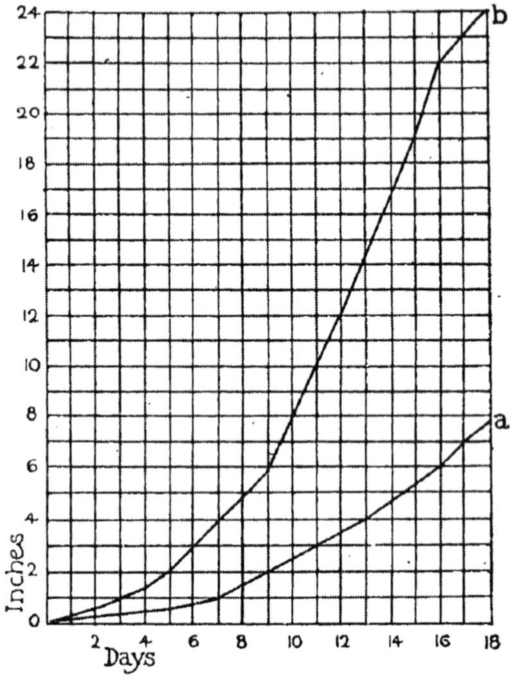

FIG. 48. GROWTH CURVE OF BEAN SEEDLINGS.
a, in the light; *b*, in the dark.

The conditions necessary for the formation of chlorophyll are light, warmth, and a supply of oxygen and iron.

We are now in a position to realize some of the most general points of difference between roots and shoots. These are, that roots grow downwards into the soil, avoid light, are not green, and do not produce leaves, while shoots grow upwards, bear thin flat leaves, and expose a large green surface to air and sunlight. We will now endeavour to find out the importance of these peculiarities of a shoot.

Carbon-assimilation

Carbon as a plant food. Photosynthesis.—Take a series of jars (1–5) similar to that shown in Fig. 49. In (1) place a little water and charge the bottle with sufficient carbon dioxide to extinguish a taper. Put in this one or two green leaves and close securely; then expose to bright sunlight. Prepare the other bottles in the same way, but in (2) place leaves that have been killed by boiling; in (3)

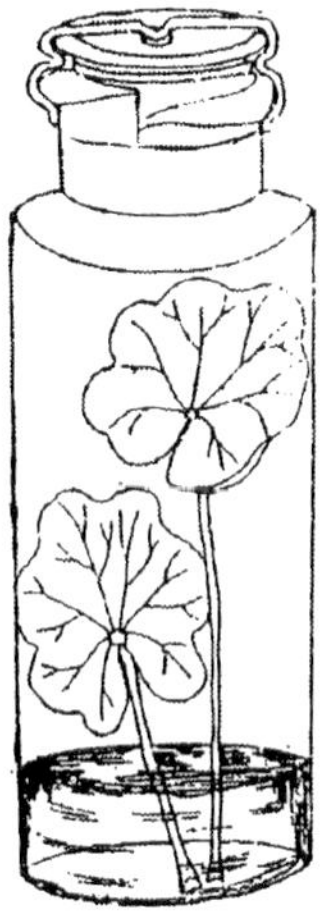

FIG. 49. EXPERIMENT TO SHOW THAT GREEN LEAVES EXPOSED TO SUNLIGHT ABSORB CARBON DIOXIDE.

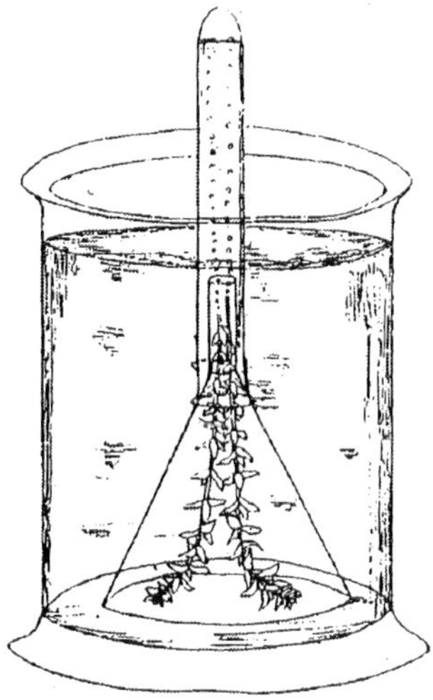

FIG. 50. EXPERIMENT TO SHOW THAT SHOOTS OF WATER-WEED EXPOSED TO SUNLIGHT GIVE OFF OXYGEN.

living non-green roots; in (4) carbon dioxide only, as a control, and in (5) the same as in (1), but let this be kept in the dark while the others are exposed to sunlight.

Leave these for a day or two, then test each (*a*) with a lighted taper for oxygen and (*b*) with lime-water for carbon dioxide. Under what conditions has carbon dioxide disappeared and oxygen taken its place?

From this experiment we learn that green leaves under the influence of sunlight are able to take up carbon dioxide and give off oxygen, whereas non-green, living parts or green leaves kept in darkness are unable to bring about these changes. This absorption of carbon dioxide by green plants under the influence of sunlight is called **photosynthesis** (Gr. *phos, photos* = light, *synthesis* = a putting together), or **carbon-assimilation**.

That plants do give off oxygen under such conditions as above described may be shown by an experiment arranged as in Fig. 50. Take a jar or beaker filled with tap-water and saturate the water with carbon dioxide. Place in the water a branch or two of **Canadian Waterweed** (*Elodea canadensis*), with the cut ends uppermost, and over these a funnel with a shortened stem. Completely fill a test tube with water and invert it over the stem of the funnel; then expose to sunlight. Set up a similar experiment, but instead of charging the water with carbon dioxide, drive off all the dissolved air by first boiling the water, and then allowing it to cool, or—simpler still—add lime-water to it so as to remove all the free carbon dioxide. Place the bottles side by side and compare.

In the second case it will be found that few or no bubbles are given off, but in the former, bubbles are given off freely and displace the water in the inverted test tube. When sufficient gas has accumulated, test it with alkaline pyrogallic acid and note how suddenly the brown colour is produced, or insert a glowing match which at once bursts into flame; these are proofs that the gas given off is oxygen. We thus see that such a plant when growing in water containing carbon dioxide, and exposed to sunlight, gives off oxygen.

Formation of starch in green leaves. Starch prints.—We have seen in a former chapter that starch is commonly present in the tissues of roots and stems. We have now to consider how this starch makes its appearance and under what conditions it is formed. For this purpose, plants of the cultivated Geranium serve very well. Keep a plant in the dark for a day, then partially cover two or three of the leaves while still on the plant with black paper, tin-foil, cardboard, or pieces of cork, as shown in Fig. 51, 1, in such a way that light is excluded from one part of the leaf while the other may be illuminated. Place the plant so prepared in sunlight for several hours or, if more convenient, on two or three successive days. At the end

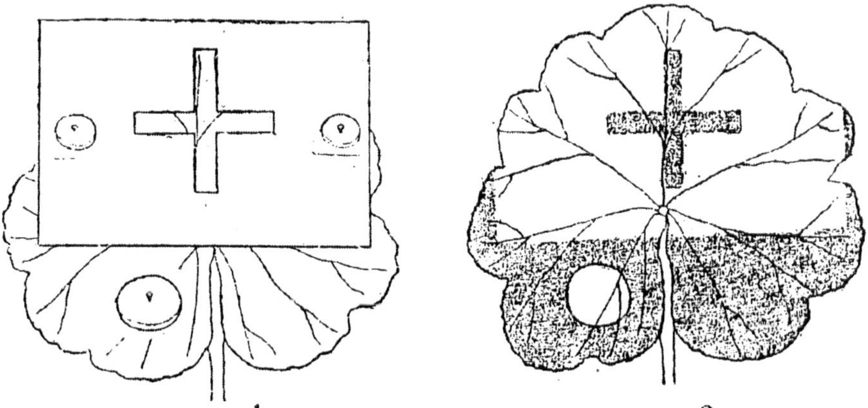

1 2

FIG. 51. EXPERIMENT TO SHOW THAT STARCH IS FORMED IN GREEN LEAVES WHEN EXPOSED TO LIGHT.—1, leaf covered with opaque paper, from which a cross has been cut; below is a cork pinned to the leaf to exclude the light; 2, the same leaf when tested later with iodine solution; the shaded areas contain starch.

of a period of several hours' illumination make the following test. Have ready some boiling water; remove the covered leaves, noting that they are still green, and plunge them at once into the water to kill them. Next place the leaves in alcohol, by which means the green colouring matter is gradually extracted. (This may be hastened by boiling in alcohol, but it must be done very carefully to prevent ignition of the alcohol vapour.) The resulting colourless leaves should be washed in water, then laid out in a shallow dish and covered with iodine solution.

Watch the effect of the iodine, and note that the parts which have been kept in the dark remain colourless or are merely stained yellowish brown, while those which were exposed to light take on a blue-black colour (Fig. 51, 2), indicating the presence of starch, which appears only in those parts to which light had access.[1]

Conditions necessary for the formation of starch.—A similar test should be made with leaves of a variegated Geranium, one with white patches on its leaves being selected. The leaves need not be covered, but after exposing the plant to sunlight as in the previous experiment, test for starch by the same methods. It will be seen that starch is formed only in the parts that are green.

With a Geranium plant that has been kept for a day in the dark arrange an experiment as shown in Fig. 52. Place a little caustic potash solution in the bottom of a bottle, tilt it as in the figure and turn into it a leaf of the plant, taking care that the leaf does not touch the liquid. Close the bottle with a split cork perforated to admit the petiole without injuring it, carefully seal the cork with vaseline, and then expose the whole to sunlight as before.

Now consider the following points: What effect will the caustic potash have on the air in the bottle? What changes take place in the composition of the air by the action of a green leaf? Test the leaf as above and determine whether starch has been formed under the conditions of this experiment? We have seen that plants absorb oxygen from the air and give off carbon dioxide; this is the process known as respiration or breathing; also that a green leaf exposed to air containing carbon dioxide

[1] If it is necessary to carry out these experiments during very dull weather, satisfactory results can be obtained by exposing the plants to artificial light, care being taken not to injure the plant by heat.

is able, under the influence of sunlight, to absorb carbon dioxide and give off oxygen, and further, that the light rays absorbed are converted into forms of energy capable

FIG. 52. EXPERIMENT TO SHOW THAT A LEAF OF PELARGONIUM EXPOSED TO LIGHT IN AIR DEVOID OF CARBON DIOXIDE DOES NOT FORM STARCH.

of bringing about chemical changes resulting in the formation of starch. But starch is found only in those parts of leaves which contain the green pigment chlorophyll, and the experiment just performed proves to us another important fact, namely, that a green leaf, working under normal conditions, but in air devoid of carbon dioxide, is unable to form starch. In addition to the above conditions it is found that the work of a leaf can only proceed at a suitable temperature and when the plant is able to obtain a sufficient supply of water. Thus moisture, warmth, sunlight, carbon dioxide, and chlorophyll are all necessary for the formation of starch in a leaf.

If we think over the previous experiments we meet with an apparent contradiction. We have just learnt that starch is formed in green leaves only under the influence of sunlight, and not in parts that do not contain chlorophyll, yet we found an abundance of starch in the cotyledons of the Bean and in the endosperm of the Wheat and other grains ; and it also occurs, as we shall see, in the non-green parts of stems and roots. It is obvious, therefore, that starch arises in more ways than one within the tissues of a plant.

Now the starch grains formed in green leaves are formed entirely from inorganic substances ; the exact method is not known, but a possible explanation is the following :

$$\underset{\text{(carbon dioxide)}}{CO_2} + \underset{\text{(water)}}{H_2O} = \underset{\text{(formaldehyde)}}{HCOH} + \underset{\text{(oxygen)}}{O_2}$$

$$\underset{\text{(formaldehyde)}}{6\,HCOH} = \underset{\text{(sugar)}}{C_6H_{12}O_6}$$

$$\underset{\text{(sugar)}}{n(C_6H_{12}O_6)} - \underset{\text{(water)}}{n\,H_2O} = \underset{\text{(starch)}}{(C_6H_{10}O_5)n}$$

In words, the carbon dioxide and water within the living chlorophyll-containing cells of the leaf, and under the conditions already enumerated, may be split up and their constituent atoms rearranged to form a compound called formaldehyde and also the element oxygen. This will account for the carbon dioxide taken in and the oxygen given out during photosynthesis. Six molecules of formaldehyde are now supposed to combine to form sugar. By further action the sugar is deprived of a molecule of water and is converted into starch. Another possible explanation is that during sunlight chlorophyll is continually breaking down and re-forming, and formaldehyde may be one of the products formed as a result of the decomposition of the chlorophyll. This building up of starch grains is intimately associated with the chlorophyll corpuscles, but when formed, they become detached and lie in the cavity of the cell. These grains, however, are solid, and, as we have learnt from our experiments in osmosis, are quite incapable of being transferred from cell to cell in this form. Obviously, if the grains are not removed, the cell will soon reach its limit of activity in this direction.

Starch digestion.—The following experiments will help us to understand how the transference of food-materials is brought about. Place a little potato starch in a beaker, add water, and note that the starch grains are not dissolved. On boiling the liquid for a few minutes, a mucilage is produced, but the starch does not completely dissolve. Dilute this with cold water. Place a little of the cooled liquid in a test tube, add saliva from the mouth, mix thoroughly, and keep it at the temperature of the body for fifteen or twenty minutes. This may be conveniently done by enclosing the tube firmly in the left hand while the following experiments are carried out. To another portion of the weak mucilage add a little diastase or malt-extract (which contains diastase), and allow it to stand at the temperature of the room for twenty minutes or more. In the meantime take another test tube and place in it a little grape-sugar, add water to it, and note how readily the sugar dissolves. Take a small portion of this, test it with iodine solution, and note that no violet coloration results. To the remaining portion add Fehling's solution[1] and boil. Note the deep orange colour produced. This reaction is characteristic of grape-sugar. Test by the same means (1) the juice of the Grape and (2) a few small pieces of Onion and compare the results. Grape-sugar occurs abundantly in each case; it is of common occurrence in plant tissues, and is a valuable and easily transported food. Now examine the starch mucilage which has been acted upon by the saliva. Note that the mucilage has dissolved. Test a small portion with iodine solution. Is starch present? As no violet coloration results, we may conclude that the starch has been converted into some other substance. To the remainder add Fehling's solution and boil. What is the new compound formed? From this experiment we learn that saliva contains a substance which has the property of converting starch into sugar. Such a body is called a ferment or enzyme, and it is by means of such ferments that we are able to digest the starch present in our food. The starch-digesting ferment in saliva is called ptyalin. Apply these tests to the solution acted upon by diastase and compare the results. In this case also the starch has disappeared. Diastase is a ferment commonly present in plant cells, and it is by means of such ferments that the insoluble starch grains are corroded and disorganized, and finally converted into sugar.

We have seen above that sugar is formed in green leaves during sunlight. Part of this is converted into starch grains within the cells of the leaf; the rest is drained away to the stem or other parts. In these organs it may either (1) be converted into starch and stored, or (2) serve for the nutrition of tissues that are growing. At night, when the

[1] Fehling's solution may be prepared and kept in two stock solutions as follows: (1) dissolve 35 grammes of cupric sulphate in 200 c.c. of water; (2) dissolve 70 grammes of rochelle salt in 200 c.c. of a 10 per cent. solution of caustic soda. When required, make a solution of equal volumes of 1, 2, and water.

influence is withdrawn, and photosynthesis is not going on, the ferment diastase is actively at work digesting the starch in the leaves, and the sugar thus formed is drained away to other organs. On the return of sunlight, the leaf is again ready to continue the work of photosynthesis. We thus see how a plant utilizes the alternating periods of day and night.

Starch formed from sugar in the dark by leucoplasts.—By osmosis the sugar solution so formed in the sap is transferred from cell to cell and carried downwards through the leaf-stalk to the stem, and even to the root; and in plants developing their seeds it is conveyed to the cotyledons, e.g. the Pea and Bean, or to the endosperm, as in the Wheat. On reaching these organs the sugar is once more converted into starch grains; thus the surplus organic food is transferred to storage organs and there laid by until required.

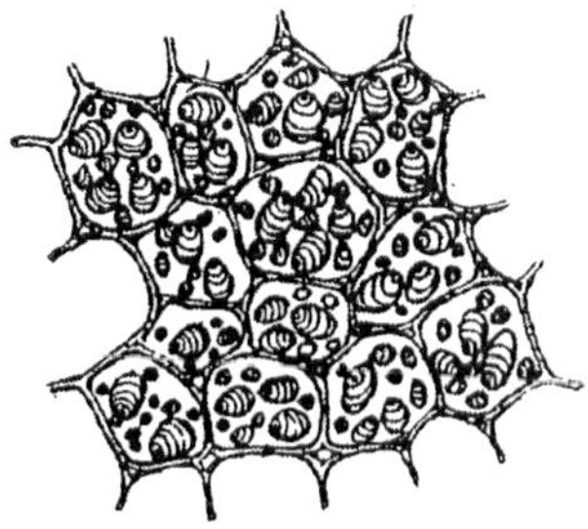

FIG. 53. CELLS OF THE POTATO CONTAINING STARCH GRAINS.

In the inner parts of plants, where light cannot penetrate, are often found small, rounded, colourless bodies having the same origin as chlorophyll corpuscles and known as **leucoplasts** (Gr. *leukos* = white). Owing to the absence of chlorophyll they are unable to manufacture starch from carbon dioxide and water, but can build up starch grains from sugar carried to them from the green parts. In this way starch grains arise in parts that grow in the dark (Fig. 53). If, however, tissues containing leucoplasts, e.g. potato tubers, are exposed to light, the leucoplasts develop the green pigment and become chlorophyll corpuscles. Thus, by the action of green corpuscles, solid food-substances are formed during sunlight. These are rendered soluble by ferments and can be transferred to organs where they may be reconverted by white corpuscles into solid food-reserves.

Chlorophyll, the green colouring matter of plants, is a very complex nitrogenous substance. As we have seen (p. 81) it may be extracted by means of alcohol, and if sections of leaves so decolorized are examined under the microscope the corpuscles will still be found in the cells (Fig. 29, 5 *ch*). In the living plant the pigment is probably dissolved in some oil, and this solution is enclosed in the meshwork of the corpuscles. Each chlorophyll corpuscle or chloroplast, therefore, consists of (*a*) a protoplasmic body or plastid and (*b*) a pigment, chlorophyll.

Light rays absorbed by chlorophyll.—An alcoholic solution of chlorophyll is fluorescent: if it is held up to the light and examined it is green, but if examined against a dark background it is blood-red. Examine a beam of light by means of a spectroscope and note the band of colours—red, orange, yellow, green, blue, indigo, and violet. This band is called the spectrum of white light. Place an alcoholic solution of chlorophyll in the path of a beam of light before it reaches the slit of the spectroscope; note the spectrum which results, and compare it with that of white light. Observe the dark bands produced and note carefully their position in the spectrum. Seven vertical bands are produced, but some of them are difficult to see. The darkest is at the red end of the spectrum; three fainter, but broader, bands occur at the blue end; the remaining three bands are much paler and occur in the yellow and green. We thus see that the rays of light falling on a green leaf do not all pass through it. Chlorophyll has the power of absorbing most of the red rays, many of the blue and violet ones, and, to a much less extent, some of the yellow and green. It is the energy thus absorbed from the sun's rays that enables the chloroplasts to carry on the constructive work of photosynthesis. We are now able to understand why starch is not formed in the white parts of leaves or in green leaves kept in darkness.

Currents in the stem.—Let us now perform a few experiments which will enable us to determine the channels along which travel the materials used by the shoot.

Place in a bottle red ink or a solution of eosin, and obtain shoots of Rhododendron, Ivy, or similar evergreen. Place the freshly-cut ends of the shoots in the coloured solution. From one of these, and above the level of the solution, cut a broad ring of tissue into the wood. Place a similar shoot in another bottle, but instead of eosin, use water to which a little finely-powdered carmine has been added. The particles are exceedingly minute and of such a nature that they remain in suspension a long time and produce a coloured liquid. Leave these for a day or two and then compare them. Are they equally fresh? Is there any difference in colour in the shoots? Cut short pieces from the lower end of each and compare the cut surfaces. How do they differ? Cut off a piece two inches long and split it longitudinally down the middle. Is the stem uniformly coloured? Are the shoots coloured similarly? Scrape off some of the bark and determine which tissue is coloured. Trace this coloured tissue upwards and determine whether it extends into the leaves. Cut the leaf-stalk and leaves across and note whether they are coloured; and, if so, how?

What do we learn from these experiments? We find that the shoots in eosin have taken up the coloured solution, and it has ascended only through the woody portions of the stem and leaves, and not through the bast, cambium, cortex, or epidermis, for in the shoot from which these tissues have been removed the eosin has ascended, notwithstanding their removal. The shoot in carmine, however, remains uncoloured. Why? Here the fine particles of carmine are in suspension, not in solution, and although a coloured solution like eosin may be absorbed, the particles, even so fine as those of the carmine, are unable to pass through the woody tissues.

Water, however, is absorbed and serves to keep the shoot fresh, while that in eosin becomes discoloured and dies. Judging from these experiments we may conclude that the wood of such a shoot is the path along which water ascends through the stem to the leaves. But what of the other tissues of the bundle, for example the bast?

If we examine the plants in a garden or park, especially shrubs or trees which have been tied for some time to a support, it will be interesting to note the mode of growth in the neighbourhood of the ligature. Fig. 54 is a sketch of a rose stem which has been tied in this way and allowed to grow for some time without further attention. Careful examination of such a shoot shows that as the stem has grown in thickness, the ligature has gripped it with increasing pressure, and the delicate tissues of the inner cortex and bast have been so compressed that substances could not pass along them, but the rigid walls of the woody tissues have withstood the pressure, and sap can still ascend as usual. The chief changes noted, however, affect the portion of the stem above the ligature. Here nutrient materials have accumulated, obviously carried from a higher level, and have become stored in an abnormal tissue which forms a swelling.

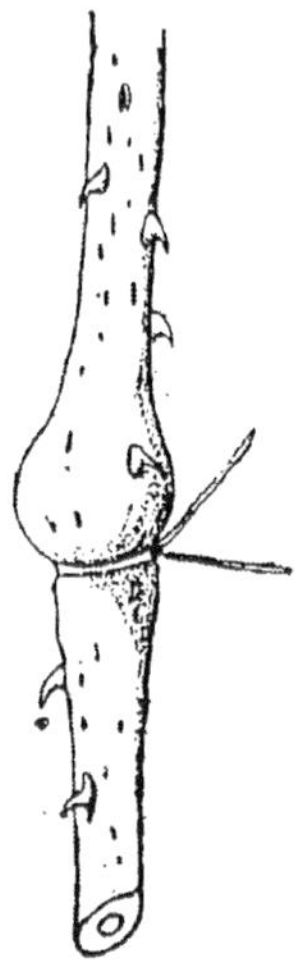

FIG. 54.
LIGATURED STEM OF THE ROSE.

Fluids are able to travel not only upwards through the wood and downwards through the bast, but there are many cross-currents as well, especially through the thin plates of tissue, the medullary rays, passing from pith to cortex. In some stems these tissues become loaded with food-materials carried to them from leaves and other parts. This is seen very clearly in the Clematis. Obtain a piece of stem about the thickness of a lead pencil and cut several slices across it. Lay them on a glass slip and place on them a drop of iodine solution. By means of a pocket lens examine the surface and note the effect of the stain. Compare the yellow-brown walls of the wood with those of the medullary rays and with the outer cells of the pith. These are crowded with dark purple-stained starch grains.

CHAPTER VIII

WORK OF THE SHOOT (*Continued*)

Transpiration.—It is a matter of common observation that, when plants are grown under a bell-jar or in a glazed case, the sides of the chamber are covered with drops of water. Where has this water come from ? Has it come from the soil or from the shoots ? If from the latter, seeing that a plant is provided in its roots with so excellent a means of obtaining water, why should so much be thrown off by its shoots ?

A simple experiment will enable us to decide the matter. Plant in a pot a single rosette of **London Pride**, take a piece of lead-foil large enough to cover the top of the pot, cut it from one side to the centre, bring the two cut edges round the stem beneath the leaves and fold them closely over, pressing the sides around the pot so as to exclude all moisture from the soil (Fig. 55). Take a small vessel, fill it with dry calcium chloride (a substance which readily absorbs water), and weigh it carefully; place this on the top of the lead-foil alongside the plant, and cover the whole with a bell-jar, sealing the edge with vaseline. Allow this to remain for a day; then examine. Remove the bell-jar and note what change has taken place in the contents of the vessel. Weigh again, and note the difference. To what is this increase due and whence has the water come? This giving off of water by living shoots is called **transpiration.** A vessel with the dry calcium chloride weighing together 16 grammes has been found by experiment to have gained 1 gramme in weight in twenty-four hours; i. e. the leaves have given off during twenty-four hours 1 gramme or 1 cubic centimetre of water.

FIG. 55. EXPERIMENT TO DETERMINE THE AMOUNT OF WATER GIVEN OFF BY A PLANT. —*a*, vessel containing calcium chloride.

If this water has come from the leaves, has it come equally from the two surfaces, and are leaves similar in this respect? To determine these points, prepare a few sheets of cobalt paper by dipping pieces of filter-paper in cobalt chloride solution and allowing them to dry. Test a small piece of this paper and note the changes in colour that occur (1) when the paper is warm, and (2) when exposed to ordinary air or when breathed upon. Obtain a dry duster, fold it several times so as to make a pad, and lay on it a few leaves, some with the lower, others with the upper surface to the duster. Place over these a sheet of dry (blue) cobalt paper, and cover immediately with a sheet of thick glass to exclude the air from the cobalt paper. Try this with several kinds of leaves, including some evergreens, and note the changes. Which surface gives off the more water? Are there any differences in this respect in the different leaves you have examined? From what we have seen in the structure of a Box leaf, is it likely that the differences noted can be accounted for by differences in the distribution of stomata?

Take two evergreen leaves, as nearly alike as you can find, and coat the upper surface of one leaf and the under surface of the other with vaseline. Carefully weigh each; then expose both to air for half an hour or more, and weigh again. What difference do you find? How does this result compare with your previous experiments? As previously seen, a cut shoot placed in water absorbs a considerable amount of liquid, and we now see that water, in the form of vapour, is given off from the leaves through the stomata.

Let us now try to determine the rate at which this water travels, and the amount absorbed in a given time. To do this, arrange an experiment as shown in Fig. 56. Take a wide-mouthed bottle provided with a tight-fitting rubber stopper with three holes. Through one is passed the tube of a funnel provided with a tap. Through the second is passed a bent, thick-walled capillary tube, passing only to the lower level of the stopper. Behind the tube is a scale marked in inches or centimetres. Through the third hole is passed a shoot which has previously stood in water (Rhododendron or Laurel will answer very well). Fill the bottle completely with water and press the stopper, together with shoot and tubes, firmly into the bottle. If the tap of the funnel is open, water rises in the tubes. Now close the tap and fill up the reservoir. By opening the tap, water flows along the bent tube and drops from the open end. Now close it, and the water stops immediately. As the shoot absorbs, water is drawn back along the tube and readings may now be made. Care should be taken that the temperature is fairly constant. Obtain the capacity of the tube and determine the amount of water absorbed in a given time.

By turning the tap, the tube may be refilled and the

experiment repeated or varied in several ways; e. g. by coating some of the leaves with vaseline (1) on their

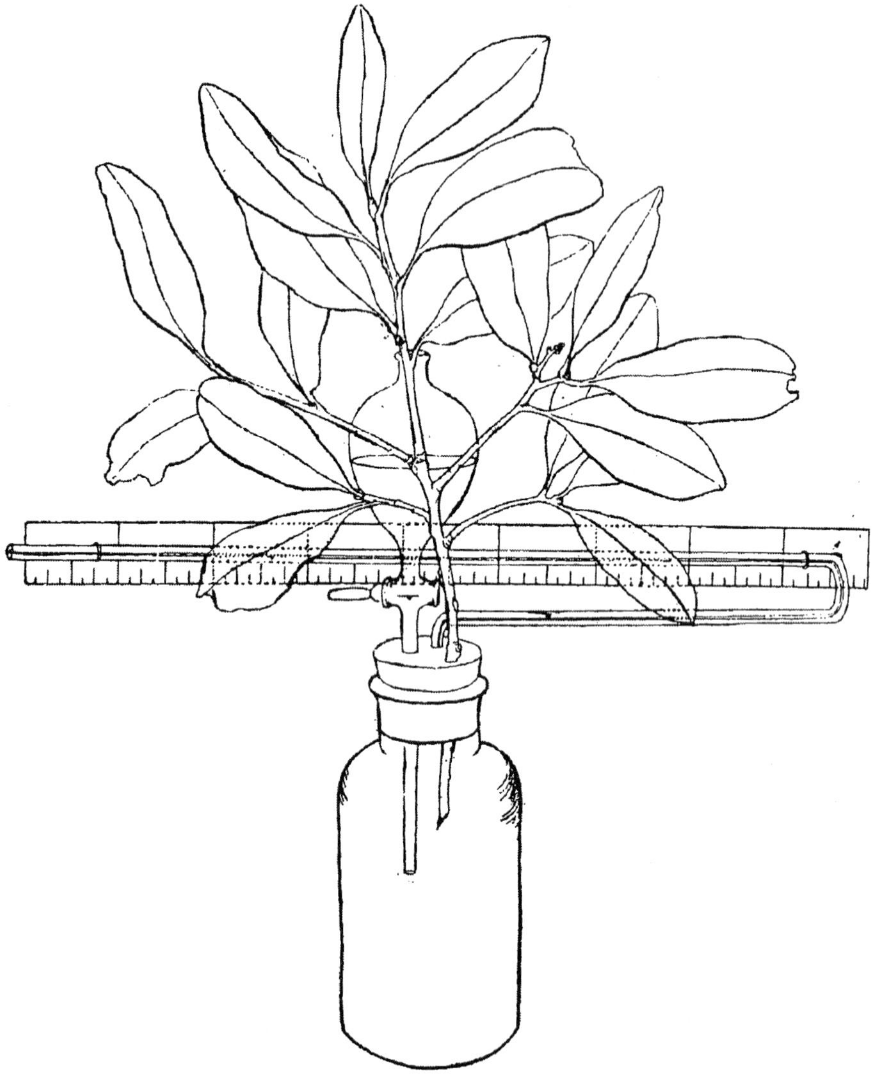

FIG. 56. POTOMETER, TO MEASURE THE RATE OF TRANSPIRATION IN A SHOOT (Farmer).

upper and (2) on their under surface; (3) by removing some of the leaves; (4) by exposing the apparatus to sunlight or (5) to diffused light; or (6) to dry air or (7) to moist air.

The food of plants, taken up by the roots, is absorbed as very weak solutions, and one of the great functions of leaves is to get rid of surplus water. The rate of transpiration varies in different plants and under different conditions; there are many peculiarities and exceptions, but generally the circumstances which favour transpiration are: (1) winds; (2) warm air; (3) height of the plant above ground, the upper layers of the atmosphere being drier than those near the ground; (4) numerous leaves, large leaf-surface, and many stomata. On the other hand, transpiration is reduced: (1) in calm weather; (2) when the air is cold; (3) in plants which are low-growing; and (4) in plants which have fewer leaves, smaller leaf-surface, and fewer stomata.

Protection of stomata.—For a shoot to perform its functions in a satisfactory manner it is important that the stomata should not become blocked by rain or dew, and it is interesting to determine how leaves are protected from this danger. Dip the leaves of several common plants in water, then take them out and examine both upper and lower surfaces. To what extent are they covered with water? Has immersion resulted in blocking a considerable proportion of stomata? What are the peculiarities of surface-structure which prevent the surface from becoming wetted? Not only is the leaf practically non-absorbent, but surface coverings such as hairs and wax render wetting difficult or impossible. Repeat these experiments with leaves of Pinks or Carnations; note the blue-grey coating of wax or 'bloom' and see how easily it may be wiped off. Dip a leaf in cold water, note the silvery-looking air-cushion on the surface; then remove it. Is it wet or dry? Dip another leaf into hot water, note what happens to the bloom; remove the leaf and compare it with a fresh one. Bring the flame of a match near an uninjured leaf and note the effect of heat on the bloom. Other leaves (and many fruits) provide similar examples, and in each case we find that the waxy coating very effectually prevents the surface from becoming wetted. Further modifications will be noted in connexion with the habitats of plants.

In the familiar process of wilting we have another good illustration of transpiration. Take two shoots and place the freshly-cut end of one in water, but allow the other to lie on the table. Compare them in an hour. The one in water is fresh and rigid, while that on the table has become limp, i. e. the shoot cut off from its water-supply

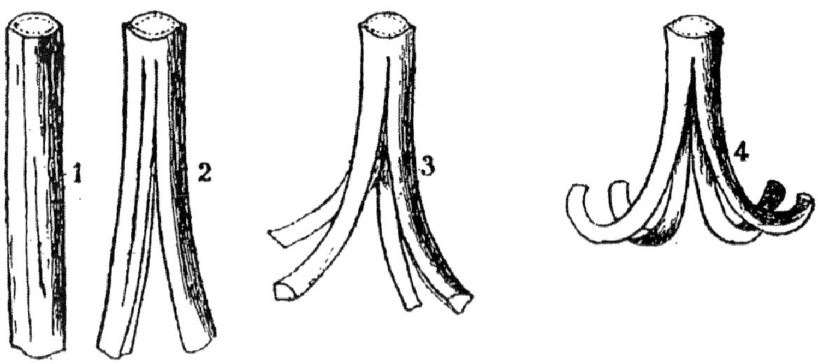

FIG. 57. EXPERIMENT TO ILLUSTRATE TURGIDITY.—1, piece of Daffodil stalk; 2, cut into two strips; 3, cut into four strips; 4, the same in water.

has wilted. Similarly, if a plant rooted in the soil be insufficiently watered it wilts, i. e. its shoots and leaves become limp and droop; but such a plant soon regains its freshness and turgidity on being watered.

Turgidity.—A simple experiment will help us to understand this. Take a piece of the flower-stalk of the Dandelion or Daffodil, about three inches long, and cut it down the middle, as in Fig. 57, 2. Does any change in shape take place? Which is now the longer side? the inner or the outer? The inner side being the convex and therefore the longer side, in what condition were the tissues of this surface before the stalk was cut? Clearly they were compressed,

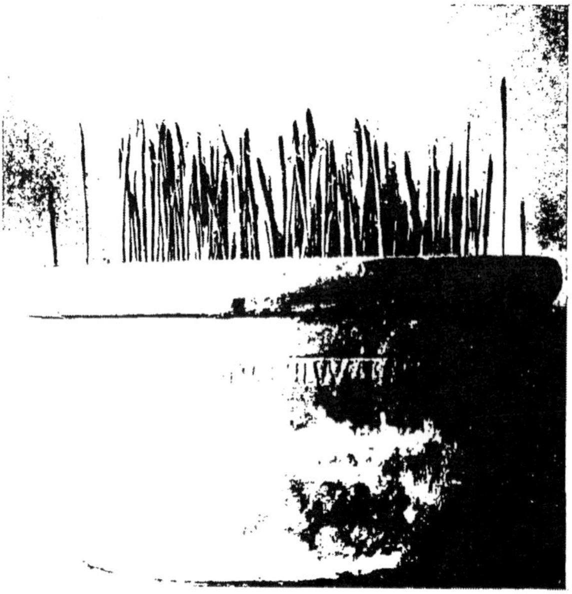

FIG. 58. EXCRETION OF WATER BY WHEAT SEEDLINGS.

FIG. 59. LEAF-ROSETTE OF A SAXIFRAGE.—The margins of the leaves are encrusted with lime from chalk glands at the ends of the teeth.

and, being released from the rest of the stalk, have expanded. Make another cut, at right angles to the first, and note the result. Now place the cut end in water and note what happens. The inner surfaces have become still more convex.

Our previous experiments in osmosis will help to explain this. The water, through the attraction of the cell-contents, has been absorbed by these cells, increasing the internal pressure and stretching the elastic cell-walls. The distension of the cell by the internal pressure of sap in plants is called **turgidity**. Next place the cut stalk in strong salt solution and note the change in shape. The salt solution, attracting water from the cells, reduces the internal pressure; the walls contract, the cells become smaller, and consequently what was before the convex side becomes now the concave side. Wash the cut stalk thoroughly in water, and allow it to remain in water a short time. Do the cells regain their turgidity and the strips resume their former shape? It is by such changes in the internal pressure of the cells of plants that shoots are at one time fresh and turgid, at another limp and wilted. The pith cells of herbaceous stems, and of quick-growing shrubs, show the same tendency to elongate, and though unable to do so, owing to the resistance of the surrounding woody tissue, they help considerably to maintain the rigidity of the shoot.

Root-pressure.—We have already seen, in our observations on Wheat seedlings (p. 33), that the roots may absorb more water than a plant can utilize in a given time, and that the excess is forced out of the tips of the leaves (Fig. 58). Drops of water may often be seen on the leaves of certain plants in the early morning, e. g. on the leaf-teeth of Fuchsia and Lady's Mantle, and at the ends of the main veins in the Garden Nasturtium. Sometimes the salts in solution are so abundant as to leave a deposit on the leaf when the water evaporates; this occurs in Wheat seedlings, and especially in some Saxifrages, where the salts form chalky incrustations on the ends of the teeth (see Fig. 59). Water is usually given off, however, in the form of vapour, especially during the day, when the stomata are open; loss of water may be so great in warm sunny weather that the plants droop. At night the stomata close, but absorption of water by the root goes on; the plants become turgid, and as the amount absorbed exceeds the amount of water-vapour transpired, water in a liquid state is forced out of the leaves. Some leaves, e.g. the Garden Nasturtium, have special stomata which are permanently open for this purpose and are called **water stomata.**

Obtain an actively growing plant of Sunflower, Fuchsia, or Dahlia, cut off the shoot about three inches above the soil, dry the cut end and examine the surface with a pocket lens. Soon water exudes from the cut surface. Arrange an experiment as shown in Fig. 60. By means of rubber tubing (*c*), attach to the stump (*s*) a bent tube (*g*). Water (*W*) now collects in this tube, and if mercury (*Q*) is placed in the bend of the tube, the column will be forced upwards. By this means the pressure of exudation may be measured. The pressure which exists in the tissues of the root and aids the upward flow of sap in the stem is called **root-pressure.**

The amount of sap which ascends in the stems of plants in spring is often very great. If, for example, the stem of the Vine is cut as the leaves are unfolding, so great is the flow of sap, that it can only with difficulty be stopped. This exudation of sap is known to gardeners as 'bleeding'.

Force of transpiration.—It is obvious that considerable force must be exerted in drawing water up a stem to the leaves to replace that given off as water-vapour through the stomata. Some idea of this force may be obtained from the following experiment (Fig. 61). Take a long, thick-walled glass tube fitted at one end with a rubber

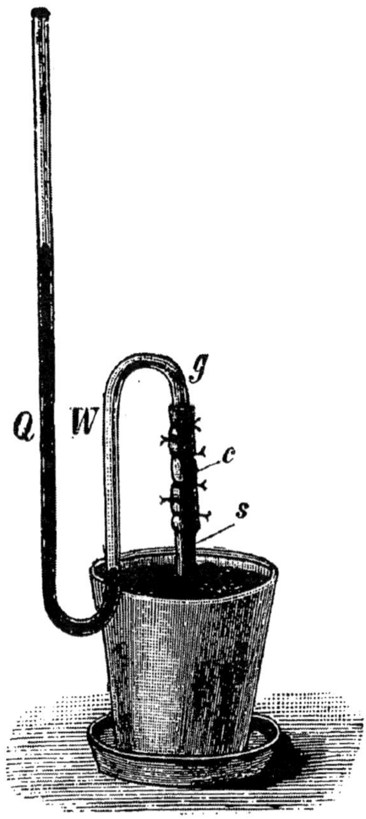

FIG. 60. EXPERIMENT TO SHOW THE PRESSURE OF EXUDATION (Jost).

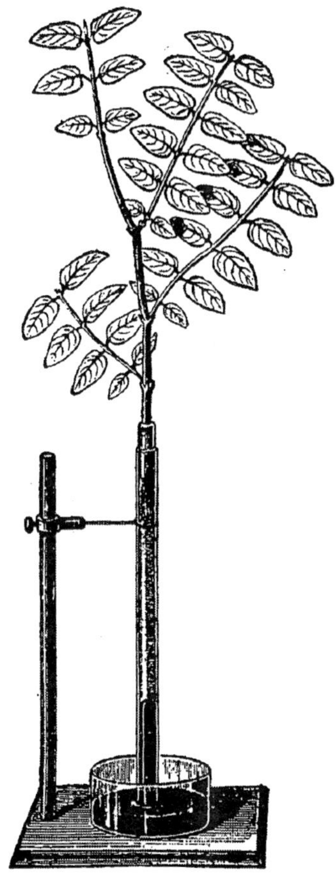

FIG. 61. EXPERIMENT TO DEMONSTRATE THE SUCTION ACTION OF TRANSPIRATION (Jost).

stopper. Through a hole in the centre push the end of a Laurel shoot, selecting one thick enough to fit tightly. Then fill the tube completely with water and, placing the thumb over the open end to prevent escape of water, put this end into a trough of mercury. Remove the thumb and secure the tube with the shoot to a retort-stand or other suitable support. Fix a scale to the tube and take readings at intervals. When the mercury has reached its maximum height, ascertain the weight of the column of mercury raised by the force of transpiration.

After a time, air will collect in the upper part of the tube. Where can this air have come from? Is it possible that the tension of the liquids has resulted in air being drawn through the stem, and that this has accumulated on the top of the column? Or did the water contain air which may have risen to the surface? The value of the latter suggestion may be tested by using water which has been previously boiled and allowed to cool. If air then accumulates it must have come from some other source. The former suggestion may then be considered. Do air-channels exist in shoots? and, if so, is it possible to draw air through them? The following experiment will help us to answer this question.

Suction of air through a shoot.—Fix a suction-pump firmly to the water-tap and connect it to a bottle by means of thick-walled rubber tubing. Fill the bottle with water and insert a rubber stopper, through the hole of which is passed the stem of a Laurel or the stalk of a single leaf, as shown in Fig. 62. Turn the tap gently, then steadily increase the flow. Note what happens at the end of the shoot. Where is the air coming from? Is a stream of air passing through the shoot?

This experiment may be reversed. Place the leaves in the water and the cut end of the shoot in the air; observe the air-bubbles coming out from the numerous tiny points on the under surfaces of the leaves and more vigorously at any broken or injured places. Suddenly turn off the water and note the change of colour in the lower surface of the leaf. Why is this? Turn on the water again and the leaf regains its colour. Repeat the experiment and notice that with the sudden back-rush of water the air-spaces of the leaf have become filled with water. How can it enter the leaf? What force has been exerted to bring about this striking result?

Importance of water to plants: water a carrier of food and waste.—The living cells of a plant are tiny chemical manufactories, and very elaborate indeed are some of the compounds

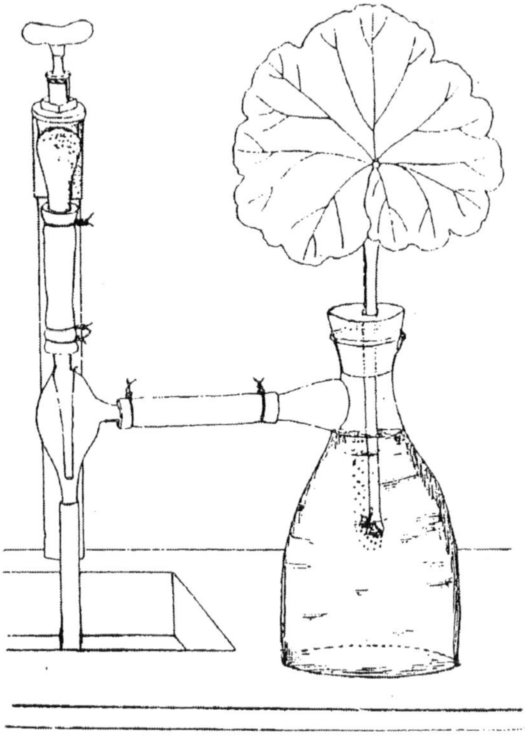

FIG. 62. SUCTION OF AIR THROUGH A LEAF.

made there. The raw materials are carried to them in the water from the soil and in the air which enters the leaves and other green parts. The latter are exposed to the influence of sunlight, and this is one of the necessary conditions for the formation of the green pigment, chlorophyll; otherwise plants would, as a rule, be sickly in colour, as they are when grown in the dark. The presence of this pigment gives to a shoot the power of utilizing the energy of the sun's rays in bringing about and carrying on many important changes in the substances entering the cells, also of rearranging their component atoms and building up new compounds from them. During these chemical changes much heat is evolved, as we have already seen in the experiment with germinating peas; but the leafy shoots of plants are always cool, and commonly cooler than the surrounding air. How is this? With all the chemical changes going on in plant-tissues, why does the temperature of the plant not rise much above that of the air, as it does in our own bodies? Some heat may be lost by radiation, but for a fuller answer we must go back to our experiments on transpiration and try to realize the enormous amount of heat required to convert the water of the cell-sap into vapour, and the large amount of vapour given off by an average leafy shoot. It is estimated that over 90 per cent. of the heat absorbed by a plant is dissipated in this way. No wonder, then, that the foliage of a plant feels cool to the touch.

But our experiments with water-cultures suggest another interesting point in this connexion. Is the soil-water (or its artificial representative, a water-culture solution) a dense, or a weak, food-solution? Is it necessary for the solution to be a weak one? and, if so, why? If a plant needs to take up an enormous amount of water in order to obtain a sufficiency of solid food, what is the consequence? The necessities of osmosis, of conduction and transmission, require a weak food-solution. This involves the absorption of an excess of water above that needed for the building up of tissue-materials. Hence we see the value of a thin, flat leaf whose exposed surface is very large compared with the amount of its tissue. Again, the spongy tissue of a leaf, with all its cells hung out, as it were, in drying-chambers, has an interesting meaning. These chambers, communicating by way of the stomata with the air outside, render the whole an admirable arrangement for getting rid of the excess of water. It seems from this that a large leaf-surface might coincide with great absorption, involving a large food-supply and consequent rapid growth. At any rate this is worth keeping in mind, and it might be considered with reference to the very different conditions under which plants grow. What differences, for example, do you find in the rate of growth and the forms of plants growing in a ditch, a hedge, on a moor, a rock, a sandy shore? Meanwhile, we see how important it is that the functions of the stomata should not be interfered with, and some of the most interesting modifications of leaves are those which concern the protection of the stomata and the economy of a plant's water-supply.

CHAPTER IX

BUDS AND BRANCHES

AT the growing end of a branch the leaves are very small and immature, and arise close together on the shoot-axis, as shown in Fig. 67. Such an undeveloped shoot is called a **bud**. In winter the leaves of the buds are often so tightly packed, and the parts are so small, that they are difficult to dissect. The essential features, however, may easily be made out from an examination of a **Brussels Sprout** (Fig. 63). Each 'sprout' arises in the axil of a leaf, like the bud of any typical plant. Remove the tightly-packed leaves one by one, noticing that they are folded, wrinkled, and arranged spirally on the axis. In the axil of each leaf a small bud will be found. How many leaves are there? How many axillary buds can you find? When you have removed all you can, cut the remainder of the bud (the 'heart') longitudinally into halves and make out by means of a pocket lens the end of the axis or growing-point, covered over by many tender, undeveloped, or rudimentary leaves. Such a bud is clearly a condensed, immature branch-system, consisting of a central axis, which bears

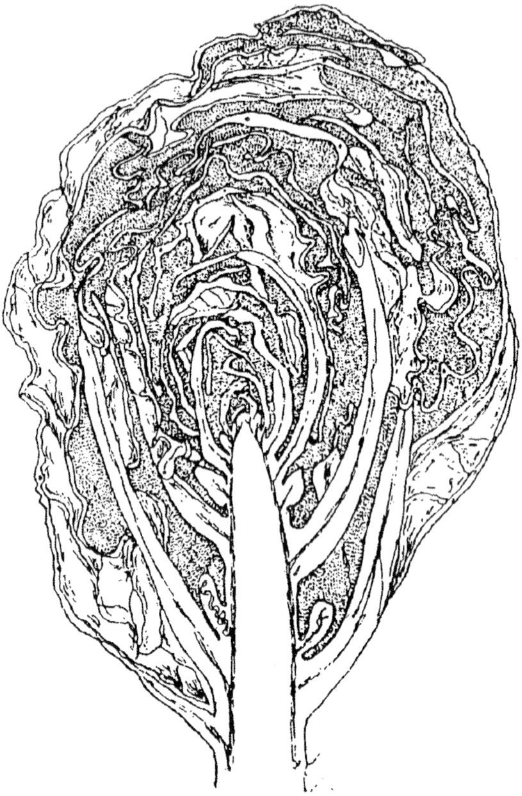

FIG. 63. BRUSSELS SPROUT IN VERTICAL SECTION.

leaves, in the axils of which buds are formed, each of these being the beginning of a new lateral branch.

If we now examine a Cabbage and Cos Lettuce we see that in such cases the axis elongates so little that, when the plants are full grown and ready for market, they still bear all the characteristics of huge buds. They are not protected on the outside, however, by tough scales, and are hence called naked buds. It will be interesting to compare with these such plants as the **Daisy** (Fig. 64), Dandelion, Plantain, Primrose, and London Pride or other **Saxifrage** (Fig. 59), where the leaves all spring close together round a short stem and quite near the ground. Each resembles a bud which has opened out its leaves and by pressing them out in a close rosette has secured a little patch of ground for itself. The leaves of such plants should be drawn and the peculiarities of outline noted. In the Daisy

FIG. 64. DAISY, SHOWING REPRODUCTION BY MEANS OF OFFSETS.

and London Pride the leaves are spoon-shaped (spatulate); in the Ribwort or Plantain they are lanceolate; and in the Primrose they are obovate and wrinkled. The margins, too, are peculiar, being even or entire in the Daisy and Plantain, wavy in the Primrose, and edged with small, rounded lobes (crenate) in the London Pride; while the margin of the Dandelion leaf has large teeth pointing backwards (runcinate), which have earned for the plant its popular name.[1] The tips of the leaves, too, vary from blunt and rounded in the Daisy to sharp-pointed or acute

[1] Dandelion is a corruption of Fr. *Dent-de-lion*.

in the Ribwort. Note the differences in the length of the leaf-stalk and the size of the blade from the lower and outer to the upper and inner part of the rosette, and observe how this prevents much overshadowing in spite of the crowding. Leaves springing from the stem near the ground in this manner are called radical leaves to distinguish them from leaves rising, like those of a Stock, on a taller stem above ground and known as cauline leaves. The habit of forming rosettes is very common in plants growing on mountains, and on rocks where the soil is liable to dry up at certain seasons. But rosette-formers are not uncommon in other habitats, especially in grassy swards. As the buds which arise in the leaf-axils of such plants also tend to form rosettes close to the parent, a large cushion is in time produced. Short lateral shoots of this kind are called offsets, and they serve as an important means of vegetative reproduction (Figs. 64 and 257).

If opportunity offers, it will be interesting to study the various rosettes of the plants growing on a rockery. You will find many forms, some compact, others lax. There will be varying lengths of offsets, and varying forms and sizes of cushions. Fleshy leaves in all grades may be found, some, like those of the Houseleek, very thick indeed and able to store much water for use in times of drought.

Observations on opening buds.—If you place winter shoots of trees in water for a few weeks you will be able to watch the opening of the buds, and it is easy to study the more important details of their structure. In such opening buds observe: (1) the number and arrangement of bud-scales; (2) their origin from leaves, leaf-bases, or stipules; (3) the transition from bud-scales to foliage-leaves; (4) the arrangement and manner of folding of the foliage-leaves; (5) whether the leaf-stalk or blade is first developed; (6) the behaviour of the leaves as they expand; (7) the differences between leaf-buds and those containing flowers.

Try to realize the great amount of work which is going on as the buds open, and to determine where the material comes from and how it is utilized. If opening buds are used instead of germinating peas in the experiment we have performed previously (p. 44), you will find that they absorb a large amount of oxygen and give off much carbon dioxide. It has been found that during this period of active respiration many of our common trees lose from 20 to 45 per cent. of their total dry weight. This helps us to appreciate the fact that respiration is a wasting or breaking-down process.

Lilac.—Quite different from the rosette type are the buds of the **Lilac.** If we watch them expanding in the spring we shall see that the leaves are not folded and wrinkled, but lie flat and edge to edge. As the shoot grows and the axis elongates, the leaves are seen to be in crossed pairs which have separated by distinct internodes (see Fig. 205). The arrangement of the leaves on the stem, and the relative positions of leaves of different sizes, stand in strong contrast with what we find in a typical rosette. Compare the leaves from below upwards, and notice the transition from small scales below, followed by larger ones, to the mature leaves with longer, grooved stalks and large, heart-shaped (cordate) blades. The bud-scales of the Lilac are thus reduced leaves, of which the lower, smaller ones fall off as the season advances, not when the bud opens, as in many trees.

Privet.—Now compare the Lilac shoot with a shoot of the **Privet** (Fig. 65). Note the small, brown scales below; their arrangement and the varying sizes and shapes, not only of the scale-leaves, but of the green foliage-leaves; also the varying positions of the blades in shoots taken from the side and from the top of the hedge. How are these differences related to the direction in which light falls on the shoot? What part of the leaf is concerned in bringing the blade into such a position? The movement occurs at the leaf-base. This is of common occurrence in plants; and the Yew, Ivy, and Virginia Creeper provide further interesting examples. By means of a pocket lens carefully examine the mode of attachment of the leaf to the stem. The three parts of a typical leaf are easily determined: (1) the swollen base, each side running as a ridge down the stem; (2) the short stalk marked off from the base by a dark transverse line, and (3) the ovate, entire, acute blade.

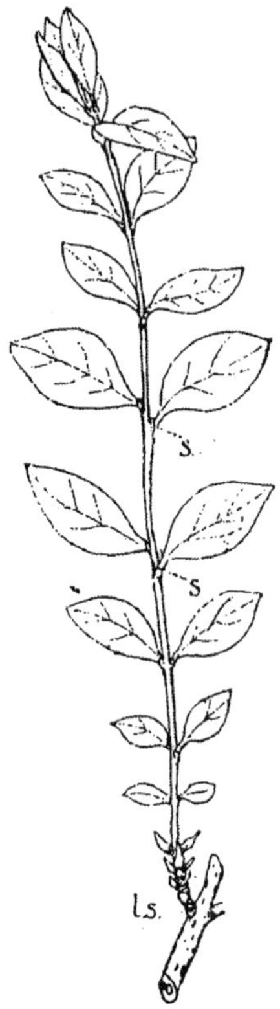

FIG. 65. SHOOT OF PRIVET. — *l.s*, leaf-scar; *s*, separation-layer.

Bend the leaf back and press the bent stalk against the stem until it snaps. Where does the break occur? Repeat this and notice that the dark line is a line of separation (Fig. 65, *s*). Examine older shoots for leaf-scars, and notice that when the leaves fall it is the blade and stalk that are thrown off, and that the base remains on the axis as a more or less prominent scar. Compare other shrubs and trees in this respect, e.g. the Common Ash.

Horse-Chestnut.—If a twig of **Horse-Chestnut** (Fig. 66) be examined we may learn much of its history. At the end is a large terminal bud, and below this are two large leaf-scars each showing seven dots, which are the broken ends of veins, while above each scar is a small bud. Lower down, at intervals, are other crossed pairs of scars and buds, the lowest of the series being frequently smaller than the rest, and below these again we find a number of small scars crowded together. Even these are in crossed pairs, the scars being the scale-scars of last year's terminal bud, and in their axils are tiny buds which, in ordinary circumstances, will not further develop but remain dormant. Thus the whole of the shoot from the rings of scale-scars to the large terminal bud has been developed during one season from the terminal bud of the previous year. The internodes between the scale-scars elongate very little, and this part of the axis remains practically in the condition in which it was formed, while the internodes between the foliage-leaves greatly elongate and separate the leaves by considerable intervals.

Place two or three shoots in water in the early spring and watch the buds as they open. We are thus able to learn a good deal about the development of a shoot. The photographs (Fig. 68, 1–9) are taken from shoots so treated. The scales of the unopened bud are covered with hairs, which secrete a sticky mucilage composed of gum and resin. This covers the surface and binds the scales together, and, with the thick scales, provides a double protection for the young leaves within.

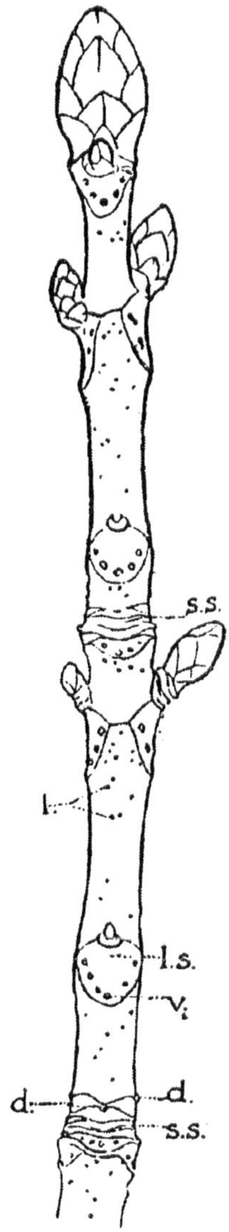

FIG. 66. WINTER SHOOT OF HORSE-CHESTNUT. — *d*, dormant bud; *l*, lenticel; *l.s*, leaf-scar; *s.s*, scale-scars; *v*, broken ends of leaf-veins.

Watch the scales as the bud opens, and follow their movements. At first they are incurved and clasp the inner leaves, later they turn outwards and backwards out of the way, the higher, bigger scales growing for some time and arching over the pleated, woolly foliage-leaves felted together with a tangle of hairs. The stalks elongate and carry the blades upwards, their tips, for a time, being fastened together by hairs, which soon fall off as the leaf grows. One by one the leaves move outwards; the leaflets expand and, growing more on the upper than the under side, bend backwards until only the upper surfaces can be seen. Growth on the under side now quickens, and the reverse process occurs, the leaflets being raised until they reach a horizontal position.

When all have unfolded, examine the shoot from above and notice that they form a closely-fitting pattern, or mosaic. The leaf-blades are mostly at the same level, though they arise at different heights on the stem. The lowest leaves have the longest stalks and the largest blades; the highest leaves have the shortest stalks and the smallest blades; those between being intermediate in these respects.

Meanwhile the scales fall off, the lowest first, leaving narrow, light-brown scars which darken with age. We can thus watch, day by day, the formation of ring- or scale-scars which indicate the beginning of a year's shoot. The uppermost scales have larger bases, are thinner, and turn green, and often remain for some time on the branch after the rest have fallen off. Then they frequently develop little blades (Fig. 69, *s*), which are different from the lower ones and are in some respects intermediate between them and the foliage-leaves. The fact that they may bear blades suggests that the true scales are really leaf-bases, the blades being usually suppressed. When the intermediate scales do fall off they leave larger and lighter scars than the others. The broken ends of their veins are well seen, and it is easy to detect their scars, even on old twigs.

Usually seven leaflets are formed, which arise from the top of the leaf-stalk, but often there are only five, and occasionally six occur, in which case the leaflets are not arranged three on each side, but there is a median one with two on one side and three on the other. When the veins

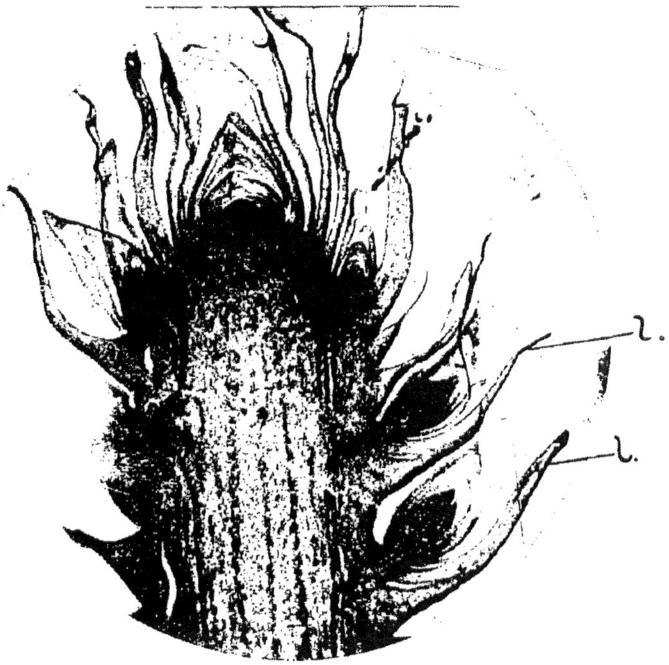

FIG. 67. VERTICAL SECTION OF BUD OF PINE.—Note that the growing point is protected by overlapping leaves; *l*, leaves which bear buds in their axils.

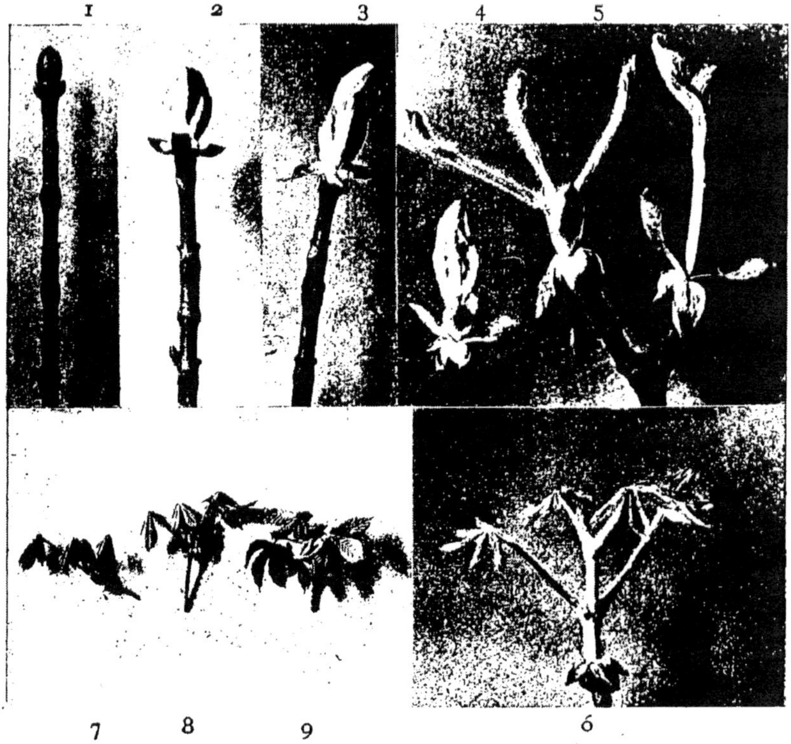

FIG. 68. OPENING BUDS OF HORSE-CHESTNUT.—The shoot (5) shows an inflorescence-scar between the branches.

FIG. 69. YOUNG SHOOT OF HORSE-CHESTNUT.—*s*, bud-scale at the end of which a leaf-blade has developed.

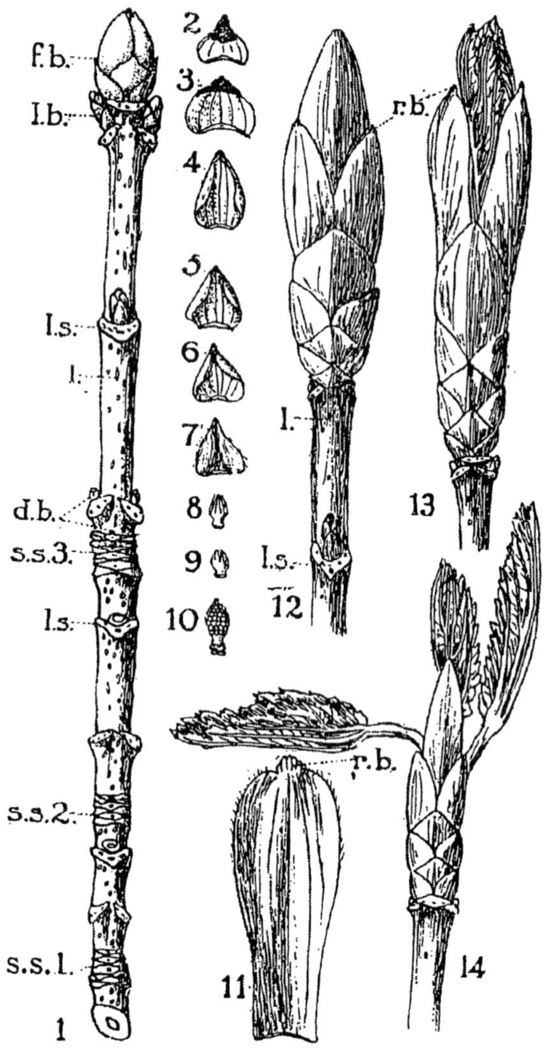

FIG. 70. SYCAMORE BUDS.—1, winter shoot of Sycamore with a large terminal flower-bud; 2–10, the parts of the bud dissected out; 2–7, one of each pair of bud-scales; 8–9, one of each of the two pairs of foliage-leaves; 10, inflorescence; 11, large scale of opening bud with rudiment of blade at the tip; 12, 13, 14, stages in the opening of a leaf-bud; *d.b*, dormant bud; *f.b*, flower-bud; *l*, lenticel; *l.b*, leaf-bud; *l.s*, leaf-scar; *r.b*, rudimentary blade; *s.s.* 1–3, scale-scars.

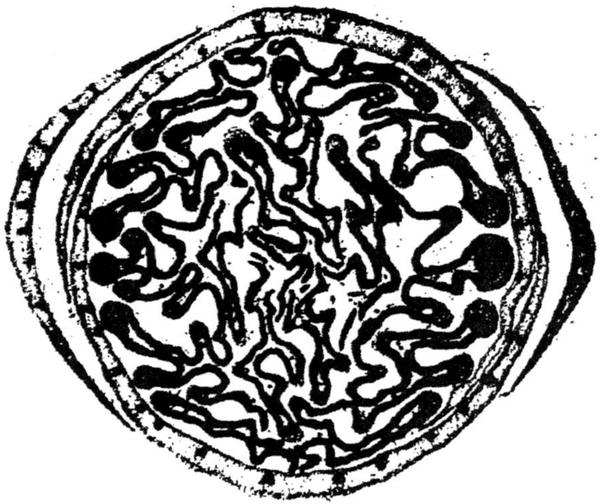

FIG. 71. TRANSVERSE SECTION OF BUD OF SYCAMORE, showing the leaves in crossed pairs, and the foliage-leaves folded fan-wise.

FIG. 72. TRANSVERSE SECTION OF BUD OF BEECH.—Outside are thin stipular bud-scales which surround the folded foliage-leaves.

of a leaf radiate from the top of the leaf-stalk the blade is said to be palmate, as it resembles the palm and fingers of the hand.

In addition to leaf-buds, still larger terminal buds may be found (see Fig. 103). These contain not only leaves but flowering branches as well.

Commonly, as in Fig. 68, 5, twigs may be found with two buds at the end. If we look carefully between them we find a distinct scar. The meaning of this is clear when a flowering branch is examined (see Fig. 103). The inflorescence is terminal, and when the work of the flowers and fruits is over, the axis which bore them dies and is thrown off in a manner similar to the fall of a leaf. A scar is thus formed at the end of the twig and further growth is continued by the next pair of lateral buds.

Thus at least three kinds of scars may be found on a branch of the Horse-Chestnut: (1) foliage-leaf scars, (2) scale-leaf scars, and (3) branch-scars or inflorescence-scars.

Sycamore.—The bud of the **Sycamore** (Fig. 70) is another useful type. It is a large one and easy to dissect, especially when expanding in the spring. Like the Horse-Chestnut, it is covered by twelve or more overlapping, boat-shaped scales in crossed pairs, the inner ones being larger and green, and often bearing at their tips rudiments of blades (Fig. 70, 11, 12, 13). These are often tucked inward and covered by hairs at the tip of the scales, but may be distinguished by means of a pocket lens. Sometimes they enlarge, and show the form and structure of typical blades. Nearer the middle of the bud are the five-lobed foliage-leaves, their blades folded fan-wise and packed closely together in crossed pairs (Fig. 71). The folding is such that the strong veins of the blade are outermost, the more delicate parts of the blade being inside (Fig. 70, 14). Examine these carefully and notice that at first only base and blade are developed, and that the stalk is the last part to appear.

These observations help us to understand the true nature of the bud-scales. The upper, large, green ones are, like those of the Horse-Chestnut, leaf-bases bearing rudimentary blades, while the lower, exposed ones are leaf-bases only. Some of the earliest buds to open, however, are larger than these and contain flowers as well as leaves (Fig. 70, 1, *f.b*).

When fully expanded the blades are seen to be in one piece (simple), the five lobes not being divided into separate leaflets as they are in the Horse-Chestnut. Note also that

FIG. 73. LEAF-MOSAIC OF SYCAMORE.

the leaves from one bud form an excellent leaf-mosaic (Fig. 73). By this means overshadowing and overcrowding are reduced to a minimum, and the leaves secure fuller advantages from exposure to air and sunshine.

Contrast the Sycamore shoot with that of the Willow (see Fig. 78), and note that in the case of the latter, similar advantages are secured in another way, viz. by long internodes and narrow blades.

Beech.—The bud of the **Beech** (Fig. 74) presents several interesting differences from those we have examined. Observe its long, tapering form and the light-brown membranous scales which are arranged in pairs. Remove

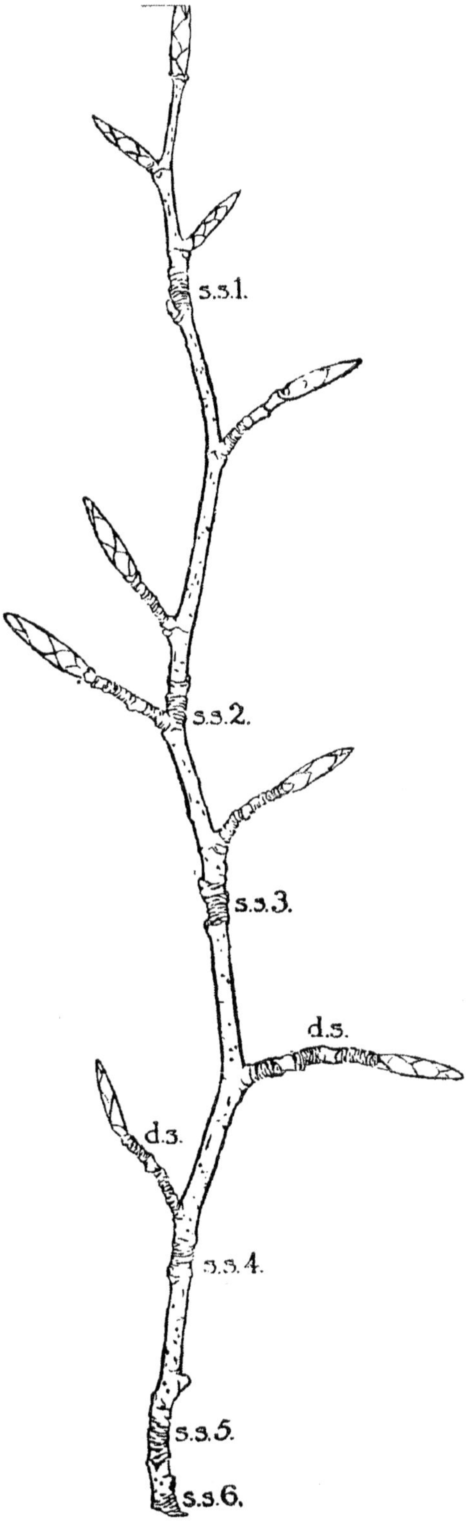

FIG. 74. SIX-YEAR-OLD WINTER SHOOT OF BEECH.—*d.s*, dwarf shoot; *s.s.*1 to *s.s.*6, scale-scars of six successive years.

the outer scales and examine the inner ones carefully. You will find between each pair a tiny green blade; the outer ones have no leaf-rudiment between them. Farther inwards are the foliage-leaves, folded fan-wise (Fig. 72), and the veins and margins are fringed with white hairs. Growing out from the base of each leaf is a pair of light-brown membraneous scales. Outgrowths of the leaf-base are called **stipules**, and a careful comparison shows that the scales of a Beech bud are such outgrowths, i. e. they are stipules. As the bud opens and the leaves mature, the scales, having served their purpose as protective structures, commonly fall off, and hence they are said to be **deciduous**. Thus the leaves, when mature, appear to be without stipules.

The bud-scales of many of our forest trees are stipules, e. g. Poplar, Oak, Hazel, and Elm. In these cases, the 'spring fall' is one of stipules, and not, as in the Sycamore and Horse-Chestnut, of leaf-bases. In the case of a few plants, e. g. the Laburnum, the scales do not fall when the bud opens, but wither on the branch. In many plants the stipules grow with the growth of the leaf, are green and leaf-like, and last as long and serve the same general purposes as the leaf; such stipules are **persistent**, e.g. Hawthorn, Rose, Pea, Violet, &c. Some plants do not produce stipules, i. e. they are **exstipulate**.

The buds of some trees and many herbs are not protected by scales at all, but are naked, e. g. Wayfaring Tree, Juniper, Barberry, Mistletoe, Ivy, Bittersweet, &c. In the Wayfaring Tree, however, the young leaves are protected by a mealy covering of star-shaped hairs, hence it is sometimes called the Meal Tree.

Watch the buds of the Beech as they open in the spring (Fig. 75) and compare the behaviour of the leaves with that of the Horse-Chestnut leaves. Note the elongation of the bud, the separation of the scales, the bright yellow-green leaves peeping above them, folded fan-wise and

FIG. 75. OPENING BUDS OF BEECH.

FIG. 76. LATER STAGE OF OPENING BUDS OF BEECH; leaves pressed downwards by greater growth of the upper surface.

clothed with silky hairs. As they emerge you will see that the leaves hang downwards in such a way as to expose only their upper surfaces (Fig. 76), thus reducing loss of water by transpiration and loss of heat by radiation, as in the Horse-Chestnut. Finally, they are raised by increased growth on the under surface and thus the blades are brought into a horizontal position.

On comparing a given shoot with that of the Lilac we find similarly that the lowest and oldest leaf is the smallest, the highest and youngest leaf the largest (see Fig. 191, 2). This difference in size, however, together with the horizontal position of the blades, results in the leaves not only forming a flat plate, but the available space is occupied with the least amount of overshadowing. In this way a leaf-mosaic is formed, but by a very different means from that of the Sycamore or Horse-Chestnut.

These features should be looked for in other trees, such as Elm (see Fig. 197), Hornbeam, and Hazel, and also in herbaceous plants. Leaf-mosaics are common in the plants of temperate climates.

Examine a winter twig of the Beech similar to that shown in Fig. 74. Note that there is a slender zigzag shoot at the tip with a bud at each bend. Below that, a series of scale-scars (*s.s*) shows the limit of the year's growth. These features are repeated as we pass backwards, six such portions being shown in the figure at 1–6. At the bends, in place of buds, are short branches with many scale-scars and terminated by a bud. Clearly these branches grow very slowly and the leaves produced by such buds are separated by very short internodes. Such short, slowly-growing branches are called Spurs, or Dwarf-shoots. Figs. 74, 75, 76 show what is produced from such a shoot in the following spring. The end bud has grown into a long, slender, slightly zigzag shoot with leaves at the bends. The next two lateral buds form the beginnings of dwarf shoots with crowded leaves, while the buds at the ends of the dwarf shoots of previous years have each developed several leaves, also crowded together, owing to the slow growth of their internodes. If you examine several twigs of the Beech, you will find that, while the end bud may grow a foot or more in a season, a dwarf shoot may have grown only three inches in ten or more years.

Scots Pine.—An excellent example of the development of dwarf shoots is seen in the **Scots Pine** (see Fig. 182). Examine a small branch, and you will find that it is terminated by a large bud, round the base of which are three or four smaller lateral buds standing at nearly the same level. Farther back the axis appears to be covered by tough evergreen needle-leaves. Carefully examine the shoot to see where and how the needles arise. Do they spring, like many leaves we have seen, singly from the axis? Remove a few of them. Do they come away singly, or in pairs? Can you find any other structure on the axis still remaining when the needles have been removed? Examine, and compare with this, part of a branch from which the needles have fallen off naturally. What are the structures producing the roughness of the shoot? Are the pairs of needles related in any way to similar structures? If any doubt as to the last-mentioned point remains, the examination of an elongating bud in the spring will make their relationship clear. The axis produces only scale-leaves. In their axils, buds arise which form very short shoots (dwarf shoots). At the base of each are several scale-leaves, and near the tip are two long, green needles. Such short shoots are called 'bifoliar spurs' (see Fig. 182). They remain three or four years on the tree, and are then thrown off. Each year new ones form, so that the tree is always green. Examine the ground under a pine tree and pick up a few of the old needles. Does the pine shed merely its leaves or its short branches also? We see that the short branches are shed as well as the needle-leaves, and the scars left on the axis are not leaf-scars, but the scars of dwarf shoots.

Vernation or praefoliation.—In the above examples we have seen how neatly the leaves are packed in the bud with the least loss of space. The manner in which leaves are thus folded and arranged is known as vernation or praefoliation, and their relationship may be seen by the examination of a transverse section, or more easily by a study of buds as they open in the spring. Determine the arrangement, manner of unfolding, and direction of greatest growth as the blades expand, in the following plants: the Dock and Rhododendron, where the leaves are back-rolled (*revolute*); the Violet, Elder, Apple, Pear, and Poplar, where they are up-rolled (*involute*); the Plum and Blackthorn, where they are rolled from one side to the other (*convolute*), and in the Ferns, where the blade is rolled from apex to base (*circinate*). When revolute leaves expand, growth is greater on the under surface; in involute, convolute, and circinate leaves growth during expansion is greater on the upper surface.

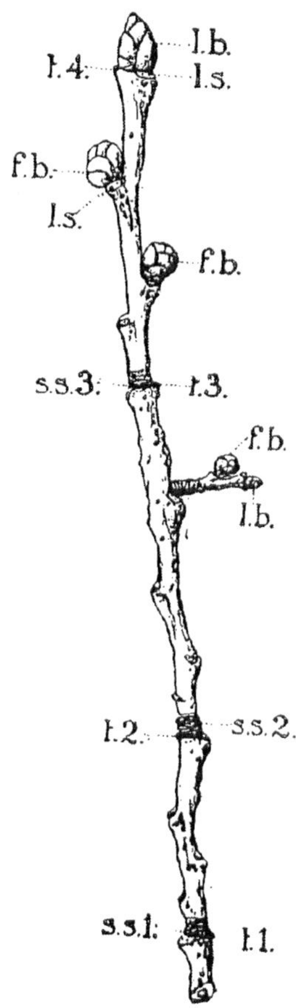

FIG. 77. WINTER SHOOT OF ELM.—*f.b*, flower-bud; *l.b*, leaf-bud; *l.s*, leaf-scar; *s.s.*1 to *s.s.*3, scale-scars of three successive years; *t.*1 to *t.*4, dead terminal buds of four successive years.

Monopodial and sympodial branching.—Trace the development of shoots from the opening buds to the formation of leafy branches and notice the variations in different plants. In some trees, like the Sycamore, Horse-Chestnut, and Pine, the terminal bud continues the growth of the main axis from year to year. A single and continuous axis is called

FIG. 78. BRANCH OF WILLOW.—*d*, dead terminal bud.

a **monopodium**. In most trees, however, growth is not so uniform; usually the end bud dies and the lateral bud arising in the axil of the next leaf below enlarges, pushes the withered bud aside, and appears to be the terminal one. A careful study of the Elm twig (Fig. 77) will make these points clear. In the following spring this lateral bud grows somewhat in the line of the original axis, the terminal bud of this branch in turn dies, and is pushed aside by the next lower lateral bud. Thus the process is repeated, and a series of branches is superposed one on another in such a way as to resemble a simple axis. Such a branch system is called a **sympodium.** In the Willows (Fig. 78, *d*), the end of the branch dies and projects as a dead stump, while in the Hawthorn a spine may result. These features are not difficult to make out in the autumn when the leaves are falling and before the true terminal bud has shrivelled and become displaced.

Long shoots and dwarf shoots.—Most of our forest trees produce two kinds of leafy shoots, but they are not always so well marked as in the Beech and Pine. It is common to find, however, that the buds on a tree do not all develop in the same manner; some grow rapidly and produce shoots with long internodes, others grow very slowly and have very short internodes. Often the leaves on a dwarf shoot vary in size and form and in the position they ultimately assume. Compare the dwarf shoots of Poplar, Birch, Beech, Elm, Hawthorn, Mountain Ash, and Laburnum. We have noticed in many herbaceous perennials that the axis remains short and a rosette of leaves is formed close to the ground. When examining the shoots of trees and shrubs observe how commonly dwarf shoots produce flowers.

Dormant buds and stool shoots. Adventitious buds.—The buds, which arise in the axils of foliage-leaves and scale-leaves, are so numerous that room could not be found for all of them to develop. The food supply, also, is insufficient for the purpose. Very frequently those buds arising in the axils of the bud-scales, and in the axils of the lowest foliage-leaves, are very small and do not enlarge in the spring, but lie dormant, although retaining their power of development for a period varying from a few years to twenty (see Figs. 70, 189, 194). If the growth of the axis above such dormant buds is arrested by injury or removal, the dormant buds begin to grow into leafy shoots. Thus old branches, and even trunks of trees, may become covered with fresh shoots.

Shoots arising on the trunks of trees, usually from buds which have long been dormant, are known as ' stool shoots ', and are common on trees with a thin bark, e. g. the Lime. They are a characteristic feature of the Elm, and occur frequently on Sycamores, Oaks, and many other trees.

New buds are occasionally formed in the outer tissues of branches and other members, e. g. on roots and leaves, and not in leaf-axils; such buds are termed **adventitious buds.** True dormant buds have a pith continuous with that of the branch, while the pith of adventitious buds is not continuous. Adventitious shoots are common on the roots of shrubs and trees, e. g. Raspberry (Fig. 27), Rose, and Poplar, also on the roots of Dandelion; while some Ferns produce adventitious buds on their leaves.

Shedding of leaves and branches.—Having now described the structure and behaviour of buds, we may conclude this chapter with a reference to the shedding of leaves and branches. The shedding of scale-leaves is a noteworthy feature in the spring, when, under Sycamores and Beeches, the ground is covered with them. In the autumn the foliage-leaves are thrown off and again the ground is covered. Thus we have at least two leaf-falls in a year: (1) A spring fall of scale-leaves; and (2) an autumn fall of foliage-leaves. From the Pine and often the Poplar whole branches are thrown off. In addition there is the fall of flowers, fruits, and their axes, so that each year a tree sheds many of its organs.

CHAPTER X

HIBERNATION; THE STRUCTURE OF MODIFIED SHOOTS

HAVING gained the foregoing knowledge as to the structure, functions, forms, and modes of growth of the vegetative organs of a few common plants, we will now pay attention to some that are peculiarly modified. In the preceding chapter we have been studying the formation of dwarf shoots, rosettes, and scales—all of which are cases of reduction. Occasionally, however, the reverse occurs, the roots, stems, or leaves becoming abnormally enlarged, in which case they usually act as storage organs, either of water, or organic food, or both. It is extremely difficult, if not impossible, to say how these modifications were brought about, but we can often suggest some useful purpose they serve when once they are formed.

Adverse conditions and their effect on growth.—With the changing seasons, plants are exposed to a great range of conditions as to temperature, moisture, and light, which greatly influence the power for work of the different plant-organs. The parts most exposed to these changes are the shoots above ground, the leaves being especially sensitive. In temperate regions the winter conditions are unfavourable for active root-absorption, and therefore for active growth, as are the dry periods of many tropical and sub-tropical countries. At the beginning of the adverse period the first changes we notice are the withering and shedding of the leaves of many shrubs and trees, and the dying down of herbaceous shoots. The strong trunks and branches of the former persist, enveloped in their coats of cork, whilst their buds are protected by tough, brown scales. But how do the more tender herbaceous plants fare? By what means do they tide over the winter? We know that many animals burrow in the ground beyond the reach of frost and cold, and lie dormant until more favourable conditions return. But do plants hibernate? and if so, how? Let us consider a few common species, e. g. Shepherd's Purse, Carrot, Daisy, Lily, Bluebell, Crocus, and Iris, and note how they pass the winter.

Annuals and ephemerals.—The **Shepherd's Purse** produces a number of seeds in the summer, but when these are shed the whole plant, roots and shoots, dies, and nothing remains but the seeds. In the following spring the seeds germinate, new plants are formed, which produce flowers, fruits, and seeds the same year; and the plants, as before, die completely in the winter.

Species which thus complete their life-cycle in one season are called **annuals**. They hibernate either as seeds or, less frequently, as fruits, and this is a very effective method. Many of our common weeds of roadsides and cornfields behave in this way, e. g. Groundsel, Chickweed, Field Pansy, Charlock, and Hemp Nettles. Some of these, like the Shepherd's Purse, may pass through their life-cycle in a few weeks if conditions are favourable, so that several generations may be produced in a season; such short-lived 'annuals' are known as **ephemerals,** and examples may often be found among the plants of a waste-heap.

Biennials.—The **Carrot** behaves differently. After the seed has germinated, the plant grows vigorously; its root enlarges considerably and becomes stored with a reserve of food-materials (see Fig. 26, 1). In this condition it passes the winter. On renewal of growth in the following spring it produces an abundance of flowers, fruits, and seeds, at the expense of the food stored in the root, which is exhausted; the seeds are shed and then the whole plant dies after two seasons' growth. To such plants the name

biennial is given. Beetroot and Parsnip are other examples. The Radish and the Turnip (see Fig. 26, 2, 3) are also biennials, but the food-material in these is stored mainly in the greatly enlarged hypocotyl. These plants, therefore, tide over one winter by means of their enlarged roots or combined roots and stems, and the following winter only their well-protected seeds remain to perpetuate the race. Just as there are variations in the life-period of annuals, so there are in that of biennials. If we nip off the flowers of biennials the plants continue to vegetate for years, and many so-called biennials, like the Foxglove and Snapdragon, often continue to grow for several seasons. Again, many plants which are annuals in the plains grow for many years in the mountains.

Perennials.—A large number of our wild plants regularly persist from year to year and may flower each season; they are called **perennials**, and in the case of certain trees may live to a great age. The ability to persist, and flower at intervals through several seasons, is termed **perennation.** In plants with shoots too tender to withstand the rigours of winter, i. e. herbaceous perennials, we meet with many interesting forms of hibernating organs.

Underground shoots: rhizomes.—In studying a plant like the Stock we receive the impression that the part below ground is mainly root, but it is not easy to decide where the root ends and the shoot begins. Pull up a plant of either the **Quick-grass** (Wicks), or the Soft-grass, and examine it carefully. What structures do you find? Is the whole of the underground part root? By what characteristics will you decide which is root and which is shoot? Do you find leaves on any of the parts? If so, what kind of leaves are they? Can you find buds arising in the axils of any of them? What do these buds become? Trace some of them. Are these structures found on some of the underground parts and not on others? What are the roots like, and where do they originate? Do these agree with the roots we found on seedling plants of the Wheat and Maize (p. 32)?

An examination of the underground parts of these plants convinces us that, (1) stems bearing leaves (scale-

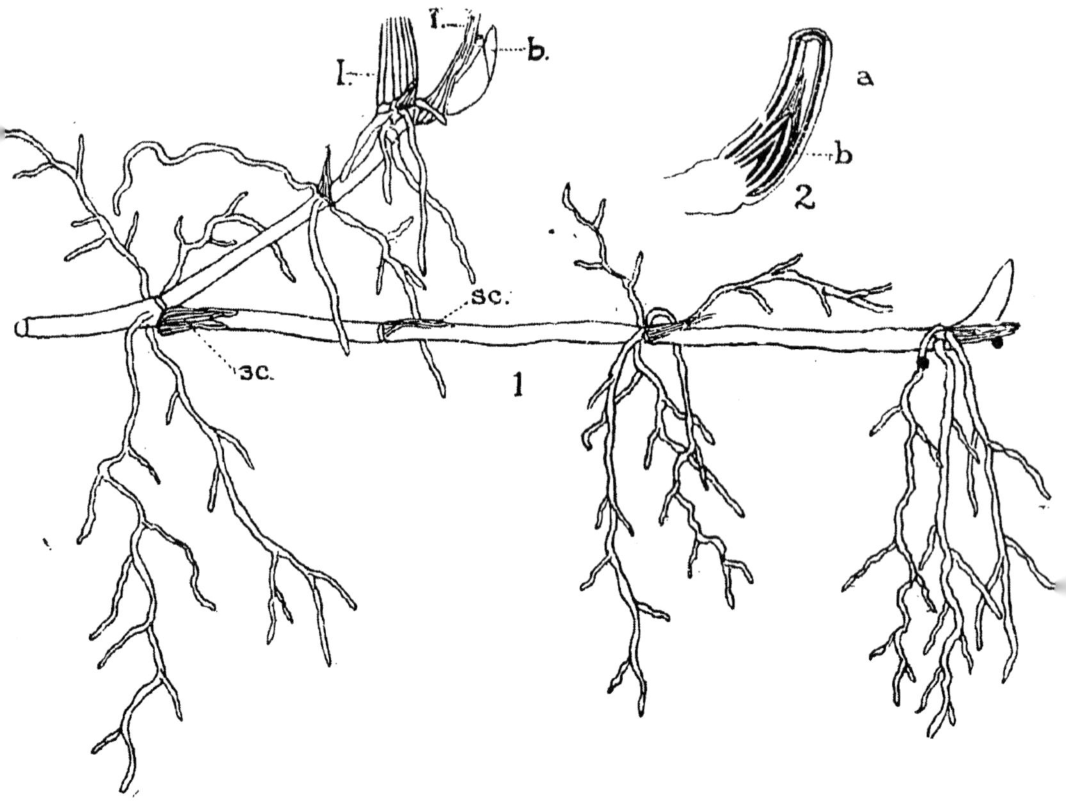

FIG. 79. RHIZOME OF LILY OF THE VALLEY.—(1) *b*, end bud emerging from the soil; *l*, withered leaf-bases; *sc*, scale-leaf. (2) section of bud: *a*, leaf-bases; *b*, axillary bud which continues growth.

leaves) with buds in their axils occur underground; (2) such plants may have not only underground stems but aerial stems also; (3) the latter are in reality branches of the former; and (4) the fibrous roots spring from the stem (often from the nodes) and are therefore adventitious.

Such underground stems are known as **rhizomes**; that of the **Lily of the Valley** (Fig. 79) is a very instructive one to study. Observe the nodes with their scale-leaves and also the branched fibrous roots springing from each node in a circle. Such an arrangement of members is termed a **whorl**. Carefully dissect a bud and compare it step by step with the parts found in other buds you have examined. Notice the different kinds of scale-leaves: the tough outer ones, forming a protective coat; further inwards some which are rather fleshy; then the foliage-leaves. If the bud is a large one, look for the inflorescence in the centre. Which of these structures come above ground? How is the further growth of the axis continued underground? Compare this mode of growth with that of the Beech, Hazel, Elm, or Willow. Is the axis monopodial or sympodial? The following examples of rhizomes should be studied and their parts compared: Garden Mint, Colts-foot, Dog's Mercury, and Wood Sorrel. In all these cases the end bud emerges from the soil, and growth is continued underground by means of a lateral bud.

Rhizomes as land-winners.—The rhizomes of some plants, such as the Marram-grass, Sand-sedge, and Horsetails, grow to a great length, often many yards, and this habit makes them useful for reclaiming our sandy shores. Fig. 80 shows how the **Marram-grass** is planted on the sands. Round the tufts wind-blown sand accumulates, and the shoots by elongating keep their leaves above the surface. Below the ground long rhizomes are formed, from the nodes of which very long, slender roots arise and grow deeply in search of water, the two producing a tangle, and serving effectually to hold the sand together. At the same time the old and decaying shoots, by adding humus to the sand, begin the formation of a soil upon which other plants can grow. Similar uses are made of rhizome-bearing plants to hold together the banks of canals and railways.[1] In these ways the underground parts of plants become valuable sand- and soil-binders and play an important part as 'land-winners'.

Thickened rhizomes.—Examine the rhizome of the **Solomon's Seal** or the **Iris**. The axis is greatly thickened and bears many branched adventitious roots; the internodes are very short, and the scale-leaves and buds are large. Cut a slice across the rhizome and examine the tissues. Outside is a layer of cork, then a thick cortex, and near the centre a number of scattered vascular bundles. Place on the cut surface a drop of iodine solution and note the large amount of starch stored in the cells. As the plant grows, the rhizome tends to rise to the surface of the ground, and if you examine Irises in a garden you will often find that the soil is washed away from the rhizomes. On plants which have thus approached the surface, it is common to find thick unbranched roots which penetrate the soil deeply, then contract and pull the rhizome downwards. Sometimes the growing end is directed downwards, and as the rhizome elongates it descends until a suitable depth is reached. Fig. 82 shows such a descending rhizome of the **Flowering Rush** as it ploughs its way through the mud in which it grows.

Stem-tubers.—The **Potato** is another strangely modified hibernating organ. On the surface, which is covered with a layer of brown cork, are small depressions, the 'eyes'. Plant a potato, or even a thick slice of potato, in a pot of soil, and as it grows you will find that shoots spring from the 'eyes', a fact suggesting that they are buds. Fig. 83 shows a plant grown in this manner. From the 'eyes' (*e*) leafy shoots have grown and the base of the stem has produced numerous branched, fibrous roots. The buds

[1] On railway banks these plants often extend their bounds, grow between the rails, and produce a weedy track very difficult to keep clean.

FIG. 80. MARRAM-GRASS PLANTED AS A 'SAND-BINDER'.

FIG. 81. RHIZOMES OF SAND-SEDGE EXPOSED BY THE WIND.

formed in the axils of the lower leaves have produced branches (*rh*) which creep on the surface of the soil, and you will observe that they are swollen at the ends to form small potatoes (*t*), bearing 'eyes' like the parent, one of which is producing a shoot (*e.s*). During the growth of

FIG. 82. RHIZOME OF FLOWERING RUSH DESCENDING INTO THE SOIL.—The thick roots are contractile.

the shoots, the potato becomes soft and wrinkled as the store of food is used up. Examine a potato and determine the arrangement of the eyes; note that at one end is a scar, left when the potato breaks from the stem. Usually the eyes are few or absent near the scar, become more numerous towards the opposite or growing end, and are arranged in a $\frac{2}{5}$ spiral. In cultivation, banking up with earth induces increased formation of rhizomes and tubers, but if left exposed to light, the tubers are small, green, and soon develop leafy shoots. A potato plant, therefore, has three

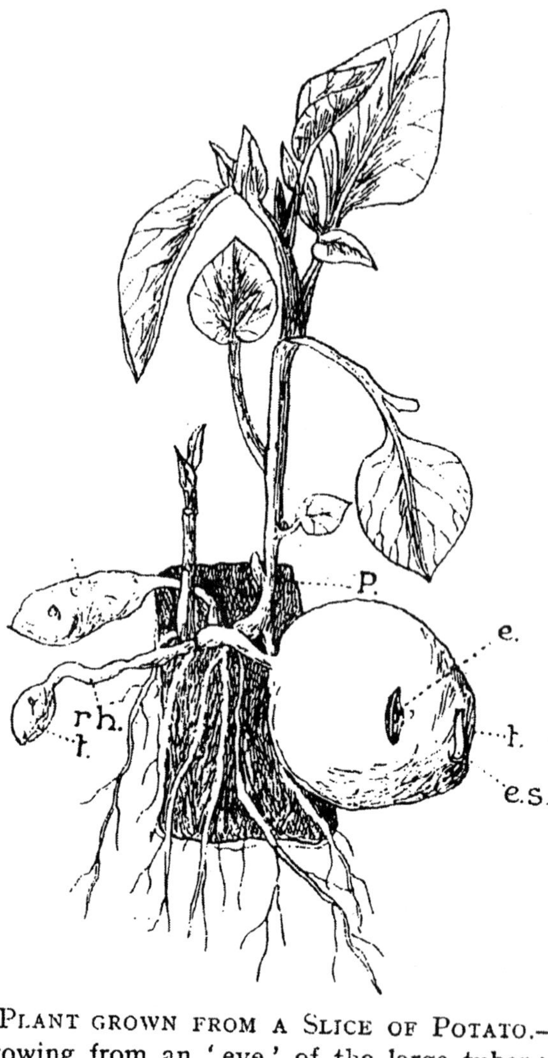

FIG. 83. PLANT GROWN FROM A SLICE OF POTATO.—*e*, 'eyes'; *e.s*, shoot growing from an 'eye' of the large tuber; P, slice of potato; *rh*, rhizome; *t*, tubers.

kinds of stems: (1) aerial stems, bearing green foliage-leaves and flowers; (2) rhizomes, bearing small scale-leaves; and (3) from the rhizomes spring greatly swollen and irregular stems, on which are reduced buds, the 'eyes'. Such swollen underground stems are known as tubers. Cut a slice from a potato and test it with iodine solution. What food-reserve is present? make a watery extract of a soft, sprouted potato and test with Fehling's solution for grape-sugar.

The **Artichoke** is another example of an underground stem-tuber. The food-reserve here, however, is not starch, as in the potato, but a substance allied to sugar, called inulin. The above must not be confused with the root-tubers of the Lesser Celandine and Dahlia.

Corm of the Crocus.—A short, thickened, underground stem, similar in many respects to those we have considered, is found in the **Crocus** (Fig. 84, 1). In this case, however, the stem is somewhat globular and surrounded by membraneous scale-leaves (*sc*); such a stem is called a **corm**. Examine a dry corm; determine the arrangement and mode of attachment of the scale-leaves, and remove them from below upwards. Examine a piece of scale with a pocket lens and observe that the fibres which form its skeleton are parallel, with many cross-connexions. Note the difference in length of the internodes from the base of the corm upwards, and the circular scale-scars (*s.s*). Are there any axillary buds? At the upper end of the corm, where the scales are crowded together, two or more buds arising in their axils become much larger than the rest. Cut the corm vertically in two (see Fig. 84, 3). Note the thick, solid axis with the veins (*v*) passing through it, and try to trace one of these to a small axillary bud. Place a drop of iodine solution on the cut surface. Of what does the food-reserve material consist?

Dissect carefully one of the large upper buds, and compare its parts with those of other buds you have examined. Note, at the base, the membraneous scales followed by four or five fleshy, cylindrical leaves or **tunics**. On removing these, we find seven or eight small, pale-yellow foliage-

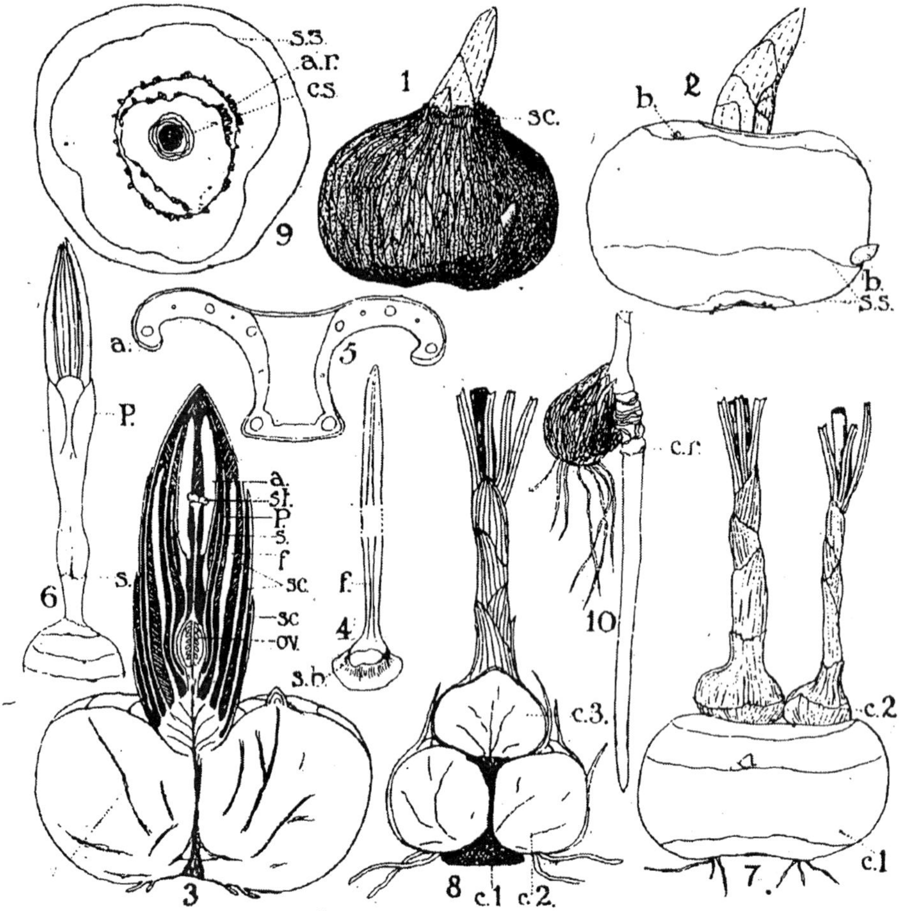

FIG. 84. THE CROCUS.—1, Crocus corm; 2, corm with scales removed; 3, vertical section of corm; 4, foliage-leaf dissected from bud; 5, transverse section of foliage-leaf; 6, flower-bud with sheath removed; 7, corm bearing two flowering shoots, a young corm forming at the base of each; 8, flowering corm in vertical section; 9, base of corm; 10, young corm with contractile root; *a*, anther; *a.r*, adventitious roots; *b*, axillary bud; *c.1*, *c.2*, *c.3*, corms of successive years; *c.r*, contractile root; *c.s*, corm-scar; *f*, foliage-leaf; P, perianth; *s*, spathe; *sc*, scale-leaf; *s.b*, sheathing base of foliage-leaf; *s.s*, circular scale-scars; *st*, stigma; *ov*, ovary; *v*, veins.

leaves. Examine these carefully with a pocket lens. Dissect off one or two leaves very carefully and note that each is attached to the axis by a sheathing ring (Fig. 84, 4). Cut a leaf across the middle and examine the cut surface, and note the back-rolled margins and thick midrib (Fig. 84, 5). The cells covering the midrib are crowded with starch grains which are used up as the leaf grows; these cells enlarge, lose their contents and become filled with air; they then reflect light from their walls and give rise to the familiar white streak of the adult leaf. In the centre are three or more flower-buds (3), each surrounded by a thin, membraneous sheath. Select one, remove the sheath, and dissect the flower (6). Outside is a short, six-lobed perianth (P), then come three stamens with short filaments and large spear-shaped anthers (*a*). The ovary is inferior and three-lobed, and above this is the long style surmounted by three large, frilled stigma-lobes (3 *st*). All the parts of the flower are present and are easily made out in the bud.

If flowering specimens (Fig. 84, 7 and 8), are examined, considerable changes will be noticed in the corm. At the base of the flowering shoots we see the beginning of a new corm (8 *c*.3), formed by the thickening of the internodes between the lower leaves. These leaves have become withered and dead, and their ring-like bases form membraneous scales around the young corm. The old corm beneath (8 *c*.2), has given up much of its food-reserve of starch, and eventually will collapse into a dead, shrivelled mass. Thus new corms arise as thickenings of the stem of an axillary bud of the old corm. In Fig. 84, 8 this relationship is shown, but in this case we can detect the collapsed remains of the still older corm at the base (*c*.1). When the old corm is cast off, a scar is left (Fig. 84, 9) at the base of the new one. This is a branch-scar, but, unlike the shoots previously examined, the new branch in the Crocus lives on as *the plant*, while the old one dies away. Round this scar and at the lower nodes numerous fibrous roots are given off in whorls (Fig. 84, 9 *a.r*). Similarly modified stems or corms are met with in the Gladiolus.

Ascent and descent in the soil. Contractile roots.—Imagine the Crocus repeating this process season after season, new corms being continually formed on the top of those of the previous year, and the mode of growth being a sympodium. What would be the position of the corm in the soil at the end of five or six years? As each year's corm is developed at a higher level than its parent, successive corms gradually approach the surface. Now it is found that many underground parts of plants have what seems to be a 'sense of depth', and if circumstances result in their being brought higher or lower than their normal depth in the soil, their behaviour is such as to raise, or lower, the young growing shoots as required.

The method adopted by the Crocus is one of which numerous examples may be found. Fig. 84, 10 shows a young corm which was developed quite near the surface of the ground; from one side a long, very thick root (*c.r*) grew, and pushed its way deeply into the firmer ground below. Its upper part then shortened and thickened, producing the wrinkles seen on the surface, with the result that the corm was pulled deeper into the soil. This process is repeated by new roots in successive seasons until the requisite depth is reached. Such roots from their behaviour are called contractile roots, and are by no means uncommon; they may be found in the Lily, Bluebell, Arum, Dandelion, and other plants.

Bulbs and droppers.—Compare the bulb of the **Tulip** with the corm of the Crocus. Cut a specimen longitudinally, as in Fig. 85, 1, and note the parts of which it is composed. On the outside are the smooth, membraneous scale-leaves (*s*.1), and about four thick, fleshy leaves (*s*.2–5), all springing from a very short and flattened stem ; then follow three foliage-leaves (*f*.1–3), surrounding a central flower (85, 2) with three outer and three inner petals (P), three outer and three inner stamens (*a*) ; and, in the centre,

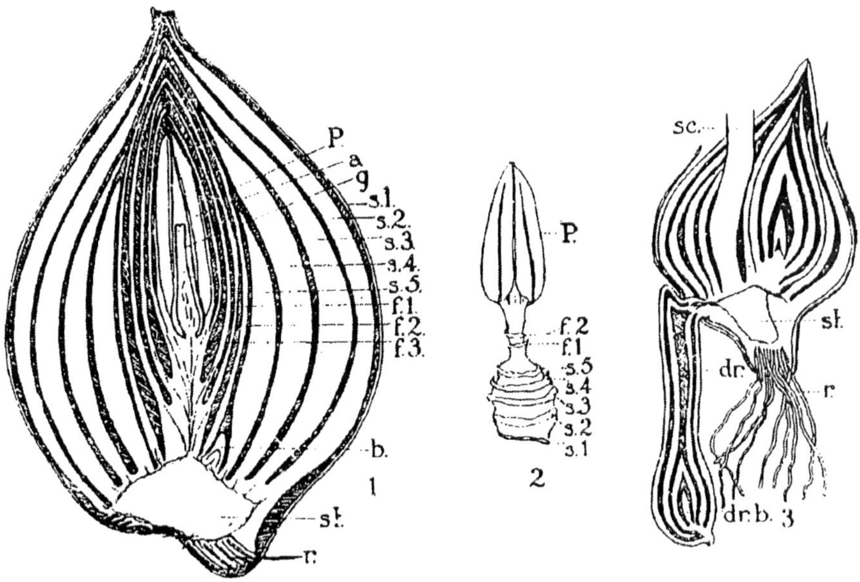

FIG. 85. TULIP.—1, vertical section of a Tulip bulb : *a*, stamens ; *b*, axillary bud ; *f*. 1, 2, and 3, foliage-leaves ; *g*, pistil ; P, perianth ; *r*, roots ; *s*.1, 2, 3, 4, and 5, scale-leaves ; *st*, stem. 2, flower-bud from bulb of Tulip : *f*.1 and 2, bases of foliage-leaves ; P, perianth ; *s*.1, 2, 3, 4, and 5, bases of scale-leaves. 3, Tulip bulb with dropper in vertical section : *dr*, dropper ; *dr.b*, dropper-bud ; *r*, roots ; *sc*, scape ; *st*, stem.

the pistil (*g*), with its three-chambered ovary, style, and the three-lobed stigma.

If a bulb is carefully dissected or cut into a series of thick slices from below upwards (Fig. 86, 1–4), it will be seen that the scales and the bases of the foliage-leaves completely surround the stem, and hence are called tunics ; while such a bulb is said to be a 'tunicated bulb'. In the axils of some of these leaves buds will be found (85, 1 *b*), which will grow and form the bulbs of another year. We thus have two kinds of leaves: (1) fleshy scale-leaves, and (2) foliage-leaves. Unlike the corm, a bulb consists mainly of leaves.

A bulb of the **Squill**, the **Snowdrop**, or the **Hyacinth** should be compared with the Tulip. In these also we have

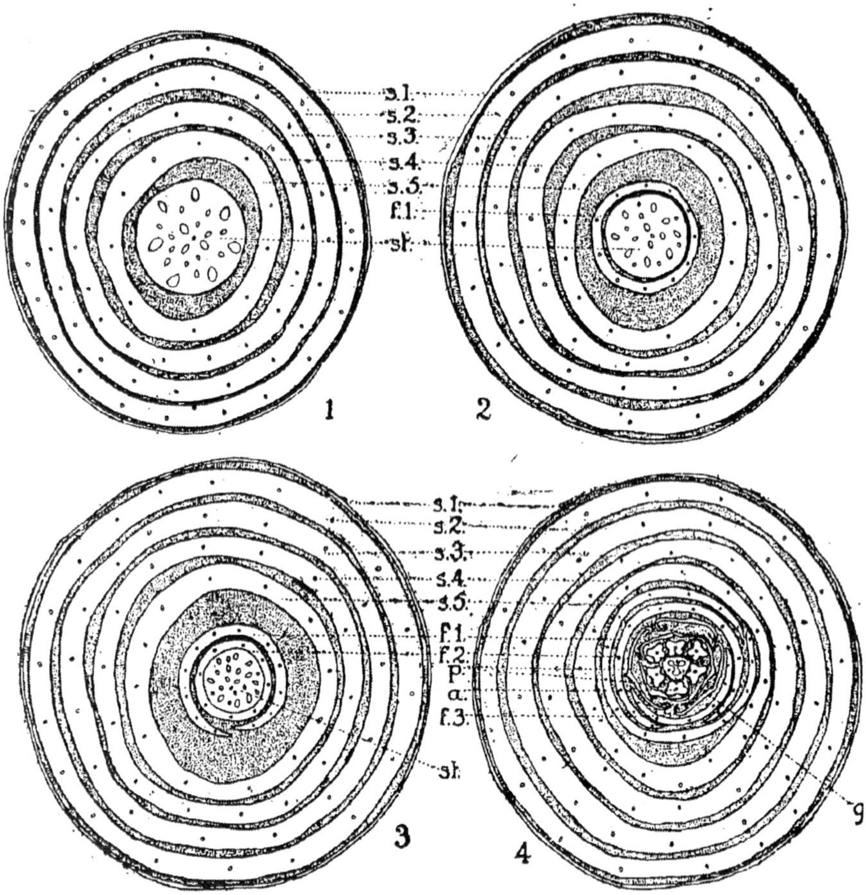

FIG. 86 (1, 2, 3, 4). SUCCESSIVE TRANSVERSE SECTIONS OF TULIP BULB.—*a*, anther; *f*.1, 2, and 3, foliage-leaves; *g*, pistil; *p*, perianth; *s*.1–5, scale-leaves; *st*, stem.

two kinds of leaves, but the bases of the foliage-leaves persist, and become swollen with food-reserves, while the green upper parts die away at the end of the season, leaving a separation scar at the top of the swollen base. In the Snowdrop, the short stem produces two narrow, green leaves and a flowering shoot, and the bases of the leaves thicken and store food, after which the green portions die away.

When digging up Tulip bulbs look out for curious forms like the one shown in Fig. 85, 3. In this case an axillary bud has pushed its way through the old outer scale and grown downwards in the form of a long, tubular, stalk-like bulb. A section through this shows a small bud at the lower end (*dr.b*). If such a bulb is carefully potted and its behaviour studied, you will find that adventitious roots are given off from the small bulb, foliage-leaves grow out into the air, and the tubular attachment to the parent bulb dies, leaving a young bulb at a lower level in the soil than the parent.

Why should some buds remain short and grow alongside the parent bulb while others elongate and push their way deeper into the soil? Seeing that the buds are axillary, and therefore produced successively at higher nodes, what would be the position of the young bulbs in the soil after several seasons' growth?

In consequence of this tendency to ascend in the soil, bulbs would eventually come too near the surface for successful development. We have seen various devices by means of which plants maintain a suitable depth in the soil, and the method adopted by the Tulip is to produce down-growing buds called 'droppers', some of which, especially in seedling plants, may be from three to nine inches long.

The means by which seedlings of bulbous plants descend in the soil and eventually reach the depth requisite for successful growth may be well studied in the **Bluebell or Wild Hyacinth**. In winter and in early spring many stages in the process can be found among the humus of the woods, and a number are shown in Fig. 87. You will find that the blue-black seeds germinate freely among the dead

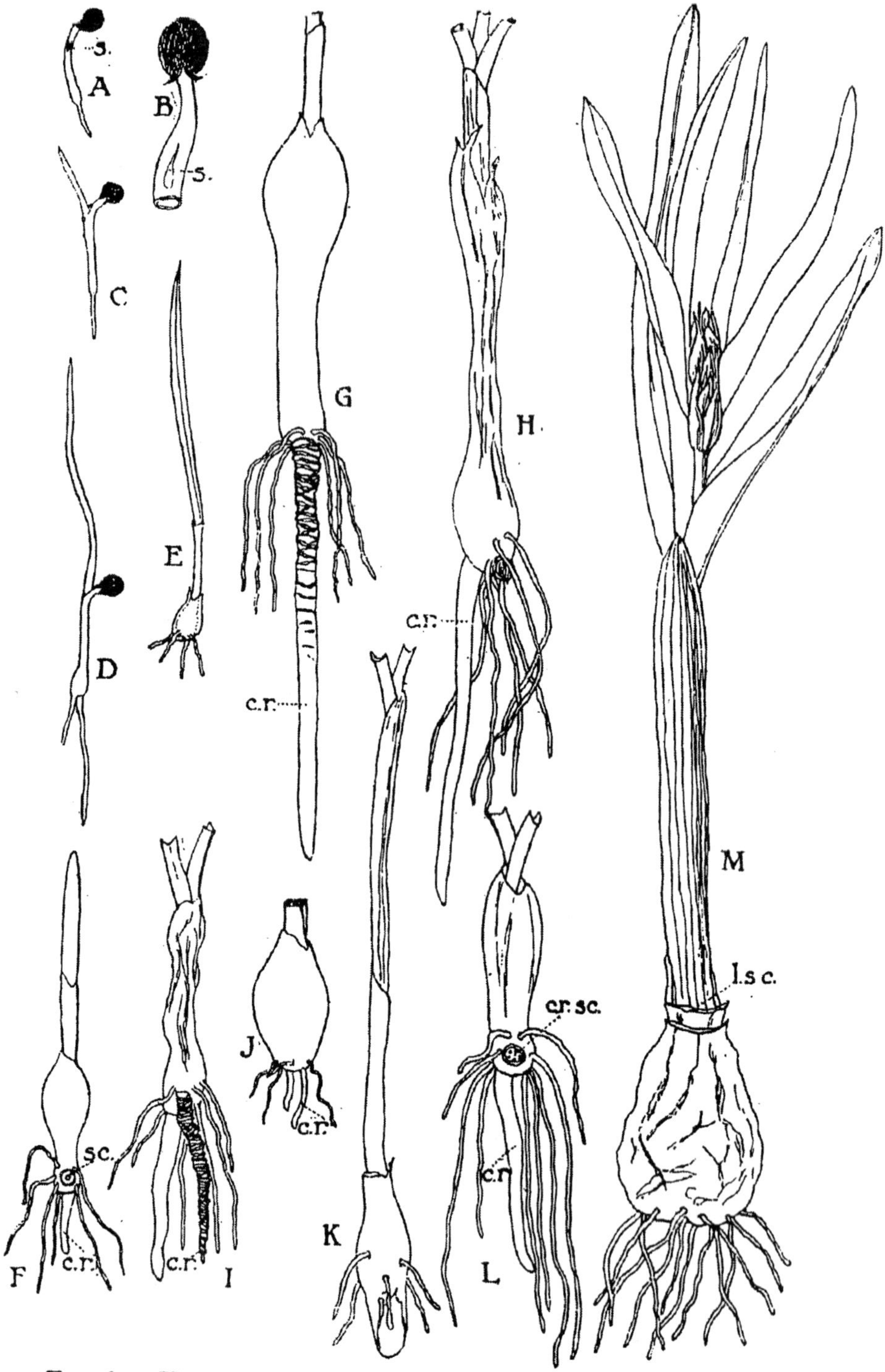

FIG. 87. HISTORY OF THE BLUEBELL BULB.—A, B, C, D, germination of the seed and seedlings; E, young bulb; F, G, H, I, L, different stages of elongation of the bulbs and development of contractile roots; J, K, bulbs again beginning to elongate; *c.r*, contractile root; *cr.sc*, contractile root-scar; *s*, slit in tubular cotyledon; M, mature flowering bulb; *l.sc*, leaf-scar.

leaves on the surface of the ground. Collect some of the seedlings and study them carefully. On germination the short radicle is carried downwards by the elongation of the single cotyledon (Fig. 87, A and B). The tip of the cotyledon is solid and remains in the seed, but it is tubular below, and has a small slit (*s*) on one side. At the base of the tube is the plumule, and when the first green leaf grows it passes up the tube and out at the slit (C, D). The tip of the cotyledon acts as a sucking organ and withdraws food from the endosperm which is passed to the growing parts below. The narrow green leaf is the first organ of photosynthesis. The food thus obtained accumulates in the bases of the cotyledon and foliage-leaf, and in consequence they become swollen and form a small bulb (Fig. 87, D E). As new leaves are formed on the stem, the bulb increases in size and begins to descend farther into the soil. It does this by the elongation of its base (F to L), and soon a curiously elongated bulb results. Often you will find on these bulbs large contractile roots (F, G, H, I, L), which aid in descent. As soon as the work of contraction is completed, a separation-layer forms across the base of each root, which then decays (Fig. 87, I) and leaves a root-scar on the bulb (F, L, *sc.* and *cr.sc*).

If several elongated bulbs are planted, allowed to grow for a few weeks, and examined at intervals, you will find that the long, outer, fleshy scale-leaves give up their food-reserve and decay (H, I). New green leaves and colourless scales are formed on the short stem within; these in turn become swollen at the base and form an oval bulb (J). At the base of each bulb roots of two kinds are formed: (1) slender, fibrous roots, and (2) long, thick roots which eventually become contractile. The processes of elongation of the bulbs and the formation and shortening of the contractile roots are repeated each season until the requisite depth is reached; then both processes cease. At each stage the bulb increases in size, and by the end of five or six years it has become a mature flowering bulb (*m*). Afterwards the Bluebell reproduces itself in two ways: (1) by means of seeds, and (2) vegetatively, by axillary buds which form new bulbs close to the parent.

In bulbs the food is stored mainly in the fleshy scale-leaves or leaf-bases. This food is used up in the spring as new leaves and flowers are formed; the old scales collapse and die, and form the dirty, shrivelled outer coverings so familiar in bulbs.

Geophytes.—The large food-store in bulbs, corms, and rhizomes provides a ready supply upon which the plant draws on the return of a favourable season for growth. It enables the plants to build up quickly new tissues and complete the growth of the young organs packed in the bud. Being situated deep down in the soil, out of reach of the frost, they are well protected, and many of these plants are among our early spring flowers. In many cases they die down early, having completed their work above ground, and after a short period of rest continue the formation of new organs in readiness for another year. Thus, much activity goes on beneath the surface and unseen throughout the greater part of the year, the actual period of rest being much less than we might suppose from a study only of the parts above ground.

Plants which pass so much of their time hidden in the soil are called **geophytes** (Gr. *gē* = the earth, *phyton* = a plant). In temperate regions, the cold season is the period of hibernation. In tropical and sub-tropical regions, hibernation occurs during the hot dry season.

Vegetative reproduction.—Underground stems of these various kinds provide very effective means of reproducing the plant and extending its range vegetatively. Those with long rhizomes are well adapted to push along and colonize new ground, like Quicks on a waste-heap or in a neglected garden, like Marram-grass on the sandy coast, or Bracken in the woods, each tending to occupy much of the ground to the exclusion of other less-favoured plants. Vegetative increase goes on not only by means of underground shoots, but to a very great extent by aerial shoots as well. For example: a plant of Silverweed (*Potentilla Anserina*) appeared in a garden and was allowed to grow. It soon produced axillary runners like those of the Strawberry (Fig. 88), and by the end of the season twelve runners

FIG. 88. VEGETATIVE REPRODUCTION IN THE STRAWBERRY.—New plantlets arising as axillary shoots on the runners.

were produced with an aggregate length of seventeen yards, and containing no fewer than 129 rooted plantlets.

Vivipary.—In the case of several Alpine plants, especially in wet autumns, a curious suppression of seed-formation occurs. The embryo, instead of passing through a period of rest in the seed, continues its growth uninterruptedly, and on the inflorescence is borne a number of small plantlets instead of fruits. These eventually drop off and reproduce the plant. Such plants are said to be **viviparous** (L. *vivus* = alive, *paro* = I bring forth).

Vivipary occurs also in species of Leek and Garlic, and the young bulb-like plantlets on the inflorescence are called bulbils'. Axillary buds sometimes drop off and form new plants; good examples are found in cultivated species of Lilium and sometimes in the Lady's Smock (*Cardamine pratensis*). Reproduction by means of axillary tubers in the Lesser Celandine has already been noticed (p. 64).

A similar mode of vegetative propagation occurs in the Wood Sorrel, and examples may often be seen in the Oxalis so common in greenhouses. In some Ferns, numerous small plantlets are produced on the fronds by vegetative budding.

Social plants.—Offsets and short axillary shoots form a very effective means of spreading and give rise to densely packed masses of plants which elbow out their weaker rivals. Plants of the same species which grow in company and cover a large patch of ground are termed **social** species.

It is by such means that the beautiful flowery cushions of rock-plants are formed on the mountains, and the tussocks of sedges and grasses which produce the monotonous Cotton-grass moors, and the grassy swards of the hills and pastures, and thus give rise to some of the most striking features in the vegetation of a country. The extensive tracts of Bracken in the woods and on the hill-slopes, the blue carpets of Wild Hyacinth, and lakes and canals choked by water-weeds, are a few examples of the spreading of plants over large areas, not by seeds, but by vegetative means.

CHAPTER XI

MOVEMENTS AND ATTITUDES OF PLANTS

MOVEMENT is one of the common phenomena of life. We usually look upon plants as stationary, and the power of movement as characteristic of animals, but this is far from a correct view of the case. Although a typical flowering plant is fixed to the soil, all its growing parts, roots as well as shoots, execute definite movements, and even in parts that are mature, definite movement occurs. As we have seen, plant organs execute movements in response to such stimuli as light, gravity, and moisture. Roots usually turn away from the light and shoots turn towards it; underground stems ascend and descend in the soil and are aided in their descent by contractile roots.

Nutation. Twining plants.—We have now to notice the movement characteristic of aerial stems. Plants growing in woods, hedgerows, and other shady places, tend to develop longer and more slender stems and thinner leaves than plants of the same species grown in open, sunny places. If the stems of these plants be observed it will be found that the growing tips move in a more or less circular orbit. This movement is called **nutation**. Plants such as the **Convolvulus** (Fig. 89) and **Black Bryony** (*Tamus communis*) (Fig. 90) develop long, slender internodes and relatively large leaves, and the stems, too weak to hold the shoot erect, lean on other and sturdier plants for support. Their growing tips describe a wide spiral, making a complete revolution in from one to two hours, and, on coming into contact with a shoot of suitable diameter, wind round it. As growth continues, the spiral so formed is drawn tighter, clasps the support firmly, becomes thicker and stronger, especially on the convex side, so that it cannot

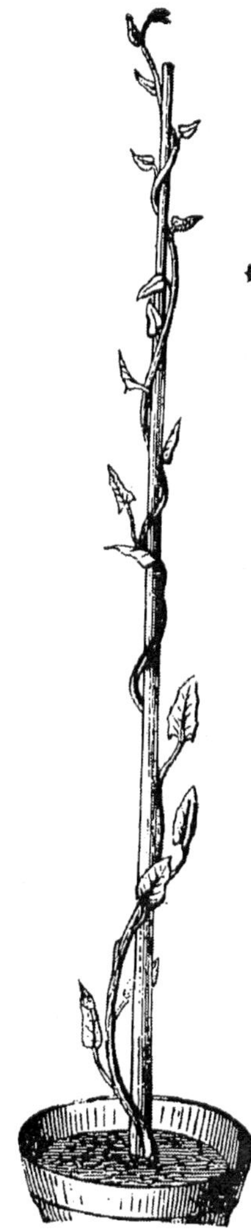

FIG. 89.
TWINING STEM OF CONVOLVULUS ARVENSIS (Pfeffer).

untwine. Observe the direction of twining in each case. The stem of the Convolvulus, looked at from above, twines from right to left (contra-clockwise). Most twining plants twine in the same direction; that of the Black Bryony, however, twines from left to right (clockwise). By this means such plants are able to climb many feet above ground, and to reach the air and sunlight without the expenditure of energy required in building up strong erect stems.

Climbing organs sensitive to contact. Tendrils.—In the **White Bryony** (*Bryonia dioica*) (Fig. 91), Clematis (Fig. 92), Wood Vetch (Fig. 220), and Sweet-Pea (Fig. 131), slender climbing organs called **tendrils** are developed, which differ from climbing stems in being sensitive to contact.

The **Passion Flower** has very sensitive branch-tendrils whose movements are easily observed. Fig. 93 shows the result of an experiment with one such tendril. At 3.10 p.m. the concave side of the tendril was gently stroked with a slender stick and records of its movements were taken, with the result shown in the diagram. If, in describing such a spiral, the tendril meets with a support, it twines round this support and clings firmly. Spiral growth continues, but being now fixed at both ends, the tendril soon develops a reversed spiral. That this form of spiral should be produced can be easily understood if you fix a piece of string at both ends and turn the middle portion: the part on the right turns in the opposite direction to that on the left. Tendrils showing the reversed spiral are also seen in the White Bryony (Fig. 91). The tendrils of the Virginia Creeper (Fig. 94), which are also modified branches, are peculiar in that they move away from the light (negatively heliotropic). At the free ends disks are formed, which, when stimulated by contact with a rough surface, become coated with mucilage and are thus cemented to the support.

Tendrils are sensitive thread-like plant organs by which

Fig. 90. Black Bryony, Stems twining from left to right, its Leaves turned towards the Light.

Fig. 91. White Bryony, climbing by Branch Tendrils.

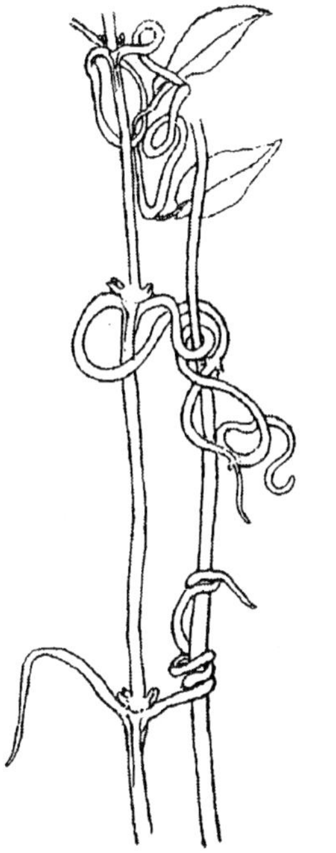

FIG. 92. LEAF-STALK TENDRILS OF CLEMATIS.

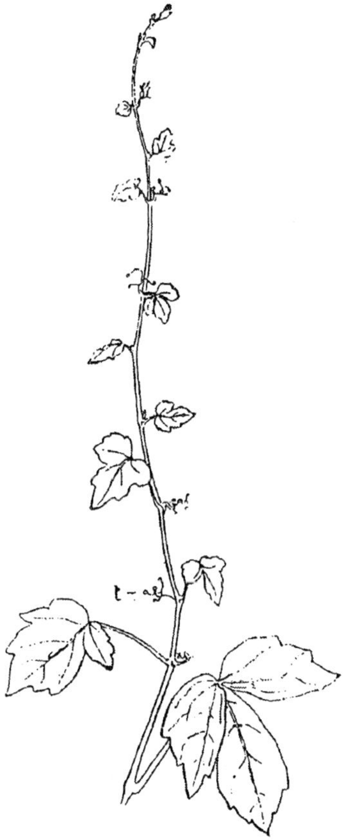

FIG. 94. BRANCH OF VIRGINIA CREEPER.—*t*, branch-tendrils with adhesive disks; the shoot shows transitions from simple to compound palmate leaves.

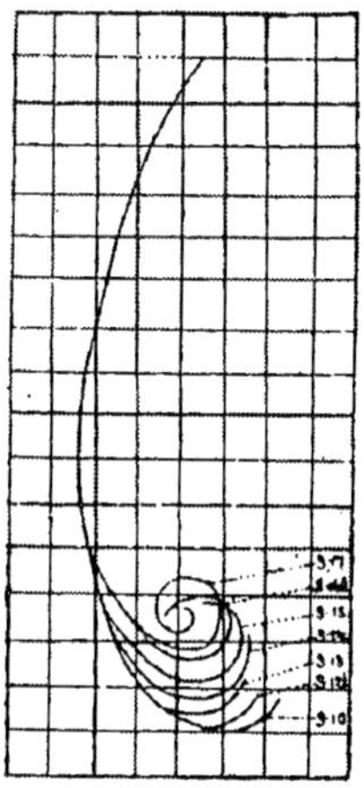

FIG. 93. MOVEMENTS OF THE BRANCH-TENDRILS OF THE PASSION FLOWER.

FIG. 95. SHOOTS OF GOOSEBERRY.—1, position of leaves in the shade; 2, in the sunlight; P, leaf-base prickles.

a plant fixes itself to a support. They may be modifications of (1) branches, e.g. Passion Flower, Vine, Virginia Creeper, and White Bryony; (2) leaves, e.g. the Chick-Pea; (3) petioles, e.g. Clematis; (4) leaflets, e.g. Sweet-Pea, Vetch, and Garden Pea; or (5) stipules, e.g. some species of Smilax.

Sun and shade positions.—If you note the differences of position of leaves in sunny places and compare these leaves with others of the same kind growing in the shade, you will see that they take up a favourable position with reference to light. Fig. 95 shows two shoots of **Gooseberry** taken from different sides of the same plant. One (1) was overshadowed by branches of an apple-tree, and exposed its leaves fully to the available light. The other (2) was not overshadowed, and its leaves moved in such a way as to expose the edges of their blades to the direction of the sun's rays. As chlorophyll is decomposed by strong sunlight, it is an advantage to a plant organ to assume a position whereby the smallest surface is exposed to the direct rays of the sun.

Fixed light position.—Some **Acacias** (Fig. 96, 1) growing in sunny regions have curiously modified leaves, which persistently turn their edges to the sky and give a characteristic appearance to the plants. The blades are not developed, but the leaf-stalks are flattened out and become leaf-like, and are called **phyllodes.** The attitude assumed by the foliage of a plant with reference to light is called the 'fixed light position'. In some plants the leaves are reduced to scales, the stems flatten out, resemble leaves, and, as in the Acacias, turn their thin edges to the sky. Such leaf-like stems are known as **phylloclades**, and occur in the Butcher's Broom. Other examples are the **Smilax** (*Myrsiphyllum*) (Fig. 96, 2) and species of **Asparagus.** Extraordinary examples occur in **Cacti** (Fig. 96, 3) and other desert-plants. The huge fleshy stems are green, store a large supply of water, and do the work of foliage-leaves. The leaves are often reduced to stiff, radiating spines (*l.s*) forming a gauze-like covering to the surface, which acts as a most effective light-screen and protects the chlorophyll of the stem from the too powerful rays of the sun. Thus, sunlight is an important factor in determining the position of stems and leaves, and there is a

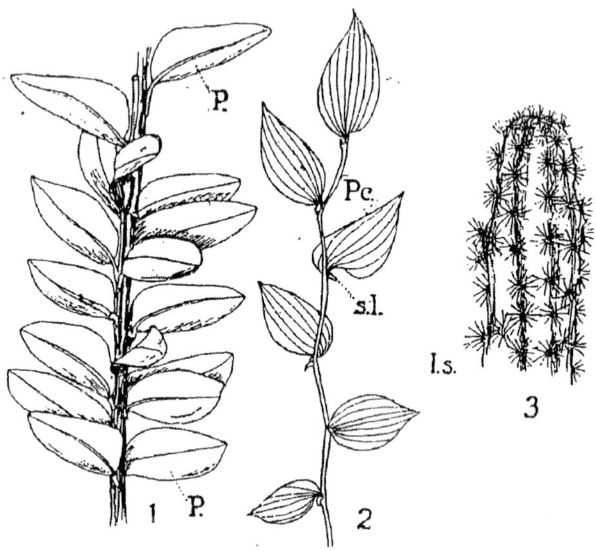

FIG. 96. STEMS WHICH PERFORM THE FUNCTIONS OF LEAVES.—1, Acacia; 2, Smilax; 3, Cactus; *l.s*, leaf-spines; P, leaf-like petioles (phyllodes); *Pc*, leaf-like stem (phylloclade); *s.l*, scale-leaf.

tendency for plants growing in dry and very sunny places to take on strange shapes and exhibit curious devices which help the plant to survive under trying conditions.

Protective movements against radiation and transpiration.—Movements which have for their object reduced radiation and transpiration are also very common. Note how young leaves emerge from the buds in spring in, e. g., the **Elm** (see Fig. 198, 1), **Beech** (see Figs. 75 and 76), **Lime**, and **Horse-Chestnut** (see Fig. 68). At first they are erect, then as they unfold and grow they bend over and hang downwards with their under surfaces applied one to another so as to expose a relatively small area to the cold and drying winds. As they expand and grow larger and stronger they assume a favourable position with regard to air and sunlight and form a mosaic.

Motile organs of hedgerow plants.—Examine the trees and shrubs in a wood or hedgerow and note how frequently the blades face the light. What part has moved to bring them into this favourable position? Privet, Yew (Fig. 97), Ivy (Fig. 222, 1), White Bryony (Fig. 91), Convolvulus (Fig. 89), Elm, Sycamore, and many other examples will be found. The bases of such leaves are swollen, and it is this cushion which is usually the organ of movement. Sometimes there is a motile organ at the upper end of the leaf-stalk near the blade. In other cases the cushions become highly specialized organs, and the leaves, and even the leaflets, are able to execute periodic, and sometimes rapid, movements. Observations should be made on a few common plants, such as the White Clover and Wood Sorrel.

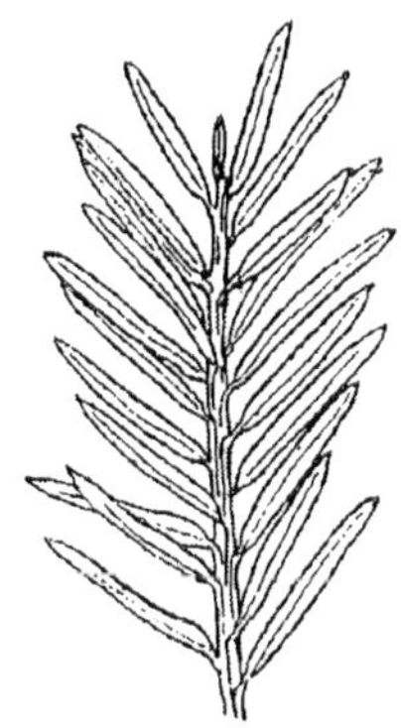

FIG. 97. BRANCH OF YEW.—Leaves arise spirally on the axis, but the blades turn to the light.

Sleep-movements in Clover, Wood Sorrel, and False Acacia.—The **Clover** leaf (Fig. 98) is borne on a long stalk with a cushion at the base, covered by a pair of stipules. During the day the blade at the end is horizontal and divided into three leaflets (trifoliate). A slight variation in the form of the cushion will cause much movement of the blade at the end of its long lever. Similar cushions are found at the base of each leaflet, and at dusk, by their aid, the two side leaflets move into a vertical position, exposing their inner edges to the sky and their outer edges to the ground. The middle leaflet now bends over them, folding the two sides of the blade downwards and exposing the back of its midrib to the sky. The amount of leaf-surface exposed to radiation is thereby greatly reduced, and in keeping with this, the lower and more exposed leaf-surfaces contain fewer stomata than the upper and more protected ones. During the day Clover leaves form an excellent mosaic, but at night, when the leaflets are tucked in, the smallness of the space they occupy is very striking.

Compare with the Clover leaves those of **Wood Sorrel**

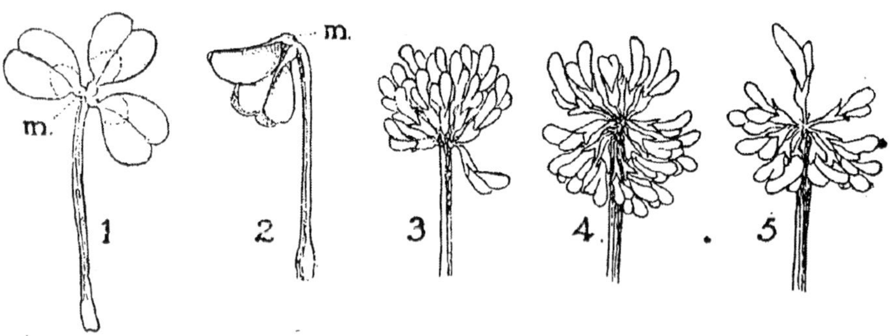

FIG. 98. WHITE CLOVER.—1, trifoliate leaf, day-position; 2, night-position; *m*, motile organs; 3, 4, and 5, inflorescences of Clover: the flowers turn downwards after pollination.

(Fig. 99) or a common garden **Oxalis**. In these plants the three leaflets droop at night, hang vertically, apply their midribs to each other, and so expose their upper and protect their under surfaces, to which the stomata are restricted. They thus secure protection against cold, but by a different method from that of the Clover. If these plants are placed in the dark at midday they do not close their leaves until the proper time; their habit of going to sleep at definite times has become fixed, and it takes some days for them to become accustomed to changed hours. Such movements are known as **sleep-movements**.

The False Acacia (*Robinia*) furnishes another example of motion in plants, and its leaf-movements should be studied. Each leaf is pinnate, i.e. with leaflets arranged on each side of the midrib like the pinnae of a feather.

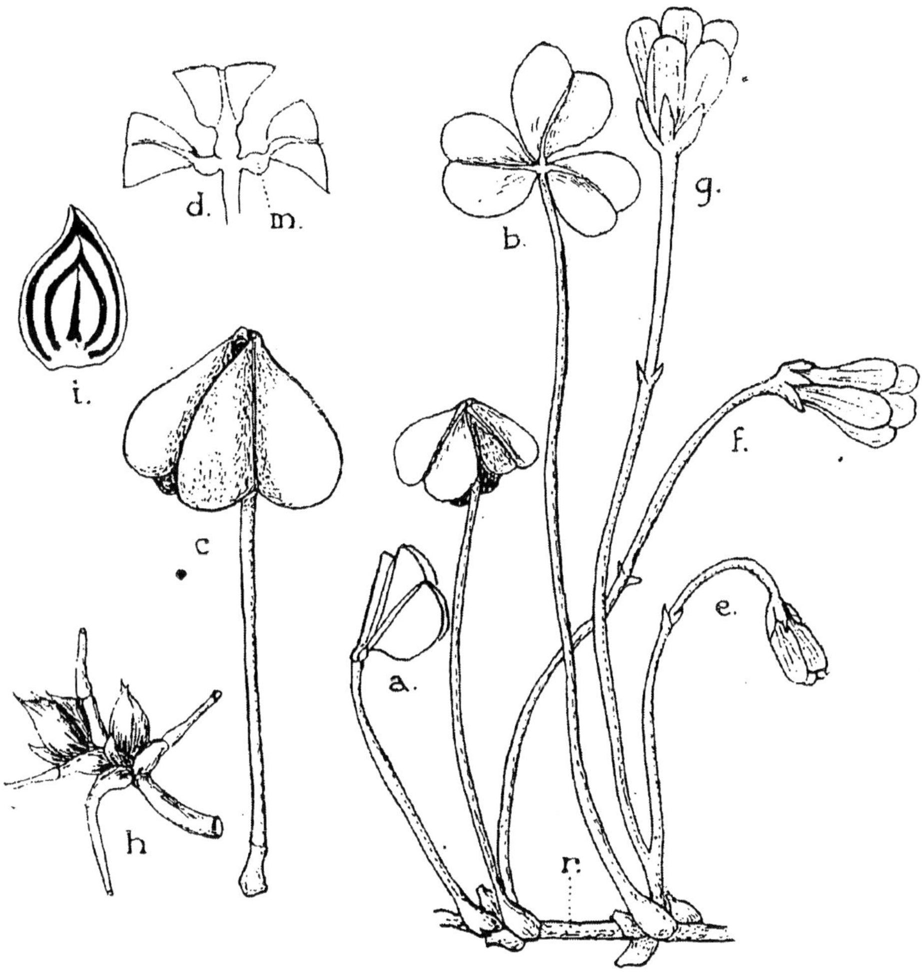

FIG. 99. WOOD SORREL.—*a*, young leaves as they first open; *b*, day-position of leaflets; *c*, night-position; *d*, bases of leaflets enlarged; *e*, *f*, *g*, movements of growing flower-stalk; *h*, bulbils in the axils of leaves; *i*, vertical section of a bulbil; *m*, motile organs; *r*, rhizome.

A leaf of the False Acacia, like most pinnate leaves, ends in a single leaflet. At the base of the leaf are two stipules transformed into spines. During the day the leaflets are horizontal, but in intense sunlight they move upwards, apply their upper surfaces each to the opposite one, and point their tips to the sky. At night they droop and apply their under surfaces together. Thus they obtain the advantages of favourable light, escape the injurious effect of intense light, and, on assuming the night-position, reduce the loss of heat from radiation.

Rapid movements in Sensitive Plants.—Some plants exhibit the power of movement to such a degree as to have earned the name of Sensitive Plants. The most familiar example is *Mimosa pudica*, the Sensitive Plant. Its leaves are bipinnate, i.e. each leaflet or pinna is again pinnately divided into similar segments or pinnules. The end of each leaflet has a pair of pinnules. At the base of each leaflet and pinnule, and also at the base of the leaf-stalk, there is an organ of movement, and the leaves exhibit sleep-movements such as are seen in the Clover and Wood Sorrel. So sensitive are the leaves, that a very slight stimulus causes the leaflets to droop in the daytime. If a lighted match be held under the end of a leaf, the heat-stimulus produces a series of remarkable changes. Not only do the heated pinnules droop, but the stimulus is transmitted from one to another, pinnules and leaflets drooping in succession, until, eventually, the stimulus reaching the cushion on the leaf-base, the whole leaf hangs down languidly. There it remains until the shock has passed off, when it gradually regains its former position. The leaves of the Venus' Flytrap close up in a similar manner, but very rapidly, in response to a contact-stimulus (see p. 366).

These movements are due to rapid changes in the turgidity of the cells of the cushions; the effect of a stimulus is to cause water to escape from the turgid cells of the cushion into the neighbouring air-spaces. Later, as the cells once more become turgid, the leaves and leaflets resume their 'awake' position.

Movements of flowers and fruits.—Flower-movements

are equally interesting and easy to observe. Sleep-movements are common and, in different species, occur at different times of the day and night. Usually flowers pollinated by day-flying insects are open by day and closed at night, while flowers visited by night-flying moths are open in the evening or at night, when they are often sweet-scented, and are white or pale yellow in colour. Some plants owe their common names to their habit of opening

FIG. 100. EVENING PRIMROSE, SHOWING THE MOVEMENTS OF THE PARTS AS THE FLOWER OPENS.

and closing their flowers, e. g. the Daisy (day's-eye), John-go-to-bed-at-noon, and Poor Man's Weather-glass. These movements are related either to the habits of the insect-pollinators, or to weather-changes, many flowers closing during cold, dull, or wet weather, and so protect their honey and pollen.

The opening of some flower-buds occurs so quickly as to be easily observable. It is quite exciting to watch the flowers of the Evening Primrose as they open on a warm summer's evening. Fig. 100 shows the stages observed in two flowers. The calyx slips down on one side and the four free tips curve back and reveal the rolled-up petals. The corolla unscrews at the base, causes the calyx to split more and more, and as the petals unroll, the mouth of the corolla opens and the stigmas appear. The calyx now splits at the bottom, the sepals suddenly turn backward, or inside out, with a distinctly audible click, and the crinkled petals unroll, slide over one another, and soon fully expand. The sound produced resembles that of two sheets of paper, one gliding over the other. If we look at the stamens we find that the stringy pollen is already hanging out of the anthers, while the stigmas, not yet ripe, spread out their four lobes well above them. As many as a dozen flowers may be seen to open in this way on one plant in half an hour.

The various attitudes that flowers assume should be carefully observed under the following conditions: (1) in bud, (2) in flower, (3) in sunshine, (4) in cloudy and wet weather, (5) during the day, (6) at night, (7) as the fruit ripens, and (8) when the fruit is ripe. Figs. 101 and 102 show the movements of the flowers and fruits of the **Wild Hyacinth.** Notice that the flowers are erect in bud; later they turn away from the axis, expand and hang downwards, the lowest and oldest opening first. After pollination they become erect again, while the fruit-stalk lengthens and becomes rigid. In the **White Clover** (see Fig. 98), the flowers are erect in bud, horizontal in flower, and after fertilization hang downwards. The **Wood Sorrel** (see Fig. 99) droops both at night and in dull, damp weather; it is erect in fine, sunny weather, droops as the seeds ripen, and is again erect in fruit.

Fig. 103 shows the movements of the parts in an opening inflorescence-bud of the Horse-Chestnut.

FIG. 101. FLOWER-MOVEMENTS OF WILD HYACINTH.

FIG. 102. FRUIT-MOVEMENTS OF WILD HYACINTH.

FIG. 103. THREE STAGES IN THE OPENING OF THE TERMINAL INFLORESCENCE-BUD OF HORSE-CHESTNUT.

FIG. 104. DANDELION.—The stalks of the open flower-heads are erect; after pollination the heads close up and the stalk bends over to the ground.

The movements of the flowers and flower-stalks of Poppies, Columbines, and Bellflowers (*Campanula*) should be compared and their differences noted.

In some cases the whole inflorescence is involved in the movements, e.g. in the **Dandelion** it is erect in bud, and when the flowers are open (Fig. 104); at night and in wet weather the flowers close and are protected by the inner whorl of bracts. After pollination the stalk grows, bends over and lies almost prostrate, becoming erect again as the fruits ripen. At this stage the disk enlarges and becomes convex, the bracts turn backwards against the stalk, the pappus-hairs spread outwards and the fruits are ready for dispersal by the wind. In the **Coltsfoot** the stalk is erect in the bud and in the flowering stages. After pollination the upper part bends over, bringing the head into a drooping position, and the fruits, protected from the rain, complete their development. As they ripen, the stalk elongates, becomes erect and rigid, and raises the fruits into a favourable position for dispersal.

The power of movement in plants is an important aid to protection. By this means, during suitable sunlight, leaves are placed in the most favourable position for photosynthesis; during intense light, the adoption of the edgewise position protects the chlorophyll against decomposition; leaves and leaflets applied to one another reduce the exposed surface and check loss of water by transpiration and loss of heat by radiation. Flower-movements protect the honey and pollen from rain and from useless insects, and finally the position of ripening fruits is a protection against rain and when ripe they are moved into the most favourable position for seed-dispersal.

PART II

THE REPRODUCTIVE ORGANS

CHAPTER XII

BIOLOGY OF THE FLOWER. DICOTYLEDONS

I. Pollination of Simple Flowers by Wind and Insects

IN the study of buds, corms, and bulbs, we have become familiar with the facts that shoots are frequently condensed; that the internodes, instead of elongating, remain undeveloped; and that, in consequence, a number of leaves spring close together from the short axis. Some of the leaves arise singly and are arranged in a close spiral, while others stand two at a level, in crossed pairs.

The flower a condensed and modified shoot.—In our examination of the Stock flower we found something very similar, viz. a condensed or dwarf shoot, with a tendency for the flower-leaves to arise close together in crossed pairs. This condensation is characteristic of flowers, and it is interesting to note how frequently flowers arise on dwarf, leafy shoots, or spurs, as in many fruit-trees. This shortening of the axis, together with the great modification that has taken place in the size, shape, colour, and function of its parts, distinguishes a typical flower from any other part of the plant.

Flowers, however, have not arisen in this simple way from a leafy shoot. It is more probable that stamens and carpels, or their equivalents, came into existence first, and that petals, and perhaps sepals, were derived from them by modification of their parts, as may be seen in double flowers like roses, and in the **White Water-Lily** (Fig. 105). Very ancient flowers had many stamens and carpels arranged spirally on the axis, but in modern flowers the parts are fewer in number and usually arranged in cycles or whorls.

Generally the flowers appear towards the end of a season's activities. In an annual they herald the closing scenes of its life-cycle and provide for the formation of fruits and seeds, which will soon be all that remain to perpetuate the race. In some cases flowers appear early in the season and

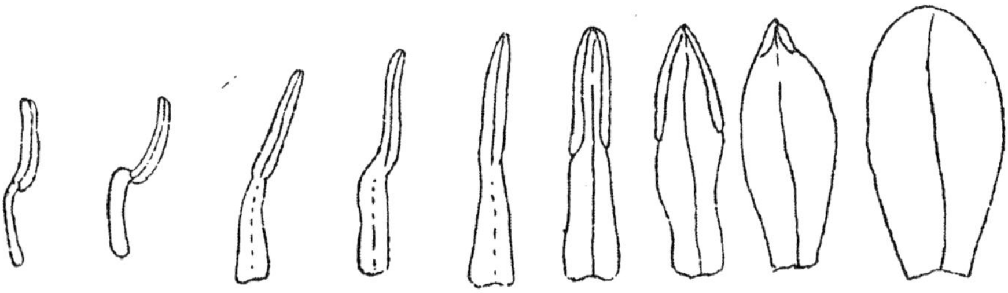

FIG. 105. STAMENS OF WATER-LILY, SHOWING TRANSITION FROM STAMENS TO PETALS.

before the leaves, as in many trees, and in the Coltsfoot, Wood Anemone, and many other herbs. In the Autumn Crocus, or Meadow Saffron (*Colchicum autumnale*), on the other hand, the leaves complete their work and die down before the flowers appear. Whether early or late, however, the chief object of the flower is to produce fruits containing seeds, which, on falling to the ground, may produce a new generation; and all the parts of a flower directly or indirectly serve this end.

Structure and functions of the parts of a flower.—The four parts of a typical flower are usually arranged in successive whorls on the short axis, which is known as the receptacle. The lowest and outermost is composed of small green sepals which form the calyx, and which in the bud completely cover and protect the other parts. The second whorl—the corolla —consists of brightly-coloured petals, which are often scented, and sometimes bear honey-secreting glands. In consequence of their colour, scent, and honey, they are attractive to insects. The two inner whorls differ from the outer ones in an important respect. They bear reproductive bodies called **spores**; such spore-bearing organs are known as **sporophylls**. The whorl of sporophylls lying immediately within the corolla is the **androecium**, and consists of small-stalked bodies, the **stamens**. Each

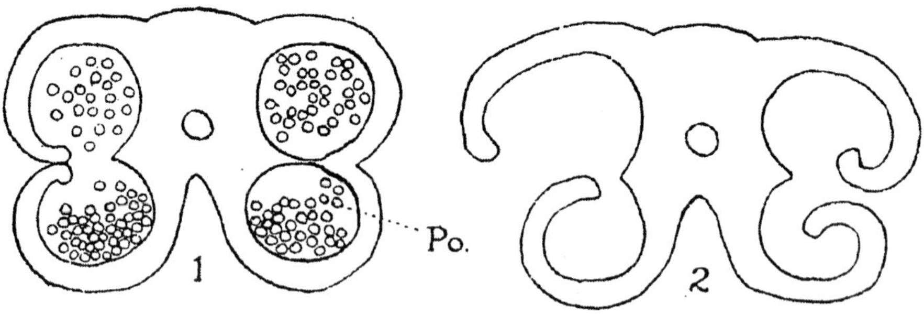

FIG. 106. TRANSVERSE SECTIONS OF ANTHERS.
1, before; 2, after dehiscence; *Po*, pollen-grains.

stamen has a slender stalk or **filament**, bearing on its free end the **anther**; this consists of four parallel **pollen-sacs** (Fig. 106), or **microsporangia** (Gr. *mikros* = small, *spora* = a seed, *angeion* = a case), within which are a large number of pollen-grains or spores (*Po*), whose production is the special function of the stamen. These minute spores are called **microspores**, and the organ which bears them (the stamen) is the **microsporophyll** (Gr. *phyllon* = a leaf). In some flowers the stamens are attractive in colour, and the pollen is an important food for bees and other insects.

The uppermost part of the axis gives rise to sporophylls of a different kind, the **carpels**, which together constitute the **gynoecium**, or **pistil**. Usually three parts of the pistil may be distinguished: (1) the **ovary**, within which the ovules are developed; (2) the **style**, growing from the top of the ovary and ending in (3), a **stigma** which, when ripe, is coated with a sticky sugary secretion. To this the pollen-grains adhere and germinate. The ovules, after fertilization by the pollen, undergo changes which result in the formation of seeds. Each ovule or **megasporangium** (Gr. *megas* = large) contains a **megaspore** (the embryo sac), and the sporophylls which bear them are known as **megasporophylls.**

The essential work of the flower is to secure the transmission of pollen-grains from the stamens to the stigmas, a process known as **pollination**, so that fertilization of the ovules may occur and seeds be formed. The flowers of different plants vary considerably in structure according to the way in which pollen is conveyed. It may be carried by the wind or by insects, or the arrangement and behaviour of the parts may be such as to transfer pollen from the anthers to the stigmas of the same flower. If we examine a number of common flowers from this point of view, we shall learn much of their structure and modifications, and also of their special use to the plant. Each flower should be carefully examined, and floral diagrams and drawings made to show the relationships of the parts, especially as seen in a vertical section.

Flowers pollinated through the agency of wind.—The flowers of the Hazel or Oak (Figs. 187 and 194) are arranged in a catkin. Each flower of the long yellow catkin is much smaller than that of the Stock; outside are five or seven green scales; within are five to twelve stamens, but no pistil and no corolla are present. The flowers in the smaller bud-like catkins have six small scales on the top of an ovary with three chambers, and there are three large sticky stigmas to catch the pollen. From these flowers both stamens and corolla are absent.

Many of our forest trees produce similarly reduced flowers. This arrangement, in which the stamens and the pistils are in different flowers but on the same tree, is called **monoecious** (Gr. *monos* = one, *oikos* = a house). Such flowers are inconspicuous and unscented, and the staminate flowers produce a large quantity of pollen which is loose, dry, and light, and easily carried by the wind to the large stigmas of the pistillate flowers. In the **Willows** (Fig. 185) and **Poplars** (Fig. 186) the two kinds of catkins are borne on different trees, and this arrangement is called **dioecious** (Gr. *di* = two). In all such cases the pollen carried to the stigma comes from another flower, and when this occurs the flower is said to be cross-pollinated. In Willows the stamens are bright yellow and numerous, and each flower contains a honey-gland or **nectary** at the base. Insects often visit these catkins and collect from them both honey and pollen, with which their bodies may become dusted. Thus the pollen may be carried to a pistil-bearing catkin and some of it deposited on the stigmas.

Fig. 107. Ripe Stigmas of Mallow curling among the Anthers.

From the abundance of fruits produced by such trees it is obvious that simple and unattractive as the flowers are, they yet contain all that is essential for fruit-production. Hence stamens and pistil are spoken of as the **essential organs** of the flower. In the flower of the Stock other parts are present, viz. the sepals, which are protective, and the petals, which are attractive. Both parts are useful, but, as we have seen, not essential, for the production of seeds.

Self-pollinated flowers.—In the flower of the Round-leaved **Mallow** (Fig. 107) there are five free sepals and five free petals; the stamens are numerous, but their filaments are all joined to form a tube round the pistil, hence called **monadelphous** (Gr. *adelphos* = brother). The pistil is superior, the carpels numerous and syncarpous. There are many long stigmas which, if not cross-pollinated, grow, curl over among the anthers, and thus receive pollen

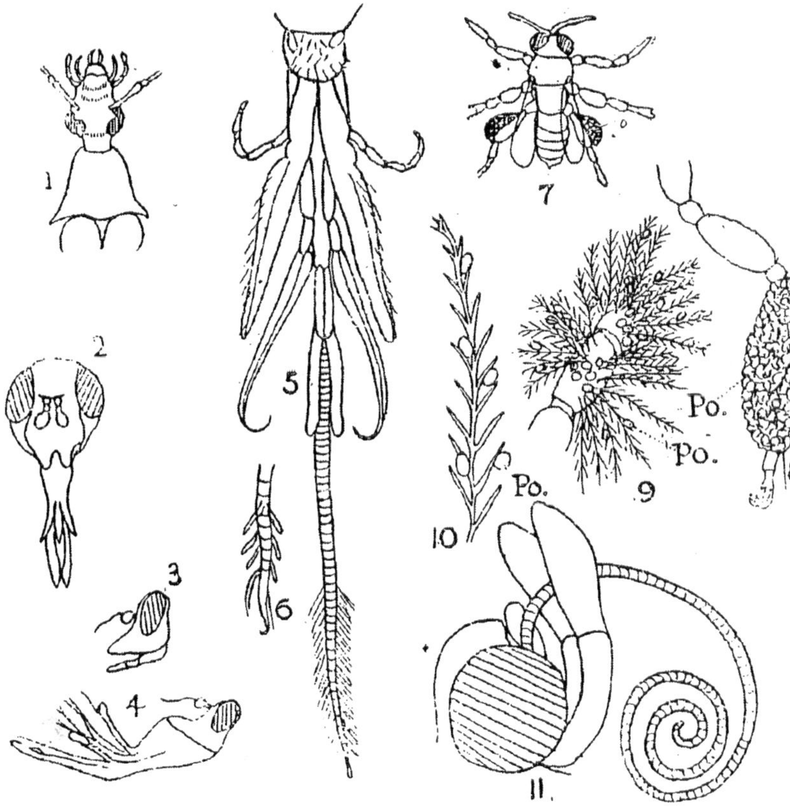

FIG. 108. MODIFICATIONS IN THE MOUTH-PARTS AND LEGS OF INSECTS WHICH COLLECT HONEY AND POLLEN FROM FLOWERS.—*Po*, pollen grains (after Sharp and Müller).

from the same flower. By this means self-pollination takes place.

Flowers attractive to, and pollinated by, insects.—Flowers which develop bright colours, scent, or honey, attract large numbers of insects, which feed on the honey and pollen;

and many of the modifications found in flowers are paralleled by modifications of the mouth-parts of insects (Fig. 108). The mouth-parts of the simpler insects, e. g. beetles and flies, are so short that they are unable to reach the honey unless it is exposed in an open, shallow flower. Honey in a deep tube is inaccessible to them, and only insects with mouth-parts elongated to form a proboscis are able to reach it. Some flowers have tubes several inches long (8 to 10 inches), and there are insects with probosces long enough to obtain honey from the bottom of them.

Insect pollinators and their mouth-parts.—Fig. 108 shows the mouth-parts of various insects which visit flowers. 1 is the head of a beetle (*Strangalia attenuata*) which can lick honey from shallow flowers; 2, the head of the Drone Fly (*Eristalis arbustorum*) with an extensible proboscis; 3, the head of a fiy (*Rhingia rostrata*), in side view; 4, the proboscis fully extended; 5, the mouth-parts of the Long-tongued Bee (*Anthophora pilipes*); 6, the end of the proboscis enlarged to show the 'honey-spoon'; 7, the Honey Bee (*Apis mellifica*) with masses of pollen (*Po*) on the hind legs; 8, the hind leg of a bee with mass of pollen; 9, collecting-hairs on the leg of a Honey Bee; 10, a hair magnified, with pollen-grains attached; 11, proboscis of a moth, which is very long and coiled up like a watch-spring.

Such insects perform unconsciously a valuable service in carrying, on their bodies, pollen from the anther to the stigma, and their habit of visiting flowers for food has probably been an important factor in the evolution of many flowers. On the other hand, the insects have themselves become modified, especially in the organs surrounding the mouth.

Insects, useful and injurious.—In the mature stage, insects are often very useful to flowers, and in many cases seeds can only be developed when insects act as pollinators. It is during this stage in their life-cycle that insects lay their eggs. In time the eggs are hatched, and the very small grubs or caterpillars have to search for food with which to complete their development. As they are mostly vegetable feeders, the caterpillars or larvae, especially of moths and butterflies, do much damage to plants; and, if they are abundant, may strip a whole forest of its leaves in a short time. Garden and field crops often suffer greatly from this cause, resulting in a loss of many thousands of pounds.

Insects of many kinds, either in their larval or adult stage, are destructive to plants, both wild and cultivated. On the other hand, some are useful. Among the more destructive are:

Injurious insects.—(1) The small Aphides or Plant Lice, and Scale Insects (*Hemiptera*), which make great ravages in both garden and field.

(2) The larvae of butterflies and moths (*Lepidoptera*), e.g. the Cabbage White Butterfly, the Cabbage Moth, Magpie Moth, and Antler Moth, which injure field and garden crops. The Lackey Moth, Buff-tip, Vapourer Moth, Ermine Moth, and several *Tortrices* affect fruit and forest trees.

(3) The larvae of many flies (*Diptera*), such as the Wheat Midge, Gout Fly, Daddy-long-legs or Crane Fly, Cabbage Fly, Radish Fly, Mangold Fly, Onion Fly, and Root Fly are pests on roots and other crops.

(4) Larvae and mature beetles (*Coleoptera*), especially the Mustard Beetle and many Weevils.

(5) Many *Hymenoptera*, such as the Gooseberry Saw-fly, Pine Saw-fly, Turnip Saw-fly, and Corn Saw-fly, are often very destructive to both herbaceous and woody plants.

Fortunately for man, these pests are the chief source of food for certain animals, and are thus kept in check. Such birds as the Fly-Catcher, Wagtail, Tits, Wren, Hedge-Sparrow, Swallow, and others, eat insects in enormous numbers. The Lady-bird Beetles, both larvae and mature beetles, live on Aphides, Scale Insects, Mites, and other pests, and are invaluable friends of the farmer and gardener. The larvae of Ichneumon flies live as parasites within the bodies of many plant-eating insects, and so destroy large numbers. It is thus important that nature's balance should not be interfered with. Man often destroys useful animals, and in consequence suffers from the depredations of pests which these animals would naturally keep in check.

As botanists, our chief interest in insects is as pollinators of flowers, and we will examine a number of common forms of flowers to determine the chief devices for securing pollination and the part played by insects in bringing it about.

Pollen-flowers : simple forms visited by insects for pollen.—The **Clematis** or **Traveller's Joy** (Fig. 109) has a calyx consisting of four greenish-white sepals which resemble petals, hence said to be **petaloid**. There is no corolla ; the stamens are numerous and arranged below the pistil, not in whorls, but spirally. The pistil consists of many carpels which are free from one another, and are hence said to be **apocarpous** (Gr. *apo* = from). The flowers secrete no honey, though they provide much pollen for their insect visitors.

In the **Wood Anemone** (Fig. 110) the flowers appear before the leaves ; but below the flower is a whorl of three large green, leafy bracts. The calyx consists of five petaloid sepals which are pinkish-white and conspicuous, and act as petals, the corolla being absent. The stamens are numerous (indefinite), and arranged spirally below the pistil, which consists of many small, spirally arranged carpels, free from one another. Examine flowers of different ages, and notice the order of ripening of the stamens and carpels. The outer stamens open first, the stigmas being covered by the inner ones. There is no honey in the flower, but it is visited for pollen by insects which alight in the centre, carry pollen from the anthers on to the ripe stigmas of an older flower, and so bring about cross-pollination. Later, the younger stamens and the stigmas are ripe together, and self-pollination may occur.

The **Marsh Marigold** (Fig. 111) is a similar flower with a large attractive calyx of five or more yellow sepals. The stamens are numerous and open outwards ; the carpels are free, and each contains several ovules. On the sides of each carpel and near the base are two shallow depressions where honey is secreted. This is an additional attraction for insects. Notice the curious stipule (*st*), which is quite

FIG. 109. FLOWERING SHOOT OF TRAVELLER'S JOY.

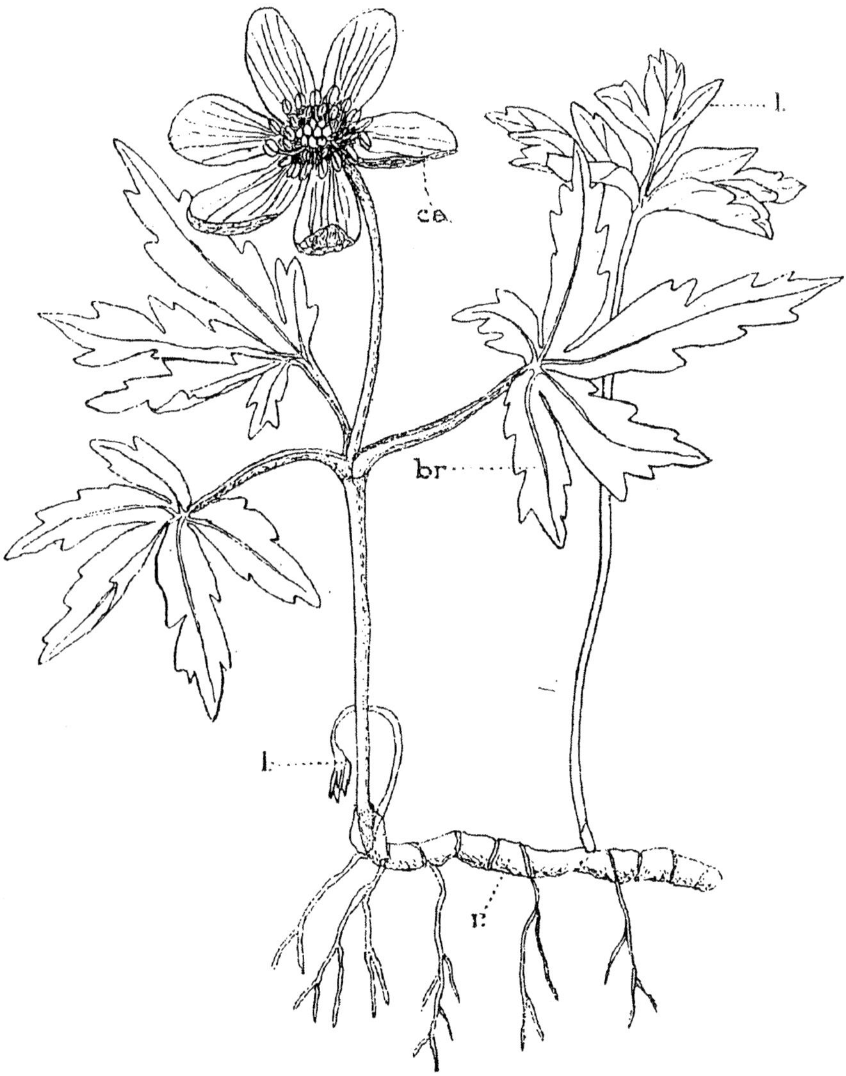

FIG. 110. WOOD ANEMONE.—*br*, bract; *ca*, calyx; *l*, foliage-leaf; *r*, rhizome.

entire in the bud and encloses the young leaf. As the leaf grows it bursts through the stipule, which remains as a thin membrane surrounding the stem.

The **Buttercup** (Fig. 112) has five free sepals (polysepalous), and alternating with them are five free petals

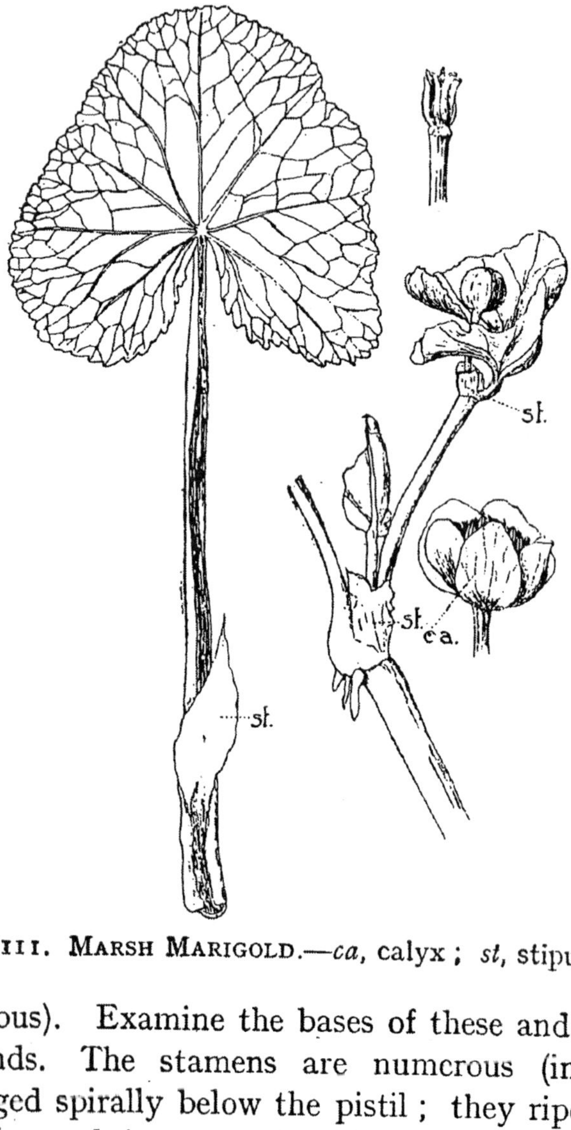

FIG. 111. MARSH MARIGOLD.—*ca*, calyx ; *st*, stipule.

(polypetalous). Examine the bases of these and note the honey-glands. The stamens are numerous (indefinite), and arranged spirally below the pistil ; they ripen before the carpels, and in succession from without inwards. The lower stamens first turn outwards and conceal the

FIG. 112. 1, TUBEROUS BUTTERCUP; 2, petal removed; 3, vertical section of flower; *a*, stamen; *br*, bract; *c*, carpel; *ca*, reflexed sepal; *n*, nectary; P, petal; *r*, receptacle; *t*, tuber.

nectaries and shed their pollen on the petals, not on the stigmas, which are not yet ripe. The pistil is in the centre, and consists of many spirally-arranged, free carpels, each containing one ovule. As in the Clematis, Anemone, and Marsh Marigold, the pistil is apocarpous and superior.

Such flowers, in which the sepals, petals, or stamens are fixed below the pistil, are said to be **hypogynous.** The Buttercup provides both pollen and honey, but in order to obtain the latter, insects must first push aside the

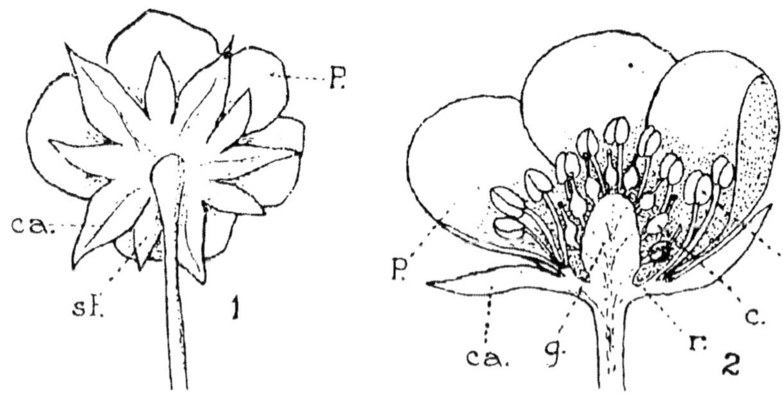

FIG. 113. FLOWER OF STRAWBERRY.—1, back of flower, showing the five sepals (*ca*) and five smaller stipules (*st*) alternating with the sepals; 2, flower in vertical section; *a*, stamen; *c*, carpel; *ca*, sepal; *g*, upgrowth from centre of receptacle bearing the carpels; P, petal; *r*, expanded and hollowed receptacle; *st*, stipule.

stamens, and in doing so their bodies become dusted with pollen. If they now visit older flowers where the stigmas are ripe, they may deposit on them some of this pollen. The different kinds of buttercups should be examined and their differences observed.

Perigynous and epigynous flowers; the simple flower-tube. —The **Strawberry** (Fig. 113) has five sepals and five sepal-like stipules, the latter forming what is called an **epicalyx.** The flower thus appears to have ten sepals. These, together with the five alternating petals and numerous stamens, are borne on the edge of an expanded and slightly hollowed receptacle, at the base of which is a fleshy, ring-like nectary. From the centre is an upgrowth from the receptacle, upon which are the numerous carpels. The stigmas ripen before the anthers, and thus cross-pollination

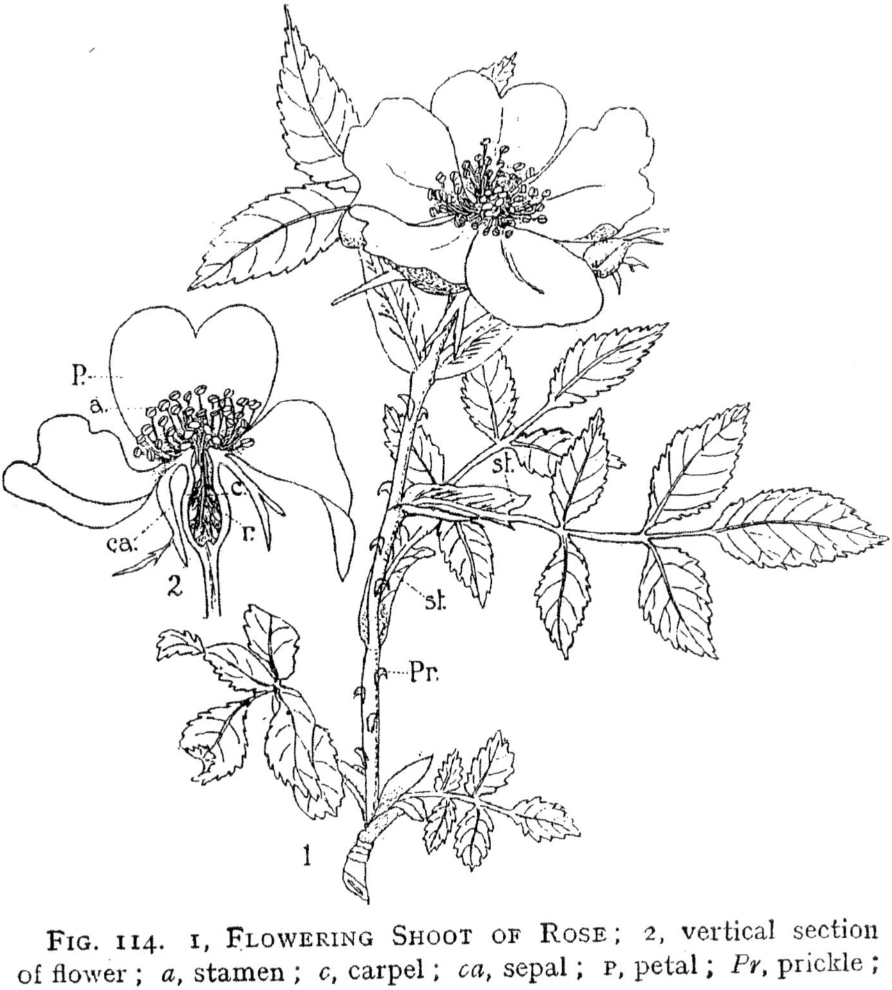

FIG. 114. 1, FLOWERING SHOOT OF ROSE; 2, vertical section of flower; *a*, stamen; *c*, carpel; *ca*, sepal; P, petal; *Pr*, prickle; *r*, receptacle; *st*, stipule.

is favoured. Flowers which thus bear their sepals, petals, and stamens on the edge of a cup-like receptacle, and therefore around the pistil, are said to be **perigynous**. The flower of the Strawberry should be carefully compared with that of the Buttercup.

In the **Rose** (Fig. 114) the receptacle is hollowed deeper still, and arising from the edge of it are five reflexed sepals, five petals, and numerous stamens, all of which are perigynous. Within the cup and fixed to the sides are several free carpels, each containing one ovule. No honey is secreted, but the stamens provide much pollen for insects. The Rose and Clematis are therefore called **pollen-flowers**.

The flowers of the **Cherry** (Fig. 115) and **Plum** have a receptacle which is hollowed, and on its edge are five sepals, five petals, and numerous stamens (i.e. they are

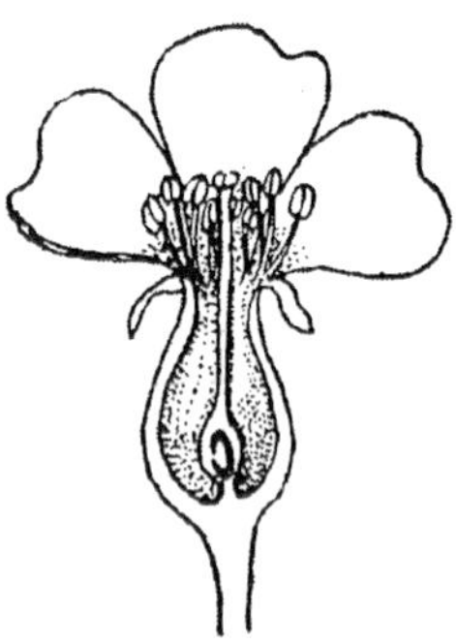

FIG. 115.
VERTICAL SECTION,
FLOWER OF CHERRY.

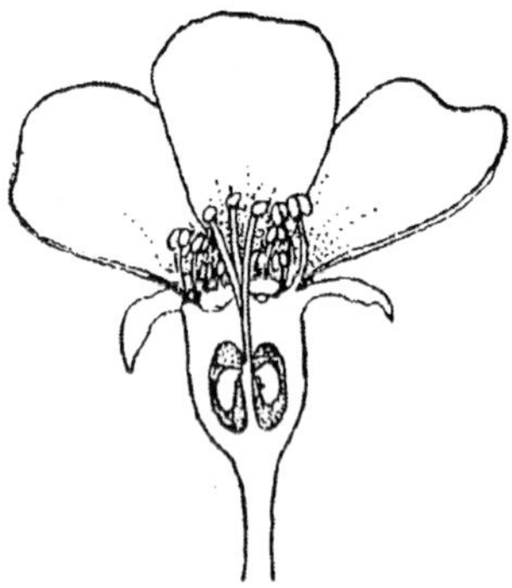

FIG. 116.
VERTICAL SECTION,
FLOWER OF APPLE.

perigynous). There is only one carpel, and this is superior to and free from the receptacle-cup. The anthers and the stigma ripen together; the anthers of the shorter inner stamens and the stigma stand at the same level, while the outer stamens are longer and overtop them. Honey is secreted by the receptacle-cup, and insects collecting honey and pollen may touch the stigma with pollen brought from another flower, and thus bring about cross-pollination; but, owing to the relative position of the anthers and stigma, self-pollination will very commonly occur. The whole of the fruit is formed from the carpel; the receptacle-cup is thrown off and does not form part of the fruit.

In the **Apple** (Fig. 116) and **Pear** the pistil consists of five carpels, which are syncarpous and closely united to the hollow receptacle; the five sepals, five petals, and numerous stamens are thus carried on to the top of the ovary (i. e. the flowers are **epigynous**). Owing to the union of the five carpels with the receptacle-cup the honey is easily obtained, and the flowers are visited by a great variety of insects. The five stigmas are prominent, ripen before the anthers, i. e. they are **proterogynous** (Gk. *proteros* = before), and so favour cross-pollination. If insect-visits fail, pollen may be shaken or may fall on to the stigmas; this is aided by the horizontal position of the flowers. After fertilization the receptacle enlarges and forms the fleshy part of the fruit. The five united carpels form the core (Fig. 155).

Examine old fertilized flowers of the Strawberry, Rose, Cherry, and Apple, and note in each case the mode of origin of the fruit and the structures concerned in their formation.

Tubular flowers with concealed honey. Devices to secure cross-pollination.—In the **Stock** (Fig. 3, 1) the four upright sepals form a narrow but split tube, which conceals the nectaries found at the base of the two short stamens. The stamens here, unlike those of the previous flowers, are reduced to six. Such a deepened flower-tube will prevent the short-tongued insects from securing the honey, but this can easily be obtained by the long-tongued insects, such as moths, butterflies, and bees. These insects are more intelligent and better adapted for carrying pollen from anther to stigma than the short-tongued insects, like beetles and flies, which may take pollen and honey from shallower flowers with less likelihood of bringing about pollination.

Those flowers, therefore, which attract the more intelligent insects possess a double advantage: (1) they need less pollen, and (2) cross-pollination is more certain. Let us see by what means these advantages are secured in some other flowers.

In the Geranium the calyx has five sepals joined by their edges to form a deep tube. A united calyx is said to be **gamosepalous** (Gr. *gamos* = union). The five petals are free; there are ten stamens, five outer and five inner. At the bases of the five outer ones are nectaries. It is interesting to watch the movements of the stamens in the **Field Geranium.** When the flower opens, the stamens lie on the petals; they then raise themselves parallel to the pistil, shed their pollen, and return—first the outer set, then the inner—to their former position. The pistil consists

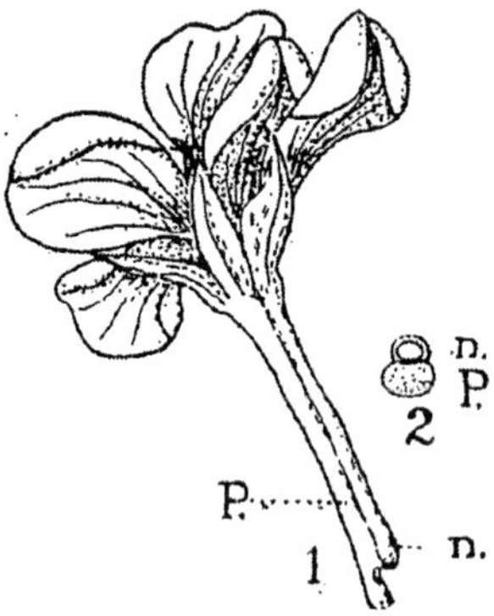

FIG. 117. 1, FLOWER OF GARDEN GERANIUM; 2, transverse section of pedicel and nectary; *n*, nectary; P, pedicel of flower.

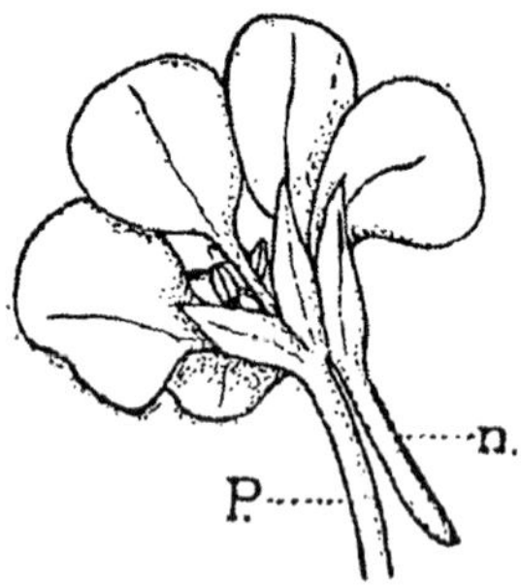

FIG. 118. FLOWER OF GARDEN NASTURTIUM.—*n*, honey-secreting spur; P, pedicel.

of five superior, united carpels. When the pollen is shed, the stigmas ripen and spread out as five lobes to receive pollen from another plant. Stamens which ripen before the pistil are said to be **proterandrous.**

Compare with this the **Garden Geranium** (*Pelargonium*) (Fig. 117). In this, do the stamens or the stigmas ripen first? In these flowers cross-pollination is secured by the stamens and the pistil, which, though existing in the same flower, ripen at different times. Look for the long tubular nectary which adheres throughout its whole length to the flower-stalk. The presence of the nectary in the Pelargonium destroys the symmetry of the flower. In the previous examples the flower can be divided into more than two similar halves. Such flowers are said to be regular, or **actinomorphic** (Gr. *aktis* = a ray, *morphē* = shape). The Pelargonium, however, can be divided into only two similar halves, and hence is said to be irregular or **zygomorphic** (Gr. *zygos* = a yoke).

The **Garden Nasturtium** should also be examined (Fig. 118). Note the long spur, which is a hollow outgrowth of the floral axis and contains the honey. The calyx, as well as the corolla, is coloured, and the fringes on

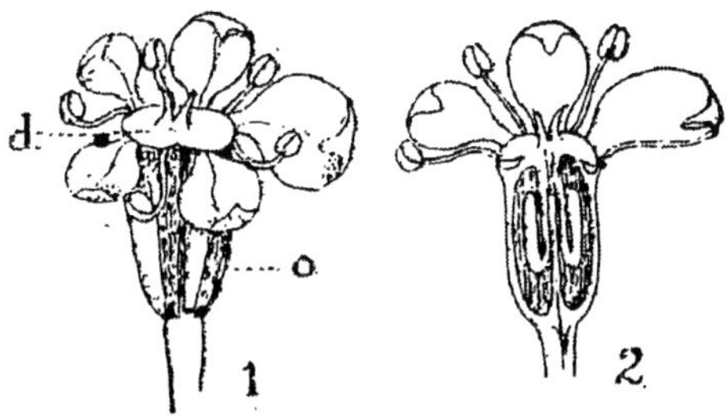

FIG. 119. 1, FLOWER OF CHERVIL; 2, vertical section of flower; *d*, disk; *o*, inferior ovary.

the three lower petals serve to keep rain out of the honey-tube. Watch the stamens as they ripen, and note how each in turn bends upwards in front of the entrance to the tube and ripens so that it will be touched by a bee visiting the flower. They then bend down out of the way, and later the stigma assumes the position previously occupied by the stamens, and is thus likely to become cross-pollinated.

The flower of the **Chervil, or Beaked Parsley** (Fig. 119), shows several important differences. The flowers are small and crowded together in a flat-topped inflorescence, called **a compound umbel** (L. *umbella* = a sunshade). The calyx consists of five minute sepals, and alternating with them are five petals of different sizes; the outer and anterior one is the largest; then follow two intermediate ones;

lastly the two inner ones, which are the smallest. Alternating with these are five stamens, which shed their pollen before the stigma is ripe. These three whorls, unlike the preceding examples, spring from the top of the ovary, and hence are said to be epigynous. The pistil consists of two united carpels inferior to the other whorls. On the top of the ovary is a honey-secreting disk (*d*), which surrounds the two stigmas, and these ripen only when the pollen of the same flower has been shed. The honey is freely exposed and liable to be spoiled by the rain, and may be obtained by short-tongued insects which commonly visit the flowers. In this case conspicuousness is due to the aggregation of many small flowers at the same level in the inflorescence, and by the outer petals enlarging at the expense of the inner ones.

The Buttercup, Stock, Strawberry, Rose, Geranium, and Chervil all agree in one important respect—their petals are not joined, i.e. the corollas are polypetalous. In most cases the sepals also are free. In the Geranium, however, they are united, and the calyx is gamosepalous. These flowers also show different methods of forming the flower-tube, namely: (*a*) by erect sepals, as in the Stock; (*b*) by a hollow receptacle, as in the Strawberry and the Rose; and (*c*) by united sepals, as in the Geranium.

CHAPTER XIII

BIOLOGY OF THE FLOWER (*Continued*)

II. Pollination of Tubular and Highly Developed Flowers

THE flowers we have considered above are generally simple in structure. Flowers pollinated by the wind have no perianth, or only a very rudimentary one; they are small and inconspicuous, and produce much pollen; and the stigmas are large, branched and sticky, to catch the pollen, much of which is wasted. Flowers possessing a perianth may have only a single whorl, and this is often petaloid. In those with a double perianth—a calyx and a corolla—the latter is usually attractive. In the lower types the parts are free, the flower-cup is more or less open, and the pollen and honey are very accessible to insects. In various ways, however, a flower-tube is developed in higher forms which protects the honey from rain and excludes the lower types of short-tongued insects.

We have now to consider a further stage in the development of the flower-tube and its relation to the habits and structure of the higher and more intelligent types of insects.

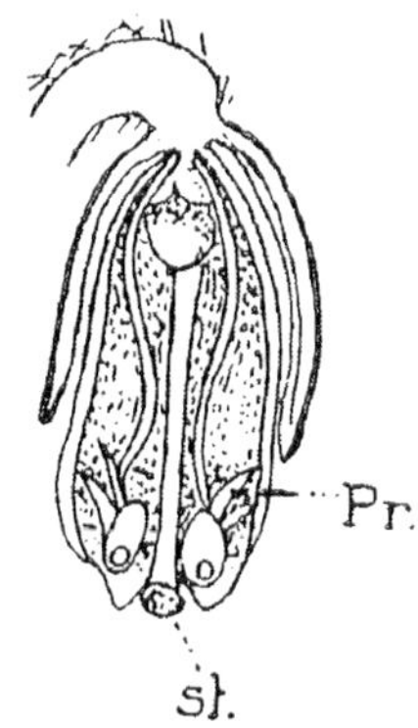

FIG. 120. FLOWER OF CROSS-LEAVED HEATH.—*Pr*, anther-processes; *st*, stigma.

Tubular flowers with united petals.—The **Cross-leaved Heath** has five united sepals and five petals joined by their edges to form a tube; the corolla is thus **gamopetalous.** Within the bell-shaped corolla (Fig. 120) are the stamens, which are peculiar. Each anther has two long processes or arms, and these project outwards towards the corolla-wall. Near the top of each anther are two pores, through which the pollen escapes when ripe. The pistil is superior and syncarpous, and around its base is a ring-like nectary; the style is long and projects beyond the anthers to the mouth of the bell. An insect visiting the flower will bring its head against the stigma and, in pushing its proboscis into the flower to obtain the honey, touch the anther-processes, which, acting as levers, will separate the anthers and cause a shower of pollen to fall on to the head of the insect. The tube is too deep for the short-tongued insects: the chief visitors are bees, lepidoptera, and long-tongued flies. The flowers of the Heaths hang downwards, so that the honey is protected from the rain.

Different kinds of flowers in the same species.—In the **Primrose** (Fig. 121, 1) and **Cowslip** (Fig. 121, 2) the five united

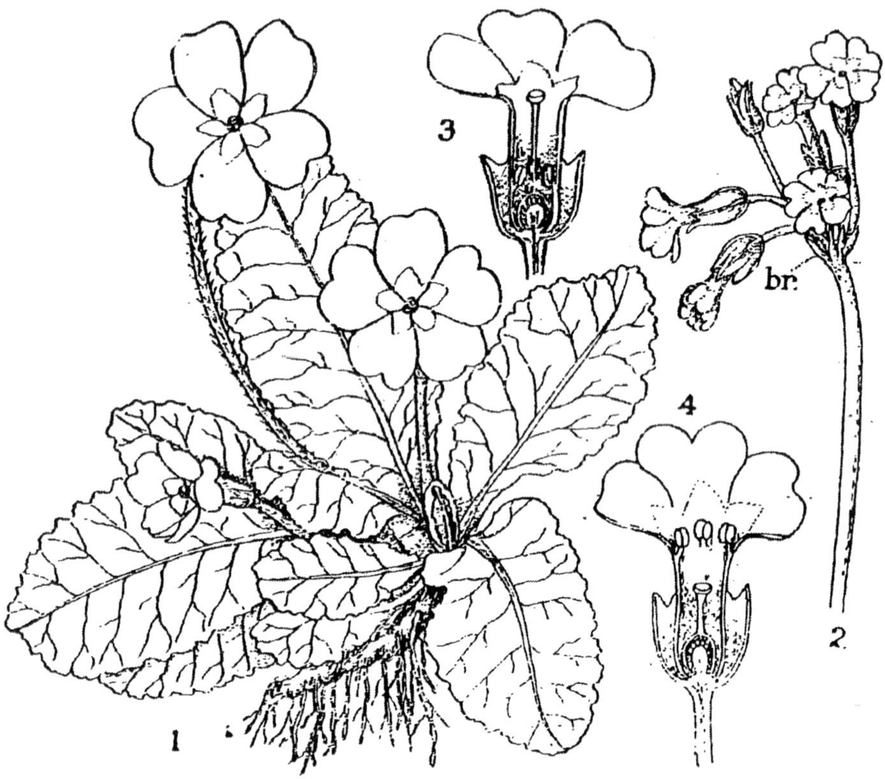

FIG. 121. 1, THE PRIMROSE; 2, COWSLIP; 3, long-styled form; 4, short-styled form; *br*, bracts.

petals form a long, narrow tube surrounded by an inflated calyx of five united sepals. If a number of flowers are examined, two kinds will be found: one with the knob-like stigma at the mouth of the tube and the anthers half-way down (Fig. 121, 3), and the other with the five stamens at the mouth of the tube and the stigma half-way down (Fig. 121, 4).

The long-styled form is known as 'pin-eyed', and the short-styled form as 'thrum-eyed'.

Species which thus produce two kinds of flowers are said to be **dimorphic**. (The Loosestrife has three kinds of flowers, and hence is **trimorphic**.)

The pistil of the Primrose consists of five carpels, superior and syncarpous, but it differs from the previous examples in that the numerous ovules are borne on placentas on a central column, which is free from the ovary-wall, and such an arrangement of ovules in an ovary is called 'free-central placentation' (Fig. 121, 3 and 4). Round the base of the ovary, honey is secreted, and an insect visiting these two kinds of flowers will receive pollen on its proboscis at two different points: at its base will be the larger pollen-grains of the short-styled form, and in the middle of it the smaller pollen-grains of the pin-eyed form. Thus pollen-grains of one form are likely to be transferred to the stigma of the other form. Here, as in the Heaths, Stock, and other plants, self-pollination may occur by pollen-grains falling on to the stigma of the same flower.

If a Primrose plant is covered with a bag to exclude insects, and its flowers compared later with uncovered ones, the latter will be found to produce more ripe seeds than the former. It was by similar experiments that Darwin and others showed how important insects are as pollinators of flowers.

Small flowers massed in heads.—The **Daisy** (Fig. 122) shows specialization similar to that already noticed in the Chervil, where small flowers, inconspicuous in themselves, are rendered attractive by being massed together in a condensed inflorescence. In the Daisy and its allies this has reached the highest stage of development; condensation has occurred to such an extent that the whole inflorescence has the appearance of a single flower. Such a head of florets (Fig. 64) is called a **capitulum**.

Outside is a series of small bracts resembling a calyx and known as the **involucre**. Within this is a series of small flowers which resemble a corolla, but on careful examination each is seen to consist below of a small inferior ovary; the calyx is absent; the corolla of five petals forms a narrow tube below, and spreads out above in the form of a white pink-tipped strap. Such a strap-shaped corolla is said to be **ligulate**. There are no stamens, but the style has two branches without hairs. These very small strap-shaped flowers are called **ray-florets** (Fig. 122, 1). The yellow disk in the centre is composed of florets of a very different type (2). The pistil, as in the ray-florets, has an inferior ovary of two carpels; there is no calyx; the corolla is long and tubular, with five teeth above. Each floret has five stamens, the anthers of which are united to form a tube round the style. Such united anthers are said to be **syngenesious** (3). These are known as **disk-florets**.

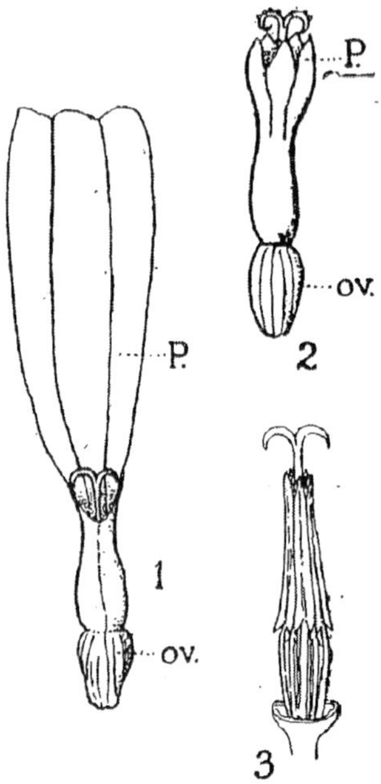

FIG. 122. FLORETS OF DAISY.—1, ligulate, female ray-floret; 2, tubular, hermaphrodite disk-floret; 3, stamens with united anthers; *ov*, inferior ovary; P, corolla.

If the disk-florets in the head of a Daisy are examined, it will be seen that the outer (lower) florets are the older, and are the first to open. The following stages should be looked for: (1) the style is short and within the anther-tube (when ripe the anthers shed their pollen into the tube and on the top of the stigma, which has two lobes, the outer face of each being provided with a brush of hairs); (2) later, the style elongates, and the stigmatic brush sweeps the pollen out of the tube; (3) the stigma-lobes then open, throwing the pollen off, and exposing their inner surfaces for pollination.

As the disk-florets open successively from outside inwards, two stages will be seen: (1) the old florets with stigmas exposed, (2) younger florets with pollen only exposed. A single insect-visit may thus readily convey pollen from one floret to the stigma of another floret. Thus cross-pollination of two degrees is possible: (1) a cross between two florets of the same head, and (2) a cross between florets on different Daisies. Honey, which is secreted by a ring-like nectary round the base of the style, rises in the tube in such quantity as to be accessible to short-tongued insects. Thus the Daisy has gained many distinct advantages: (1) conspicuousness, due to aggregation of many small flowers; (2) a large supply of honey protected from rain by narrow tubes; (3) accessibility of pollen and honey to a great variety of insect visitors; (4) pollen presented so as to secure cross-pollination in the case of insect-visits, or self-pollination in their absence.

The Coltsfoot.—The bright yellow flower-heads of the **Coltsfoot** may be found in January or February, when insects are rare. The flowers appear before the leaves, and from a single plant many flowering shoots of different ages arise which prolong the flowering-period. Growth is maintained at the expense of food stored in its thick and often long, underground stem, on which are the young leaves. Each flowering shoot is covered by small, very hairy bracts, and bears above a single capitulum. The bracts of the involucre are in a single row and protect the florets in the bud. Cut a capitulum vertically in two and note the flat disk; determine the different kinds of florets in the head, and the interesting division of labour they show (Fig. 123).

The outer, strap-shaped florets are the most numerous

(about three hundred). Each has an inferior ovary, and above it is the calyx, consisting of a whorl of hairs. Such a calyx is said to be **pappose.** Within this is the ligulate corolla. The style has a two-lobed stigma, and also a brush of hairs, which, however, is of no use, as these florets have no stamens, i. e. they are all pistillate ; no honey is secreted, and they are ripe before the inner florets. About forty tubular flowers will be found in the centre, differing from the outer ones as follows: they are smaller and less attractive ; the corolla is tubular and five-toothed ;

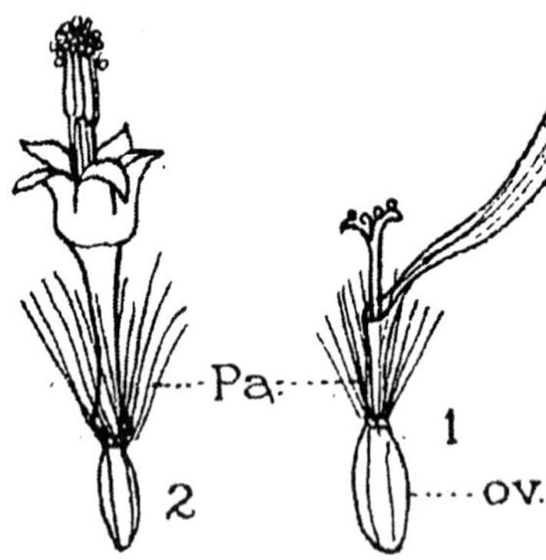

FIG. 123. FLORETS OF COLTSFOOT.—1, ligulate, female ray-floret; 2, tubular, male disk-floret; *Pa*, pappus; *ov*, ovary.

at the base, honey is secreted ; the ovule in the ovary is abortive ; there are five stamens with joined anthers, and the style has a pollen-brush, but not a functional stigma. As the outer female florets are ripe before the inner male ones, self-pollination cannot take place ; they are therefore dependent on insect-visits. The chances of pollination are increased by the prolonged flowering-period.

The Dandelion.—Compare the flower-head of the **Dandelion** (Fig. 104) with those of the Daisy and Coltsfoot, and notice that the hollow stalk is devoid of leaves.[1] The

[1] All parts of the plant contain a milky juice or *latex*, a fluid which consists partly of waste substances, and to some extent of nutritive materials. The latex is contained in irregular channels known as laticiferous vessels.

bracts of the involucre are in two whorls. The small outer ones become recurved when the bud opens, and the inner, larger ones stand erect and protect the florets. These are all alike and ligulate, and, though small, are very conspicuous when massed together in the capitulum; the corollas are yellow, but the outer, more exposed ones are often brown on the back. Cut the capitulum longitudinally and note that the florets spring from the expanded, flattened end of the inflorescence-axis, the outer ones being the oldest. Examine a floret (Fig. 124) and note that the ovary is inferior, and above it is a short neck on which is the calyx, represented by numerous hairs which form the pappus. The corolla is irregular, tubular below and strap-shaped above, ending in five small teeth representing the corolla-lobes. The five stamens are fixed to the corolla, the anthers are joined into a tube round the style, and the pollen is shed into the tube while the style is yet short. Honey is secreted by a ring at the base of the corolla and rises high in the tube, thus being accessible to many kinds of insects. When the style elongates it is seen to have two stigma-lobes and to be covered with hairs, which brush the pollen out of the anther-cylinder. The stigmas curl outwards, and their upper surfaces are covered with papillae which receive pollen brought by insects from another flower-head; or self-pollination may occur by the back-rolled stigmas coming into contact with pollen from the same flower.

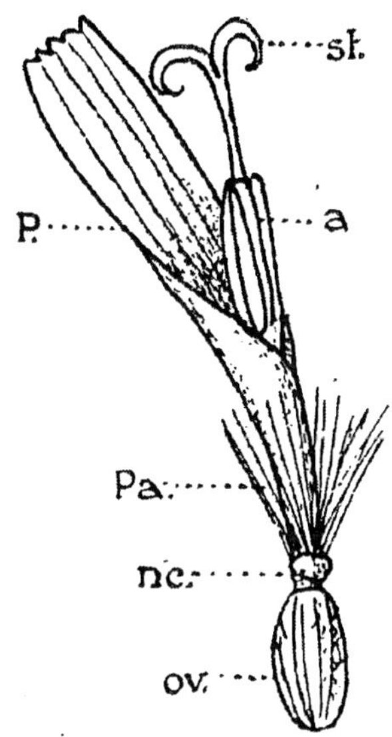

FIG. 124. FLORET OF DANDELION.—*a*, united anthers; *nc*, neck; *ov*, ovary; P, corolla; *Pa*, pappus; *st*, stigma.

The Dandelion is visited by a great variety of insects which are able to obtain both pollen and honey very readily. Its flowering-period is a long one, but if the flowers are out too early or too late for insect-visits self-pollination is possible.

Flowers of the Potato and the Woody Nightshade or

FIG. 125. 1, WOODY NIGHTSHADE; 2, vertical section of flower; 3, stamen; P, pore.

Bittersweet agree with Composite flowers in one respect, viz. the anthers are joined to form a tube (syngenesious).

The **Woody Nightshade** (Fig. 125) is frequent in hedgerows and is a lax climber. The flowers (Fig. 125, 2) are small and in irregularly-branched inflorescences called panicles. The calyx of each flower has five united sepals; the corolla has five purple, united petals, upon which are five stamens alternating with the corolla-lobes; the large anthers are joined into a tube round the style and form a conspicuous yellow cone, above which projects the two-lobed stigma. When ripe the anthers dehisce by pores at their free ends (Fig. 125, 3). The pistil is superior, and consists of two united carpels placed obliquely in the flower (Fig. 173); the ovary is two-celled with many ovules on axile placentas. The flower secretes no honey, but is visited by bees for pollen.

Irregular and specialized flowers. — The **Germander Speedwell** (Fig. 126) is pollinated mainly by drone-flies, and shows several interesting modifications. The flower (2 and 3) has four sepals, the fifth posterior one being absent. The blue corolla has four petals, but the large posterior one really represents two fused petals. All are joined into a short tube. There are only two stamens, and these spread out horizontally. The pistil consists of two united carpels, and the style projects over the anterior petal. A fleshy disk below the ovary secretes honey, which is protected from rain by hairs on the corolla.

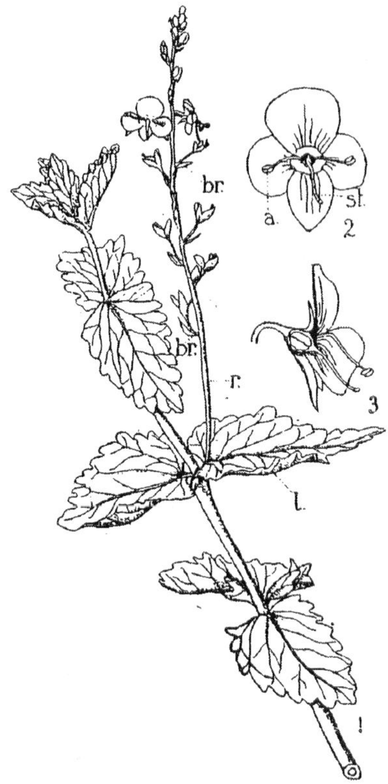

FIG. 126. 1, GERMANDER SPEEDWELL; 2, flower showing the position of the stamens and style; 3, flower in vertical section; *a*, stamens; *br*, bracts; *l*, opposite decussate leaves: *r*, raceme; *st*, style.

As the fly alights, it first touches the stigma, then grasps the two stamens, pulling them to the sides of its body,

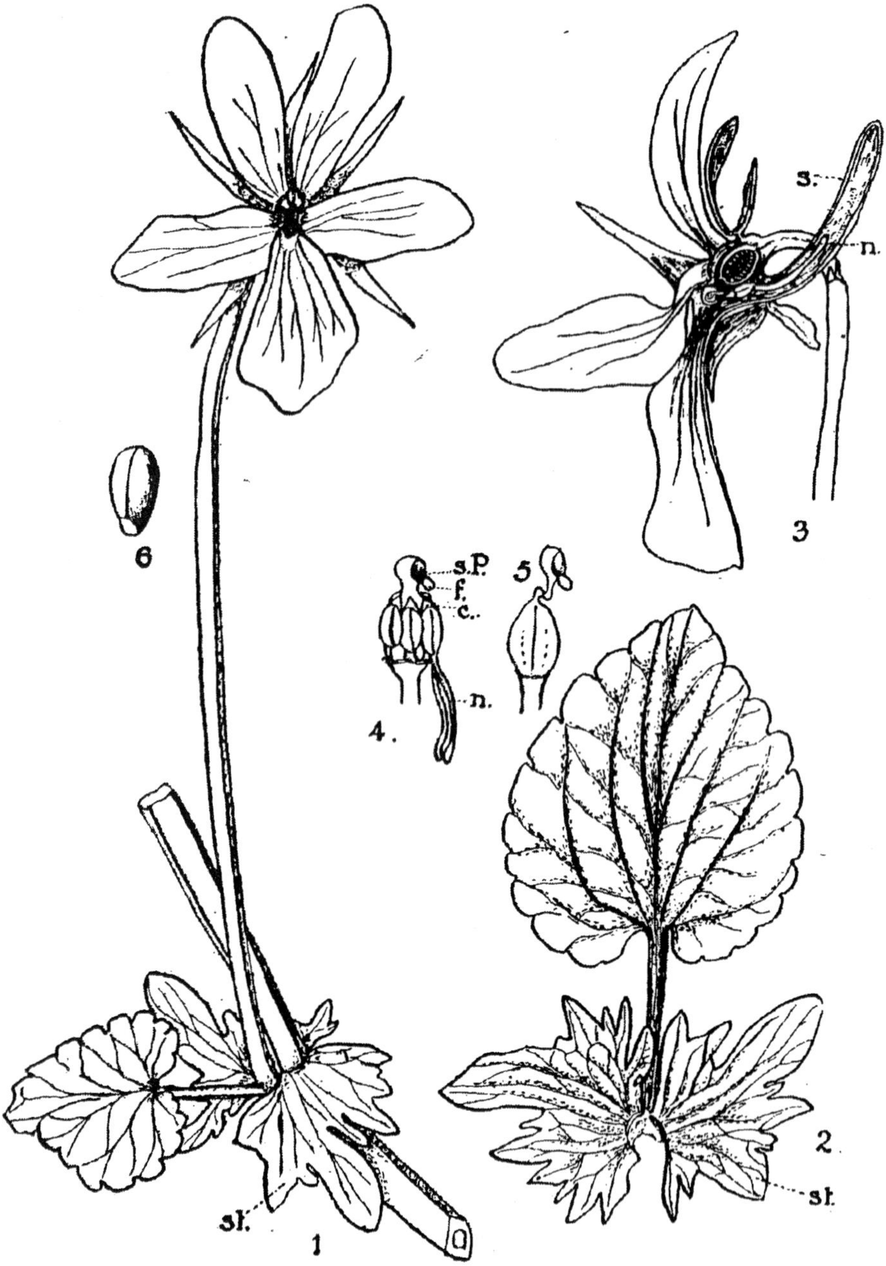

FIG. 127. FLOWER OF VIOLET.—1, solitary axillary inflorescence; 2, a leaf; 3, flower in vertical section; 4, stamens and pistil; 5, pistil removed; 6, seed; *c*, membraneous prolongation of anther-connective; *f*, flap of stigma; *n*, nectary; *s*, spur; *s.P*, stigmatic pit; *st*, foliaceous stipules.

which thus becomes dusted with pollen. On visiting another flower the ripe stigma becomes dusted with pollen from the body of the insect.

Violet and Pansy. Cleistogamous flowers.—The flowers of the **Violet** and **Pansy** are curiously modified in all their parts, and these should be carefully examined (Figs. 127 and 128). Note that the five free sepals are prolonged downwards below their points of attachment. The corolla consists of five free, dissimilar petals, the lower anterior one being produced into a tubular spur. There are five stamens attached by very short stalks and bearing orange-coloured membraneous outgrowths at the ends of the anthers (Fig. 127, 4 *c*). Notice that the two stamens opposite the spurred petal have each a long fleshy nectary projecting into the spur (Fig. 127, 3 and 4). These secrete honey which collects in the spur. The lines on the petals point towards this receptacle, direct the insect to the honey, and hence are called **honey-guides.** The pistil is superior and consists of three united carpels. The style is bent at the base and terminates in a rounded knob (Fig. 127, 5). Look for the little pit on the lower surface of the stigma; on the edge of it is a little flap. The pollen is shed on to the spurred petal, and an insect, visiting the flower to obtain the honey, becomes dusted with pollen. On visiting another flower it brushes the flap, leaving pollen on it and, on quitting the flower, presses the flap into the stigmatic pit, and so effects cross-pollination.

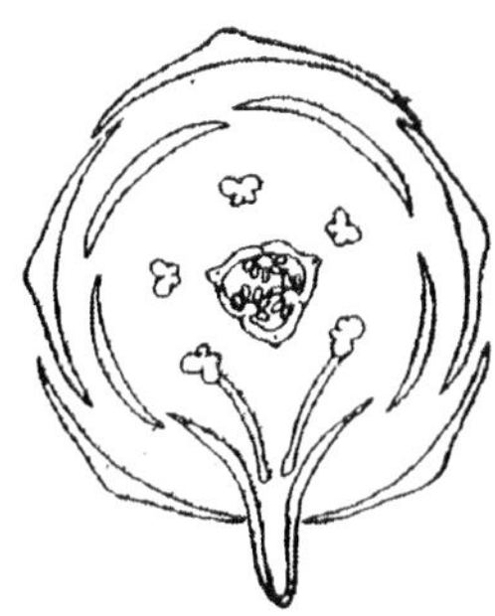

FIG. 128. FLORAL DIAGRAM OF VIOLET.

In the Sweet Violet the stigma is not globular.

Violas and Pansies both belong to the same genus *Viola*, but if the flowers are compared, differences will be noted in the arrangement of the petals. The lateral petals of

Violas are horizontal (Fig. 129, 1), while those of Pansies are directed upwards (Fig. 129, 2). These differences may be somewhat masked in cultivated forms with very large petals. Irregular (zygomorphic) flowers, as those of the Violet, are well adapted to the structure and habits of bees. They not only provide honey protected in tubes or spurs, but are frequently scented, and in the highest forms have a blue colour.

Sweet Violets grown in poor soil and in a shady place often cease to develop the typical showy flowers, yet ripe capsules full of seeds may be formed. Careful examination will reveal a few very small inconspicuous flowers, resembling small flower-buds, at the base of the plant and overshadowed by the leaves. These flowers never open; they regularly fertilize themselves and produce an abundance of ripe seeds. Sometimes both kinds of flowers occur on the same plant. Similar closed flowers occur in the Wood Sorrel and Henbit - Deadnettle (*Lamium amplexicaule*). Such flowers are called **cleistogamous** (Gr. *kleistos* = closed). Cold, absence of sunshine, and a poor soil, favour their development.

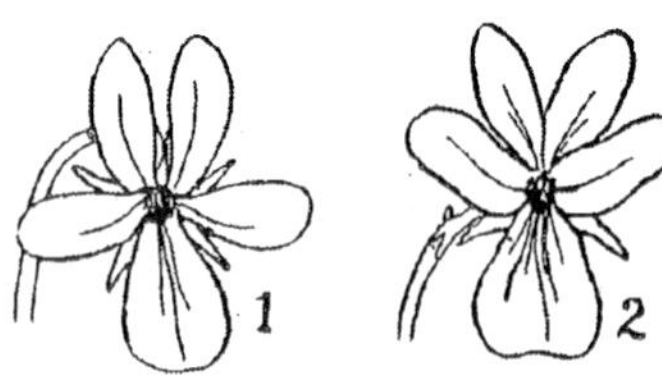

FIG. 129.
1, FLOWER OF VIOLA.
2, FLOWER OF PANSY.

Columbine, Monkshood, and Larkspur.—The **Columbine** (Fig. 130) is a humble-bee flower. Its sepals are coloured; its five free petals are prolonged into large spurs whose curved and fleshy ends secrete honey. The stamens are indefinite, but the inner ones do not produce pollen. Such barren stamens are called **staminodes**. In the centre are the five carpels slightly joined at the base. The stigmas ripen later than the stamens. The flower is pendulous, and the bee, in order to obtain the honey, has to cling to the base of the spur and also to the column of stamens and carpels. It thus becomes dusted with pollen, which it carries to older flowers where the stigmas are ripe.

It requires an insect with a long proboscis to obtain the honey from so long a tube. Often, however, the humble-

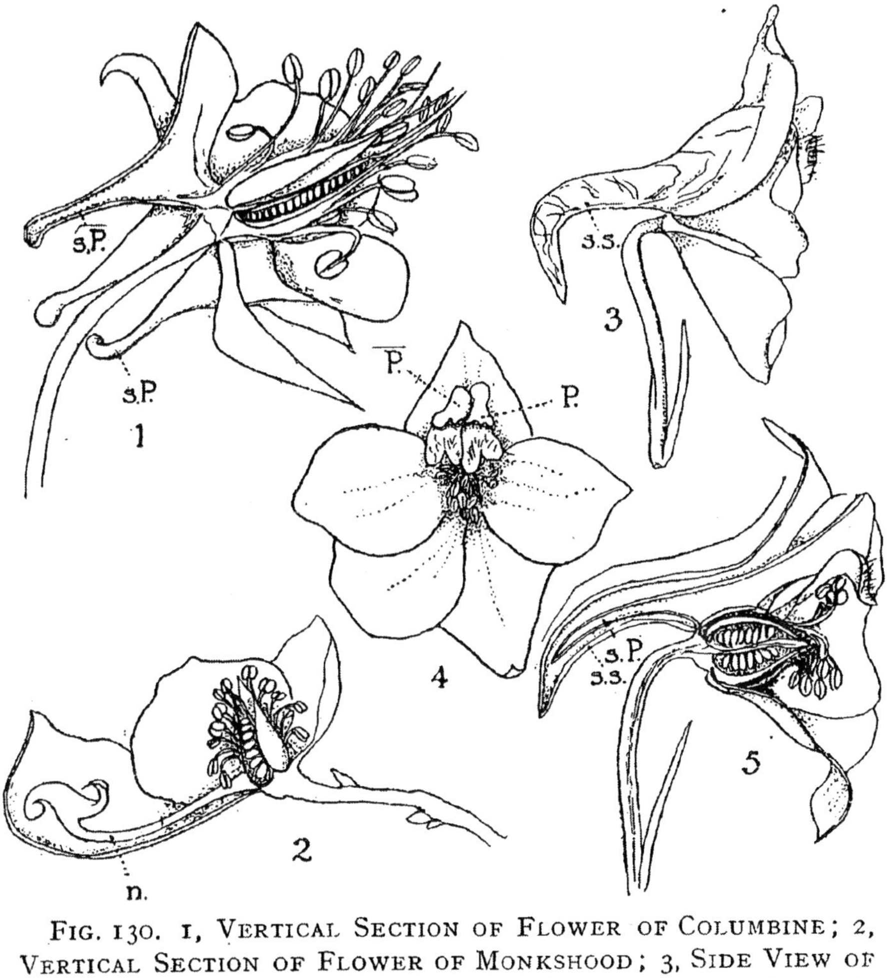

FIG. 130. 1, VERTICAL SECTION OF FLOWER OF COLUMBINE; 2, VERTICAL SECTION OF FLOWER OF MONKSHOOD; 3, SIDE VIEW OF FLOWER OF LARKSPUR; 4, front view of the flower; 5, flower in vertical section; *n*, nectary; P, petals; *s.P*, honey-secreting spur of petal; *s.s*, spurred sepal.

bees bore holes in the spurs and so obtain the honey without effecting pollination. These holes may then be used by other insects which do not themselves pierce flowers.

The **Monkshood** (Fig. 130, 2) is another humble-bee

flower, and the plant grows wild only where humble-bees are found. It has five blue sepals, the posterior one forming a large hood which protects the anthers and nectaries. The two posterior petals are modified to form long, clawed nectaries or honey-leaves (*n*) and the other petals are usually absent. The stamens are indefinite and they ripen before the stigmas (i. e. they are proterandrous).

The **Larkspur** (Fig. 130, 3, 4, 5) has five sepals and two spurred petals ; the posterior sepal is prolonged into a membraneous spur which encloses the honey-secreting spurs of the two posterior petals. The flower of the Columbine is regular, while those of Monkshood and Larkspur are irregular.

Sweet-Pea, Scarlet-Runner, and Gorse.—The **Sweet-Pea** (Fig. 131) is a complex bee-flower. Its calyx has five united sepals, two above and three below. Its corolla has five petals curiously shaped and known by distinctive names (Fig. 131, 2). The large posterior petal is the *standard* (*s*), the two lateral ones are the *alae* or *wings* (*a*), and the two anterior ones, which are slightly joined, form together the *carina* or *keel* (*k*). Note carefully how these are related to one another, especially at the base, observing that depressions in the wings correspond to bulges in the keel. Enclosed in the keel are ten stamens, nine of them being united by their filaments to form a stamen-trough around the pistil, the posterior one lying free over the slit (Fig. 131, 3). When the stamens are united by their filaments into two sets they are said to be **diadelphous.** The filaments and style are bent upwards at the end of the keel. Honey is secreted by the bases of the stamens, and collects in and is protected by the stamen-trough (*s.tr*).

A clever insect like the bee is required to open the flower and obtain the honey. It uses the wing-petals as an alighting stage ; but these being articulated with the keel, both are depressed by the weight of its body. When the

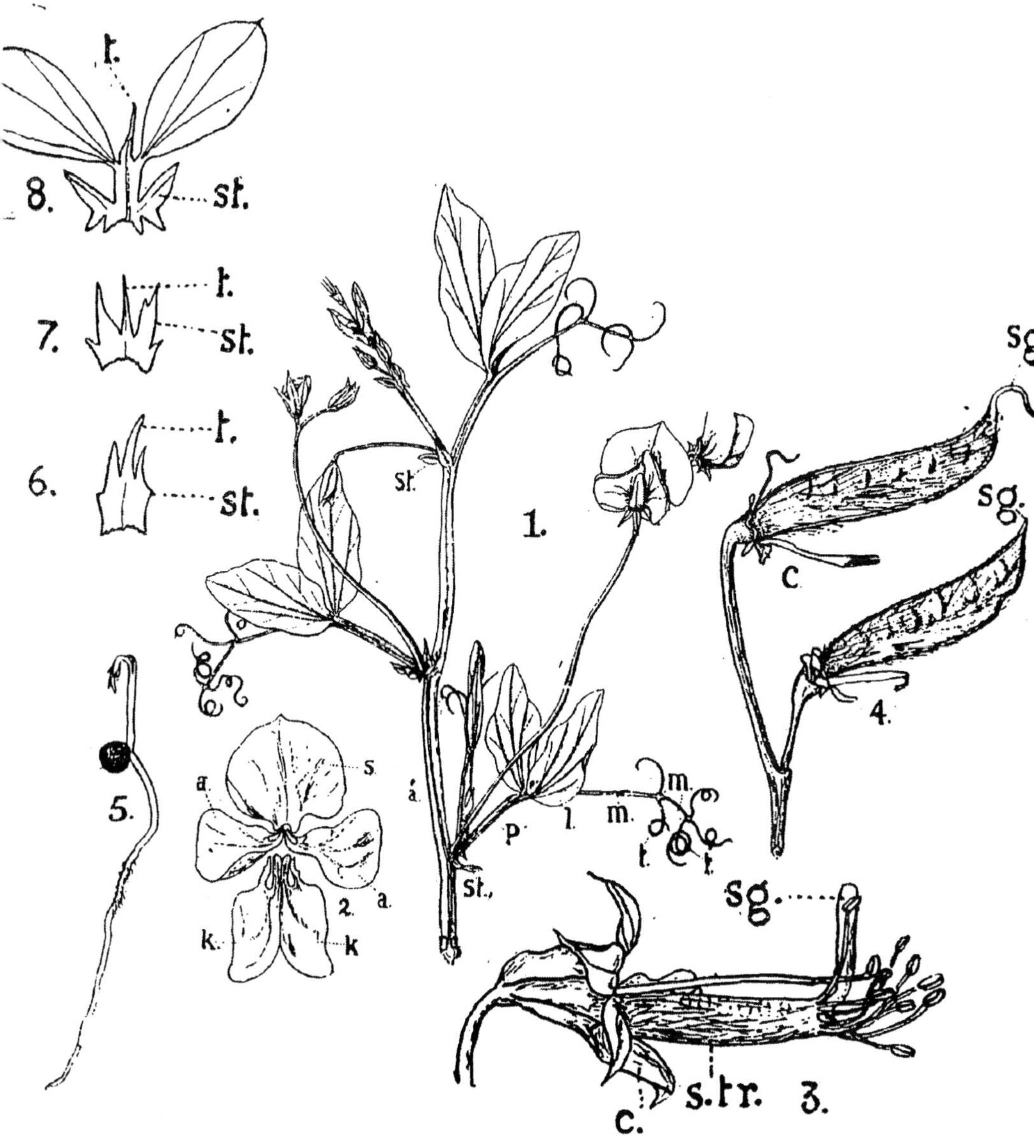

FIG. 131. 1, SWEET-PEA; 2, corolla; 3, flower with corolla removed; 4, fruits; 5, seedling; 6, 7, and 8, the first, second, and third foliage-leaves; *a*, alae; *c*, calyx; *k*, keel-petals; *l*, leaflet; *m*, midrib; P, petiole; *s*, standard; *sg*, stigma; *st*, stipule; *s.tr*, stamen-trough; *t*, leaflet-tendril.

bee pushes its proboscis into the tube, it obtains the honey through two openings, one on either side of the base of the free stamen. The stamens and style are thus exposed; the pollen is swept out by a brush of hairs at the end of the style and dusts the under surface of the bee. When the bee flies away, the keel springs back to its former place. If at this time the stigma is ripe, it may be touched with pollen as the keel returns to its place; self-pollination would thus occur. As the bee flies from flower to flower it may deposit pollen on the stigma of another flower of the same species and so effect cross-pollination.

The pistil (Fig. 131, 3) consists of one carpel (apocarpous), and is covered with hairs. Look for the small ovules within the ovary.

Other pea-like flowers should be compared with the Sweet-Pea, and their characteristics noted. The Vetch, Broad Bean, and Scarlet-Runner are similar to the Sweet-Pea: the style is provided with a pollen-brush and the flowers may be visited a number of times. The Scarlet-Runner, however, is peculiar in that the keel and style are spirally coiled. The flower of the Garden Pea is regularly self-pollinated. No European insect is strong enough to open the flower and pollinate it. The White Clover produces much honey, and its short flower-tube does not exclude short-tongued bees. When an insect alights on the wing-petals, the stamens and stigma emerge from the flower. After the visit they return to their former position within the keel.

The **Gorse** (Fig. 132) differs in several respects from the above. All the stamens are united to form a tube; no honey is secreted, and the flowers are visited by bees and other insects for the sake of the pollen. The stamens ripen and shed their pollen into the tip of the keel. A bee, resting on the wings and pressing its head beneath the standard, bursts open the keel and the under side of its body is first touched by the stigma and then receives a shower of pollen. The wings and keel are dislocated; they hang vertically downwards, and cannot return to their former position. If the bee is dusted with pollen from another flower, cross-pollination will occur; if not, self-pollination may take place as the insect leaves the flower. Imitate the action of the bee by depressing the keel with a pencil and watch the

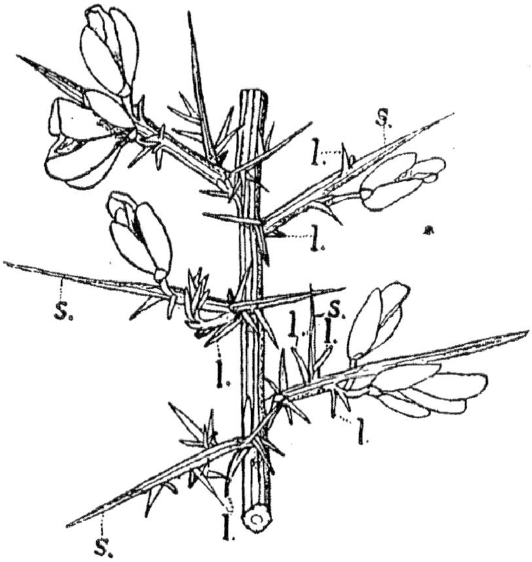

FIG. 132. GORSE.—*l*, leaf; *s*, spine.

explosion. Such explosive flowers can profit only by one visit.

Summary of the various grades of flower-structure.—Flowers may be regarded from two points of view:—(1) the Biological, which is concerned with the functions of the flower and its relations to the outside world, especially wind and insects. In the series of flowers which we have studied, we find an increase in complexity of structure from simple wind-pollinated flowers to more complex insect-pollinated flowers, and in the latter, commencing with flowers in the form of an open cup, and accessible to the lower orders of insects, we find a tendency to form a deeper flower-tube, a union of sepals and petals, production of honey and scent, and at the end of the series we arrive at special forms with irregular flowers. These are so constructed as to exclude the lower forms of insects, but are attractive to and visited by the higher and more intelligent forms, e. g. bees, moths, and butterflies.

Colour-changes follow somewhat similar lines of development. The simpler flowers are small, green, and inconspicuous, a stage higher the petals are larger and yellow or white, while the tubular and more complex flowers are often red, blue, or violet.

(2) The Morphological point of view, which regards the structural differences, forms, and relationships, of the parts of the flower. If we summarize the chief structural differences we find that the small and simple flowers of the Poplar and Willow have no calyx or corolla. Their flowers are unisexual and arranged in catkins, the male and female catkins occurring on different trees, i. e. they are dioecious. In the Oak and Hazel both male and female catkins occur on the same tree, i. e. they are monoecious. The Anemone, Buttercup, Marsh Marigold, Columbine, Monkshood, and Larkspur are hermaphrodite, and though differing much in form and colour, all agree in possessing a free calyx and corolla, numerous hypogynous stamens, and a superior apocarpous pistil. The perianth is whorled or cyclic, but the stamens and carpels are arranged spirally on the axis. The Stock agrees with the above in that the sepals and petals are free, but there are only six stamens, four long and two short. The ovary consists of two united carpels, divided by a central plate. All the whorls are cyclic. The Violet, while having free sepals and petals, has an irregular corolla, one petal being spurred. The pistil consists of three united carpels and a one-celled ovary.

Important differences from the above are met with in

the Strawberry and the Rose. While the sepals and petals are free and the stamens indefinite, as in the Buttercup, these parts are developed on the edge of the hollow receptacle; they are therefore around, and not below, the ovary —thus being perigynous. The carpels are numerous and apocarpous.

The irregular flowers of the Sweet-Pea and Gorse also have perigynous stamens, but only one (apocarpous) carpel.

The Chervil flower is also irregular, but the sepals, petals, and stamens are developed on the top of the ovary (epigynous), the inferior ovary being syncarpous.

In all these cases the petals are free (polypetalous).

In the remaining examples the chief difference we noticed was that the petals were united to form a tube (gamopetalous).

In the Heath, the corolla is bell-shaped; the anthers possess small appendages or horns, and the pollen escapes through pores. The ovary is superior. In the Primrose the five petals are joined, forming a long, narrow tube to which are attached five stamens. The superior ovary has only one chamber with free-central placentation. The corolla of the Speedwell is also gamopetalous, and on it there are two stamens; the ovary is superior and two-celled. In the Daisy and Dandelion, the flowers, though small, are massed together into a head, or capitulum. The petals are joined, and the five stamens on them have their anthers united round the style. The ovary is inferior, and when mature has only one chamber containing a single ovule.

It is upon such characters in the structure and arrangement of the parts of the flower that the classification of plants depends; and it is by comparison of these characters on the lines indicated that we gain an insight into plant relationships. All the plants we have just considered agree in a few broad characters. The vascular bundles of the stem are arranged in a ring; the leaves are net-veined; and the parts of the flower are usually disposed in fours or fives. Further, the seeds have two cotyledons, and they therefore all belong to the same great class of flowering plants called Dicotyledons.

We now proceed to a few examples belonging to the other great class, the Monocotyledons.

CHAPTER XIV

BIOLOGY OF THE FLOWER (*Continued*)

MONOCOTYLEDONS

Hypogynous flowers, with three parts in each whorl.—Another type of flower, constructed on a different plan from the preceding, is represented by plants such as the Bluebell and the Daffodil. The flowers of the **Bluebell** arise on the upper part of a long leafless axis or scape (Fig. 133, 1). Each flower (Fig. 133, 2) is attached by a stalk or pedicel, at the base of which are two long, narrow, coloured bracts which cover the young flowers when in bud (*br*). An inflorescence of this kind with stalked flowers, of which the oldest is at the bottom and the others are in succession younger as we near the top, is called a **raceme**.

The parts of the flowers are arranged very regularly in threes (trimerous). On the outside are three sepals, then three petals, and the six are slightly joined together at the base. Here the sepals and petals are all similar and coloured, and the name **perianth** may be used to indicate the two whorls when they are not differentiated into calyx and corolla. There are six stamens, three outer and three inner, fixed below the pistil, which is in the centre and consists of three united carpels. The ovary is superior and divided into three chambers, and above it is a long style, terminating in a three-lobed stigma. Cut the ovary across, and note that the ovules in each chamber are attached to the axis in double rows, i.e. the placentation is axile. Honey is secreted by glands in the ovary-wall between the carpels, and is much sought for by bees. The three outer stamens ripen before the stigma, and the anthers (at first vertical)

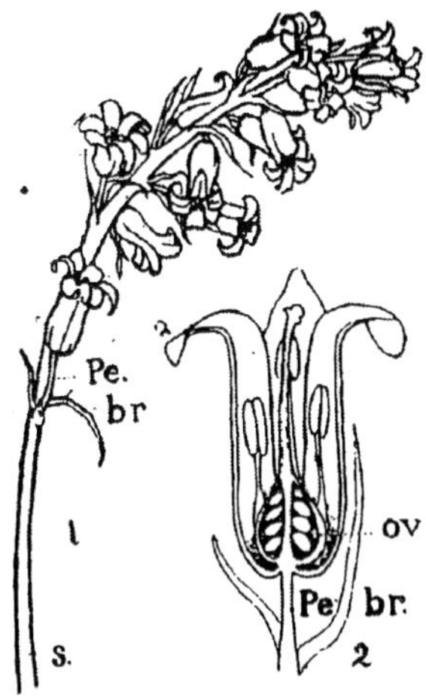

FIG. 133. WILD HYACINTH. —1, racemose inflorescence; 2, vertical section of flower; *br*, bract; *ov*, ovary; *Pe*, pedicel; *s*, scape.

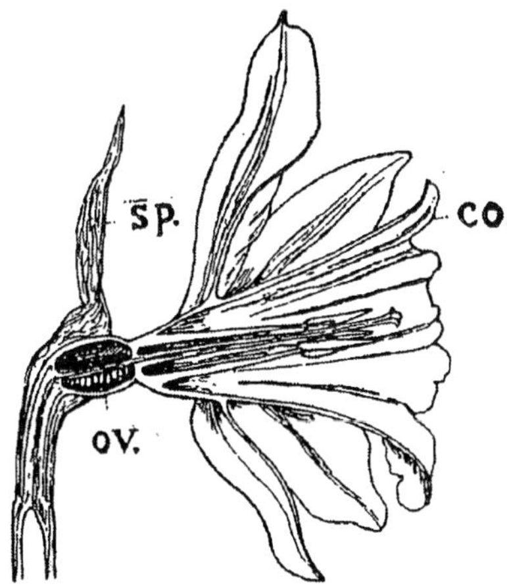

FIG. 134. VERTICAL SECTION OF DAFFODIL FLOWER.—*co*, corona; *ov*, ovary; *sp*, spathe.

dehisce inwards (introrse dehiscence), then turn horizontally and almost close the entrance to the flower. A bee visiting the flower presses its head into the corolla, forces apart the deeply-divided lobes, and in doing so becomes dusted with pollen. The style now elongates, the stigmas ripen and may become cross-pollinated. At this time the three inner stamens dehisce, and if cross-pollination does not occur, self-pollination is certain, as the three stigma-lobes when ripe come into contact with the anthers.

Epigynous trimerous flowers.—The **Daffodil** has three

outer and three inner perianth-leaves united into a tube, an outgrowth from which forms a large bell-shaped corona (Fig. 134, *co*). Interesting modifications of this may be seen in different species of Narcissus; in some the corona is in the form of a small, brightly-coloured ring, and gradations may be found from this up to the large corona of the Daffodil. The six stamens, in two whorls of three, are attached to the perianth. The pistil, however, differs from that of the Bluebell, in that the ovary is inferior. The style is long, reaching to the mouth of the tube, and has three stigma-lobes. The pistil consists of three carpels and the ovary is three-celled, with axile placentation. The bract of the Daffodil is in the form of a dry membraneous sheath or **spathe** (*sp*).

Examine flowering specimens of the **Crocus**, and note, when closed flowers are brought into a warm room, how quickly they open, the lobes often becoming strongly reflexed. The flowers and leaves are surrounded by colourless sheaths, which keep together and protect the inner parts while they are growing through the soil. Remove the sheaths, and note how limp and slender are the organs within, and how dependent they are upon the support of the sheathing cylinders. An interesting example of division of labour is thus afforded. Note the scale-leaf in the axil of which this flowering shoot arises, and observe that the base is already enlarging. Thus we see a young corm forming as a branch upon the old one (Fig. 135).

Remove the foliage-leaves and note their mode of attachment (see p. 133 and Fig. 84). The flower is surrounded by a thin, colourless sheath (Fig. 135, *sh*), and is supported on an under-ground cylindrical stalk—the scape (*sc*). The perianth-leaves are in two whorls of three each, and are united to form a narrow tube three to four inches long. There are three stamens fixed to the top of the tube; the large arrowhead-shaped anthers (*a*) dehisce outwards (extrorse dehiscence), and are ripe before the stigmas. Look for the ovary (*ov*) and note its position in relation to the ground-level (*g*). A section across it shows it to be three-celled and to contain many ovules on axile placentas. The ovary is inferior and lies one to two inches below the surface of the soil.

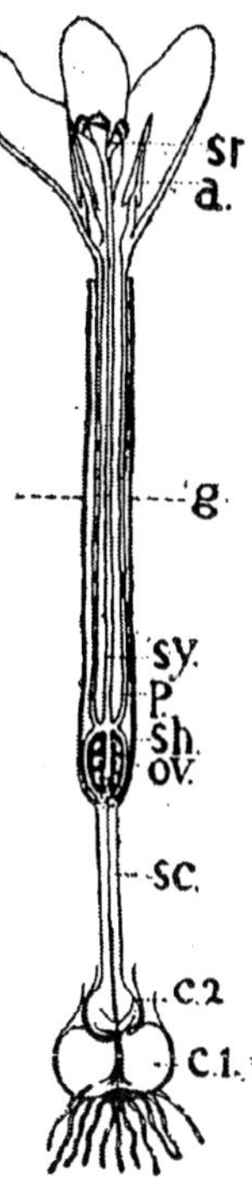

FIG. 135. VERTICAL SECTION, CORM AND FLOWER OF CROCUS.—*a*, anther; *c*.1, old corm; *c*.2, young corm; *g*, ground-level; *ov*, ovary; P, perianth-tube; *sc*, scape; *sh*, sheath; *st*, stigma; *sy*, style.

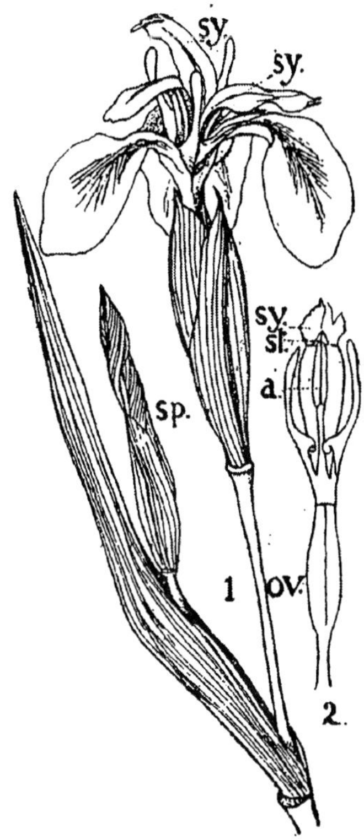

FIG. 136. 1, INFLORESCENCE OF IRIS. 2, flower with outer perianth-leaves removed; *a*, anther; *ov*, ovary; *sp*, spathe; *st*, stigma; *sy*, petaloid style.

Arising from the top of the ovary, and nearly filling the perianth-tube, is a very long style which divides at its free end into three large, frilled stigmas (*st*). The ovary secretes honey which rises high in the tube and so comes within reach of the long-tongued bees, but only insects with very long tongues (such as the hawk-moths) are able to extract the whole of it. Insects visiting the flower for honey will at first become dusted with pollen, which may be transferred to an older flower with ripe stigmas. If insects fail, the stigma-lobes curl outwards between the anthers and become self-pollinated—an advantage in a species flowering early before many insects are about. As the fruit develops, the scape elongates and carries the ripening capsule above ground, where it dehisces and the seeds are scattered.

The flower of the **Iris** (Fig. 136) has three large outer and three small inner perianth-leaves, but has only three stamens, the inner whorl being suppressed, and, as in the Daffodil, they are above the ovary, i. e. they are epigynous. The three styles are transformed into large petaloid lobes, against which lie the stamens. On the under side of these lobes and near the tip is a little flap covering the stigmatic surface. As in the Daffodil, the flowers are enclosed in a large spathe. The petaloid style arms (*sy*) of the Iris are often applied to the perianth-leaves in such a way as to form a split tube, at the base of which are honey-glands. The difference in size of the perianth-leaves, together with the large petaloid styles, gives to the Iris a striking, and at first rather puzzling, appearance.

The Orchid flower.—Still more puzzling is the flower of the **Purple Orchis** (Fig. 137). Each flower arises in the axil of a bract and appears to be stalked, but the stalk consists of the inferior ovary and is twisted (2, *ov*). Hence the flowers are sessile and the oldest are below as in a raceme.

Such an inflorescence of sessile flowers is called a **spike**.

The perianth is epigynous, consisting of six perianth-leaves, one of which (the **labellum**) forms an alighting stage (Fig. 137, *l*), while the others form a hood, covering in the stamens and stigmas. From the base of the labellum is produced a tube or spur in which honey is secreted. There is only one stamen (*a*), the remainder being suppressed. Two of them, however, are represented by small, barren stumps, called **staminodes** (*st*), one on either side of the stamen. Just below the single stamen, which lies under the hood, is a sticky disk—the **rostellum** (*r*)—and on either side of it is a stigmatic surface.

Imitate the action of a bee by inserting the point of a pencil into the throat of the flower, and in doing so press it against the rostellum. Now remove it, and if the stamen

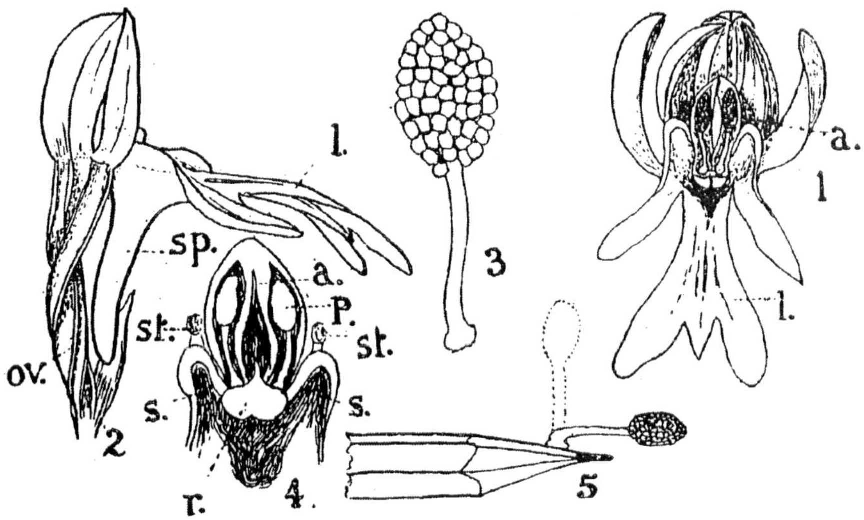

FIG. 137. PURPLE ORCHIS.—1, front view of flower; 2, side view of flower; 3, pollinium; 4, anther- and stigma-lobes; 5, pollinium removed from a flower and bending horizontally; *a*, anther; *l*, labellum; *ov*, ovary; P, pollinia; *r*, rostellum; *s*, stigma; *sp*, spur; *st*, staminodes.

is ripe, notice that two stalked, club-shaped masses of pollen adhere to the pencil (Fig. 137, 5). These are called **pollinia** (P), and are the masses of pollen from the anther-lobes. Watch them for a moment and note that they bend into a horizontal position, turn outwards a little, and are therefore suitably placed for coming into contact with the stigmatic surfaces wh n again inserted into a flower. Try to brush the pollinia off the pencil and you will find that the secretion glues them so firmly to the pencil that some force is required to remove them.

The Orchid illustrates great modification in a monocotyledonous flower. The perianth is irregular (zygomorphic); the posterior petal of the inner whorl projects as a lip or labellum, and serves as an alighting stage for insects. Below, it is prolonged into a honey-secreting spur. The other petals form a hood, protecting the pollen and honey. The essential organs are borne on a prolonged outgrowth of the axis, called the **column**, on the top of which are one fertile and two barren stamens, and two stigmas, also the rudiment of a third stigma—the rostellum (*r*)—on which are developed the two sticky bodies which glue the pollinia to the bee's head. The ovary (*ov*) is inferior, stalk-like, and twisted. When ripe, the fruit contains a large number of minute seeds.

Reference to the floral diagram (Fig. 179) will help to make the various relationships clear. Six stamens, in two whorls of three each, ought to be present, but of the outer whorl only the anterior one is present, and it is over the rostellum. The other two are suppressed, and their position is represented in the diagram by a cross ×. Of the inner whorl, the posterior stamen is suppressed, and the two lateral ones are reduced to short, barren stumps called staminodes. The anterior stigma is transformed into the rostellum; the two lateral ones are functional and lie below the stamen, one on either side of it. As the flower develops, the stalk-like inferior ovary twists through 180°, and carries all the parts of the flower round, so that the posterior lip comes to be anterior in the open flower. The diagram (Fig. 179) represents the parts before twisting occurs.

Flowers of Grasses.—The flowers of **Grasses** are small, and the parts can be made out only by careful observation. The inflorescence is usually either a compound spike or a panicle; and what appears to be a single flower is a group of sessile flowers, or spikelets (Fig. 138, 1). A dissected spikelet of Wheat is shown in Fig. 138, 3, and the arrangement of the parts on the axis is shown in the diagram Fig. 138, 2. At the base of it is a scale (*g*), the outer glume; imme-

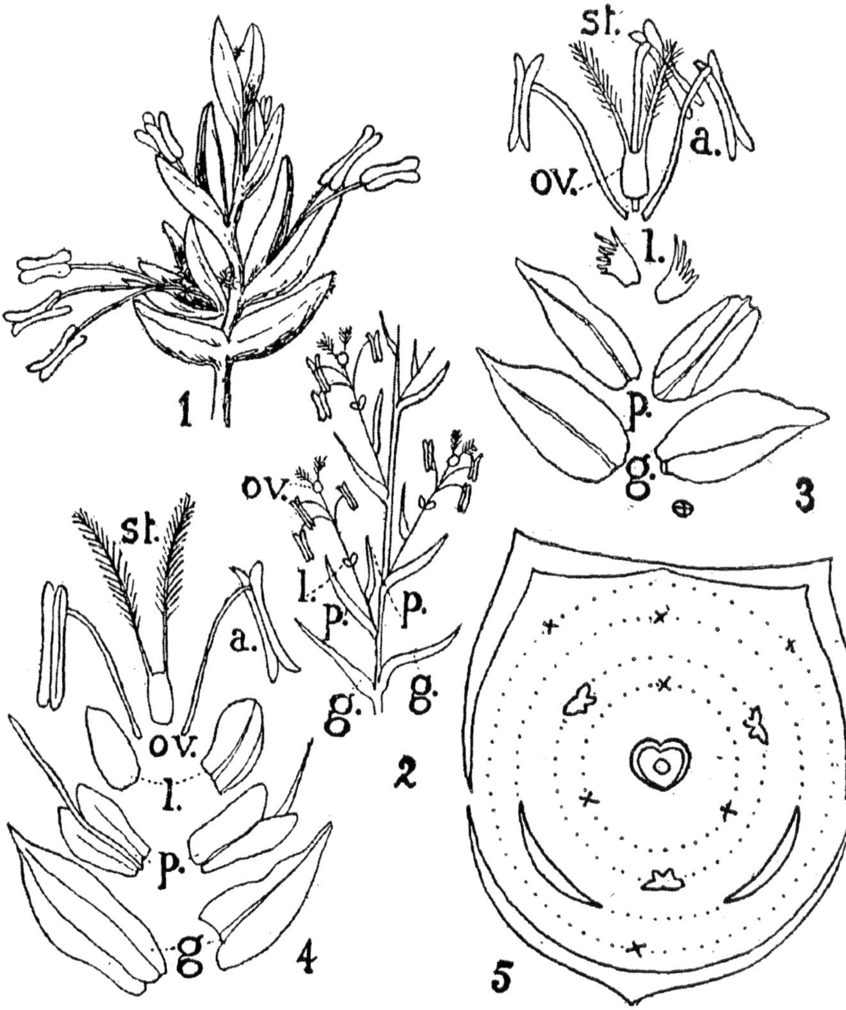

FIG. 138. FLOWERS OF GRASSES.—1, spikelet of Meadow Poa; 2, diagram of Grass spikelet; 3, parts of a Wheat spikelet dissected; 4, parts of flower of Vernal-grass dissected; 5, floral diagram; *a*, anther; *g*, glume; *l*, lodicule; *ov*, ovary; *p*, pale; *st*, stigma.

diately above it is another scale, the inner glume; and higher still is a smaller and thinner scale, the outer pale, in the axil of which a flowering branch arises. Low on the flower-stalk is a fourth scale, the inner pale; then follow two minute scales called **lodicules** (*l*), immediately above which are three stamens with long, slender filaments and large, easily moved (versatile) anthers. At the end of the flower-axis is the pistil, consisting of one carpel. The ovary contains one ovule and above it are two large, feathery stigmas. There is no perianth unless the two lodicules may be so regarded, but this is doubtful, and the flower is said to be naked. Sometimes there are two pairs of pales, and the lower ones may bear bristles, or awns, as in the Vernal-grass (Fig. 138, 4), in which flower there are no lodicules and only two stamens. In most cases the stamens ripen and shed their pollen before the stigmas are ripe, but in the Vernal-grass the stigmas are ripe first. The long, slender filaments and large, easily moved anthers, the great amount of pollen and the big, branched, feathery stigmas, are excellent devices for wind-pollination; while the fact that the anthers and stigmas of the same flower are not ripe at the same time secures cross-fertilization.

To summarize: The flowers of Monocotyledons show a series of modifications, from simple types pollinated by the wind to complex forms adapted to the habits of special insects. The Grasses, like the Willows, have no perianth, and the essential organs are enclosed by bracts called glumes and pales. The three stamens with their slender filaments and large versatile anthers, and the branched stigmas, are adapted to wind-pollination. The ovary is superior and contains one ovule. The flower of the Bluebell is cyclic, and the parts are in five whorls of three each; the perianth is petaloid; the syncarpous ovary is superior, three-celled, and contains many ovules. The Daffodil differs from the Bluebell in that its ovary is inferior and the perianth has a corona. The Crocus also has an inferior ovary, but has only three stamens. The flower has a long tube, and the stigma-lobes are large. The flowers are attractive in colour and secrete honey, but, failing insect-visits, self pollination commonly occurs, an advantage in early flowering species. In the Iris the styles are petaloid and large enough to be conspicuous. The greatest specialization occurs in the Orchids, which have irregular flowers, long honey-secreting spurs, stamens reduced to one, rarely two, and united to the style (gynandrous). They are usually incapable of self-pollination. The capsules produce an immense number of minute seeds.

CHAPTER XV

POLLINATION, FERTILIZATION, AND THE ORIGIN OF SEEDS

From our study of flowers we learn that all the parts of which they are composed serve directly or indirectly to secure the production of seeds. The modifications are very numerous, but in most cases they are definitely related to pollination, which precedes fertilization. Except in rare cases, ovules must be fertilized before seeds can be developed.

Advantages of self- and cross-pollination.—The flowers of most plants are developed in air, and the pollen-grains have to be carried through air from the anther to the stigma. The chief means by which this is secured may be summarized as follows:

1. Self-pollination, where the pollen falls on to the stigma of the same flower. This occurs in flowers which never open, like some of the Violets and Wood Sorrel, also in certain species of Deadnettle and Vetch. These are known as **cleistogamic flowers** (see p. 186). Self-pollination is very common in flowers where the pollen and stigma are ripe at the same time, e. g. in the Buttercup, Dwarf Mallow,

Scarlet Pimpernel, Stock, and Crocus. It is an effective means of securing a crop of seeds if cross-pollination fails.

2. Cross-pollination, where the pollen is carried to the stigma of another flower on the same plant or to the stigma of a flower on another plant of the same species.

Devices which favour or necessitate cross-pollination are very common, and seeds resulting from such a cross are often more numerous and produce better and healthier plants than when self-pollination occurs.

The more important means of securing cross-pollination are:

1. Stamens and pistil occurring in different flowers (**diclinous**: Gr. *di* = double, *kline* = a bed).

(*a*) Staminate and pistillate flowers on the same plant (**monoecious**), e. g. Pine, Hazel, Oak, and Birch.

(*b*) Staminate and pistillate flowers on separate plants (**dioecious**), e. g. Willow, Poplar, Red Campion, Dog's Mercury, Crowberry.

2. Stamens and pistil occurring in the same flower (**hermaphrodite**), but are not ripe at the same time (**dichogamous**: Gr. *dicha* = in two parts), though the male and female stages usually overlap, at which time self-pollination may occur.

(*a*) Stamens ripen and shed their pollen before the pistil is ripe (**proterandrous**), e. g. Daisy, Dandelion, and other Compositae; Mallow, Wood Sorrel, Meadow Crane's-bill, Chervil and other umbelliferous flowers.

(*b*) Pistils ripen before the stamens (**proterogynous**), e. g. Field Wood-rush and Plantains.

3. Anthers and stigmas are so situated that the pollen does not fall on to the stigma, e. g. Pansy, Buttercup (see pp. 167–8).

4. Different forms of flowers occur in the same species (**heteromorphic**, Gr. *hetero* = different).

(*a*) Long- and short-styled forms (**dimorphic**), e. g.

Primrose, Cowslip (see p. 176), and some species of Sorrel (*Oxalis*).

(*b*) Long, intermediate, and short-styled forms (**trimorphic**), e. g. Purple Loosestrife (*Lythrum Salicaria*), and some species of Sorrel (*Oxalis*).

As the pollen-grains are unable by their own efforts to reach the stigmas, they must be carried by external agents, the chief of which are the wind and animals. Flowers differ in several important respects, according to the agent employed in the transference of pollen.

1. Flowers pollinated by the Wind (**anemophilous**: Gr. *anemos* = wind, *philos* = loving) have usually the following characteristics:

The flowers are small, not showy, unscented, and without honey; the anthers are large and on long slender filaments; and sometimes, as in catkins, the whole inflorescence is easily shaken by the wind. The pollen is abundant, dry, and powdery. The stigmas are large and feathery, and expose a large surface to catch the pollen; but much of the pollen never reaches the stigmas and is wasted. Many of our forest trees are wind-pollinated, e. g. Pine, Larch, Poplar, Hazel, Oak, Birch, Beech; also the Grasses and Sedges, Docks, Plantains, Stinging Nettle, and Crowberry.

2. Flowers pollinated by Animals (**zoophilous**: Gr. *zōon* = an animal) are the more familiar and attractive species. The most important pollinators are Insects, and the flowers so pollinated are said to be **entomophilous** (Gr. *entomon* = insect). Flowers possess several features which render them attractive to insects. They are usually brightly coloured, often scented, and have nectaries. The pollen is sticky, readily adheres to the bodies of insects, and is often collected by them as food. The stigmas are small and frequently placed in a position favouring pollination by the insect visitor. Small animals like snails may bring about pollination when crawling over certain flowers,

and some, often scarlet exotic flowers, are pollinated by humming-birds.

Positions of honey-glands.—Nectaries occur, as we have seen, on many different organs, but usually they are on some part of the flower :

(*a*) On the receptacle at the bases of the short stamens in the Wallflower and Stock ;

(*b*) On the sepals of the Mallow and Coronilla ;

(*c*) On the petals of the Buttercup and Lesser Celandine, and in the spurred petal of Orchis ;

(*d*) On the stamens of the Violet and Pansy, the spur acting as a honey-receptacle ;

(*e*) On the carpels of the Marsh Marigold and Bluebell.

Nectaries sometimes occur on leaves (extra-floral nectaries) (Fig. 219, p. 340). These attract numerous ants, which in turn keep off the caterpillars that would eat the leaves.

Fertilization and the Origin of Seeds

Structure of the pistil.—The pistil is the inner essential organ of the flower. In a typical and simple case like that of the Pea it may be regarded as an up-rolled leaf (Fig. 139), bearing on its margins small rounded bodies called ovules (*o*). If we suppose the turned-in edges to meet and fuse, so enclosing the ovules in a box, we can form some idea of its structure. The ovule-bearing portion is called the ovary (*ov*), and that part of the edge from which the ovules spring is called the placenta (*pl*). Its tip is prolonged and known as the style, and at the end of it is a portion which receives the pollen, called the stigma. Such an ovule-bearing leaf is called a carpel, and in the case of the Pea the pistil consists of one carpel only.

In the Stock the pistil is composed of two united carpels ; in the Clematis, Buttercup, Marsh Marigold, and Rose, it consists of many free carpels. When the pistil consists of one or more free carpels it is said to be apocarpous; and when of two or more carpels united together, it is syncarpous. If it arises above the other parts (calyx, corolla, and stamens) it is superior, and if below, it is inferior.

Sometimes botanists use the word 'pistil' in a different sense. When the carpels are free each consists typically of an ovary, style, and stigma. Then the flower is said to have many pistils, e. g. the Buttercup; but when the carpels are joined, as in the Stock and Tulip, the flower is said to have only one pistil. The name **gynoecium** is given to the central part of the flower, whether consisting of one carpel or many, free or joined.

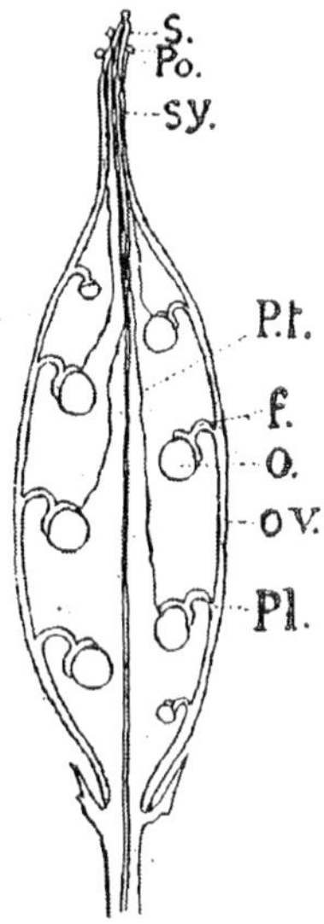

FIG. 139. DIAGRAM OF OPENED PISTIL.—*f*, funicle; *o*, ovule; *ov*, ovary; *Pl*, placenta; *Po*, pollen-grains; *P.t*, pollen-tube; *s*, stigma; *sy*, style.

The ovule and the embryo-sac.—When the ovule first appears on the placenta it consists of a small outgrowth of tissue, the **nucellus** (Fig. 140, *n*). Around the base of this, two coats grow upwards and cover the nucellus, with the exception of a minute pore, which remains at the end and forms the **micropyle.** Within the nucellus a large cell arises, called the embryo-sac (*em*), within which several cells are formed, one of these, near the micropyle, being called the egg-cell.

Fig. 140, 1–3, illustrates these points. In these the ovule is represented as a straight one, but this form is not common. More usually the ovule during its development becomes bent and often inverted, so that the micropyle is brought down to the base of the ovule stalk or funicle. Such an inverted ovule is said to be **anatropous** (Fig. 141, 2) (Gr. *ana*, denoting inversion, *trepo* = I turn). A straight one is **orthotropous** (1) (Gr. *orthos* = straight); when curved it is **campylotropous** (3) (Gr. *kampylos* = curved); and when at right angles to the funicle, **amphitropous** (4) (Gr. *amphi* = on both sides).

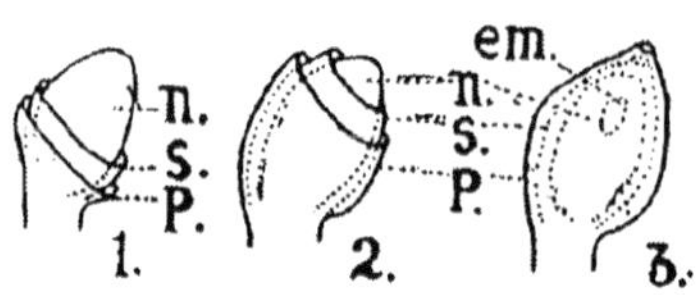

FIG. 140. DEVELOPMENT OF AN OVULE.—*em*, embryo-sac; *n*, nucellus; P, outer integument; *s*, inner integument.

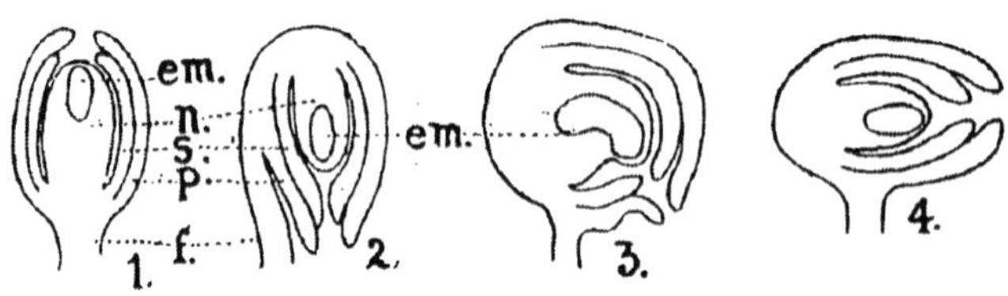

FIG. 141. FORMS OF OVULES.—1, orthotropous; 2, anatropous; 3, campylotropous; 4, amphitropous; *em*, embryo-sac; *f*, funicle; *n*, nucellus; *p*, outer coat or primine; *s*, inner coat or secundine.

When the ovule has developed thus far it is ready for fertilization. Before this can take place, however, pollen-grains of the same species must be deposited on the stigma. This, when ripe, is covered with a sticky, sugary secretion, in which the pollen-grains germinate.

Germination of pollen-grains, and fertilization.—By means of a simple experiment a good idea may be obtained of what occurs. Place a drop of 10 per cent. solution of cane sugar on a glass slip and put into the solution a few pollen-grains of the Sweet-Pea. If examined after an hour or two, short, delicate tubes called pollen-tubes will be seen emerging from the grains, and into these the contents pass as a living stream (Fig. 142).

When pollen-grains are placed on the stigma of a pistil, growth of this kind occurs. The pollen-tubes grow downwards through the style, enter the cavity of the ovary, and reach the micropyle of the ovule. Into this the tube passes, and the essential part of its contents enters the embryo-sac and there accomplishes the process of fertilization by fusing with the nucleus of the egg-cell. It is not until such fertilization has taken place that the ovule develops into a seed; for if the stamens are removed from a flower before

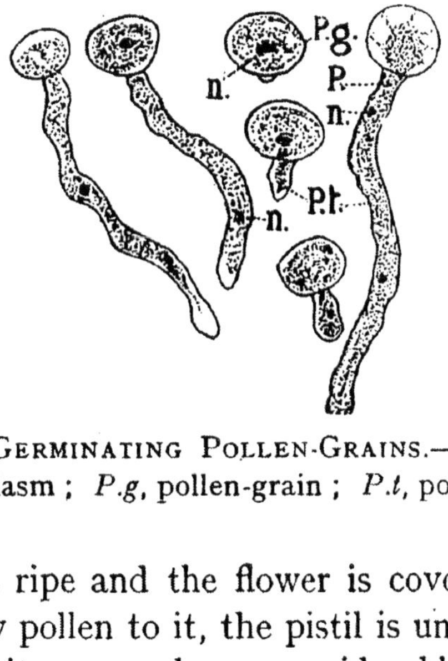

FIG. 142. GERMINATING POLLEN-GRAINS.—*n*, nucleus; P, protoplasm; *P.g*, pollen-grain; *P.t*, pollen-tube.

the anthers are ripe and the flower is covered so that no insect can carry pollen to it, the pistil is unable to set ripe seed, although it may enlarge considerably and in some cases become fleshy.

Changes produced by fertilization.—After the egg-cell is fertilized, the rest of the embryo-sac becomes filled with tissue called **endosperm**. In the Pea and the Bean the fertilized egg-cell grows at the expense of the endosperm, develops into an embryo, and at the same time the nucellus is absorbed. Thus, when the seed is ripe it contains no endosperm.

The differences between ovule and seed are important and may be stated as follows :

Ovule.		Seed.
Before fertilization.	*After fertilization.*	
2 ovule-coats	2 ovule-coats	Seed-coat
Nucellus	Nucellus	[Nucellus and
Embryo sac	Embryo-sac	endosperm used
	Endosperm	up as food for]
Egg-cell	Young embryo	Embryo.

Sometimes only part of the endosperm is used up ; the embryo then is relatively small and more or less endosperm persists round the embryo in the seed. Examples of this occur in Wheat, Maize; Common Ash, and Castor Oil. Occasionally some of the nucellus persists, when it is called **perisperm.**

Thus food-reserve may be stored in various regions in a seed : in the cotyledons of the embryo (Bean), in the endosperm of the embryo-sac (Wheat, Ash), and in the perisperm or persistent nucellus (Water-lily, Pepper).

The ovule is not the only part affected by fertilization ; many surrounding parts are affected also :

In the Pea, the carpel grows enormously and forms the pod. In the Stock, the two carpels elongate greatly. In the Strawberry, Rose, Apple, and Pear, the receptacle becomes not only very large but fleshy. In the Winter Cherry (*Physalis*), the calyx becomes much inflated and brightly coloured. In the Fig, Mulberry, and Pine-apple, the whole inflorescence or parts of it become fleshy, fuse together, and form a very complex aggregate fruit.

The changes that take place as a result of fertilization are, therefore, very great. The union of the two elements, a nucleus from the pollen-grain and the nucleus of the egg-cell, results in a stimulus to growth which is the starting-point in the life-history of a new plant. The growth-stimulus produces the changes we have outlined both within the ovule and in adjacent parts, and carries them on till the seed is ripe and ready for an independent existence.

But some of the changes may go on, and fruits form even in the absence of fertilization, as in fruits like the Banana, seedless forms of Orange, Grape, and others. These, however, do not contain ripe seed capable of germination.

CHAPTER XVI

STRUCTURE OF FRUITS

THE fruit is the structure produced from the pistil as a result of fertilization. In the Pea and Bean, this consists of one carpel, but in the Stock there are two united carpels. Commonly, however, the pistil consists of several carpels, sometimes free as in the Buttercup, sometimes united as in the Violet and Crocus.

In common edible fruits the fleshy part which is eaten, often consists of structures other than the pistil. These are known as **false fruits**, while those formed from the pistil only are known as **true fruits.**

In all cases the object of their formation is the production of seeds, containing a young plantlet capable of growing into a new plant. In some fruits the fruit-coat or pericarp is dry and does not split until the seed within germinates; in others, the fruit-coat splits and the seeds are scattered. Others, again, have a succulent or fleshy fruit-coat, and most of our edible fruits belong to this class. A few specimens of each kind should be obtained and carefully studied.

Dry indehiscent fruits.—The **Hazel-nut** (Fig. 187, 8) is enclosed in a cup consisting of large, leafy bracts. Break open the hard, dry shell and note the single seed within (sometimes two may be found). Look for the seed-stalk and notice how it is attached. The seed-coat is thin and brown; and surrounds an embryo consisting, as in the Bean, of two fleshy cotyledons, a radicle, and a plumule.

Compare with this the **Acorn** (Fig. 143). Here the bracts are numerous, small, and form a compact cup. The smooth fruit-coat encloses one seed, the embryo has two large fleshy cotyledons. Such hard, dry, one-seeded fruits

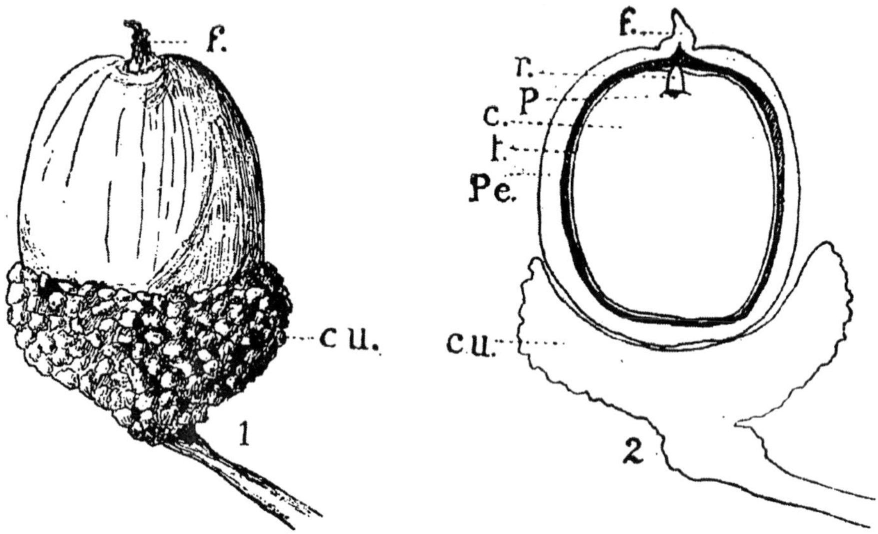

FIG. 143. FRUIT OF OAK.—1, Acorn with cupule; 2, vertical section of same; *c*, cotyledon; *cu*, cupule; *f*, remains of flower; *p*, plumule; *Pe*, pericarp; *r*, radicle; *t*, testa.

are called **nuts**. The Sweet Chestnut and the Beech are other examples.

The fruit of the **Buttercup** (Fig. 144) consists of many small, dry, one-seeded fruits, each the product of a separate carpel, and attached to a somewhat swollen receptacle. Each one is called a **nutlet** or **achene**.

The **Strawberry** (Fig. 145) is very similar, but the receptacle is slightly hollowed, and from its centre grows a large, fleshy structure, in which the achenes are embedded.

The **Rose-hip** (Fig. 146) differs from the above in that the

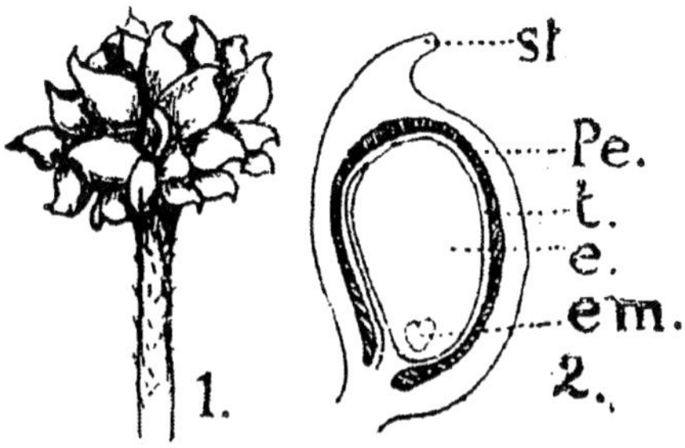

FIG. 144. FRUIT OF BUTTERCUP.—1, aeterio of achenes; 2, section of an achene; *e*, endosperm; *em*, embryo; *Pe*, pericarp; *st*, stigma; *t*, testa.

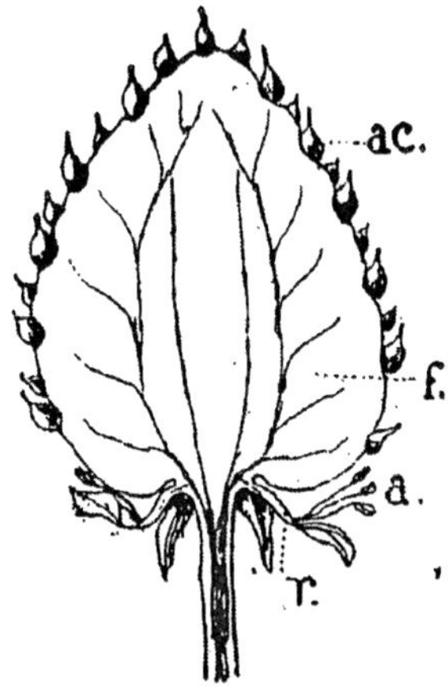

FIG. 145. VERTICAL SECTION OF A STRAWBERRY.—*a*, remains of stamens on edge of receptacle; *ac*, achenes; *f*, fleshy outgrowth of receptacle; *r*, receptacle.

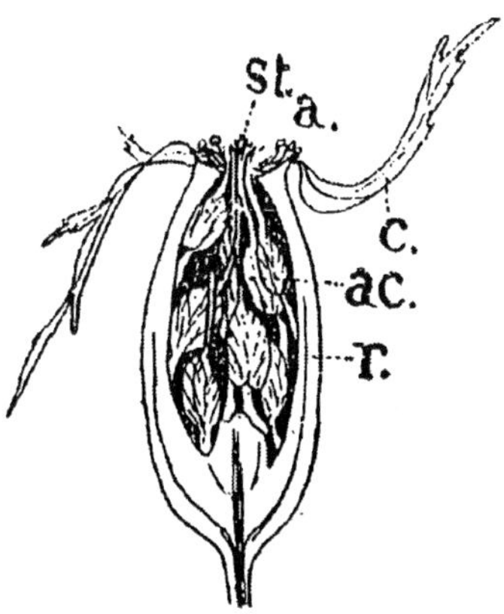

FIG. 146. VERTICAL SECTION OF HIP OF ROSE.—*a*, remains of stamens on edge of receptacle cup; *ac*, achenes; *c*, calyx; *r*, hollow receptacle; *st*, stigmas.

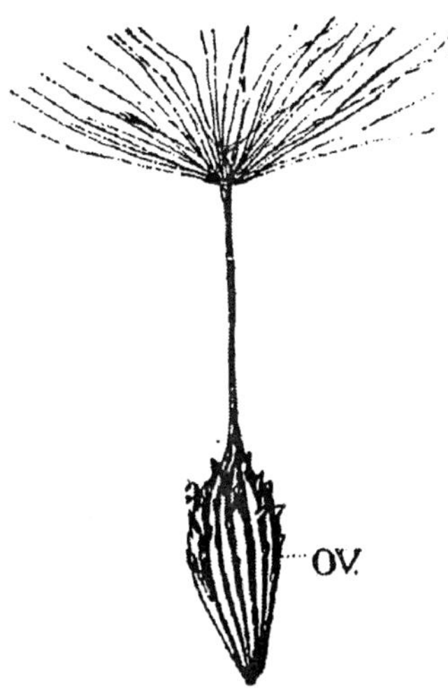

FIG. 147.
FRUIT OF DANDELION.
ov, ovary.

receptacle is hollowed and bears the achenes on the inner surface of the cup. Both in this and in the Strawberry the fleshy part is the receptacle, and these fruits are therefore succulent fruits containing many dry indehiscent nutlets.

Fruits like those of the **Dandelion** (Fig. 147), Coltsfoot, Thistles, &c., are achenes; but they are formed from an inferior ovary of two carpels, only one of which matures, and that contains but one seed.

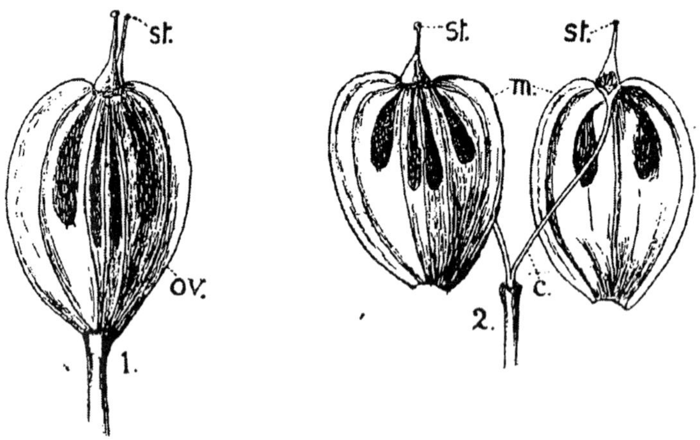

FIG. 148. 1, CREMOCARP OF HOGWEED; 2, mericarps separated; *c*, carpophore; *m*, mericarp; *ov*, ovary; *st*, stigmas.

The fruits of Grasses like Wheat, Maize, and Oat differ, as we have seen, in that the fruit-coat and seed-coat closely adhere.

The **Sycamore key** (Fig. 202, 5) consists of two or more carpels, which, when ripe, separate into part-fruits (mericarps) but do not scatter the seeds. Each part contains one seed, and is provided with a wing. Cut open a fruit; observe the thick fruit-coat lined with a felt of hairs and the seed within covered by a thin brown testa. Remove this and examine the embryo carefully; see how the cotyledons are rolled up. Make a paper model to show clearly how they are folded and rolled (Fig. 202, 6). Such winged fruits are called **samaras**. Other examples are the Maple, Ash, Elm, and Birch.

Another common type which splits into two half-fruits is found in the Hogweed, Chervil, and other umbelliferous plants. A ripe fruit of the **Hogweed** (Fig. 148) is easy to dissect. It consists of two flattened carpels, which readily separate into two half-fruits (mericarps), and remaining attached by a slender stalk, the **carpophore** (*c*), each

FIG. 149.
FOLLICLES OF COLUMBINE.

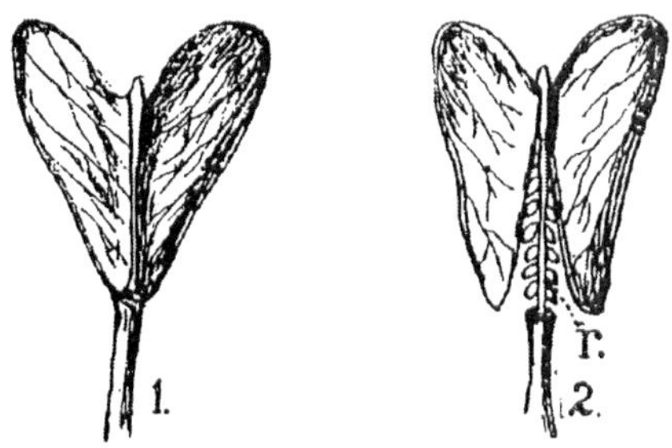

FIG. 150. 1, SILIQUA OF SHEPHERD'S PURSE; 2, siliqua dehiscing; *r*, replum with seeds attached.

mericarp containing one seed. Such fruits are called **cremocarps.** Caraway 'seeds' are half-fruits of this kind.

Dry dehiscent fruits.—In all the above, when the fruit is dispersed, the thick protective coat does not burst until germination begins, such seeds usually having only a thin testa. Dry fruits of a second class have a fruit-coat which splits, and so allows the seeds to escape. These, not having the protection of the fruit-coat, are usually surrounded by a thick testa. They include many common fruits, and several should be examined, and the different modes of splitting (or dehiscence) compared.

The **Columbine** (Fig. 149), Monkshood, and Marsh

Marigold have a pistil of several free carpels. Each is pod-like and contains many seeds; but, unlike a pod, splits along the inner seam only. These fruits are called **follicles.** The **legume** or pod of the Pea and the Bean consists of one carpel and splits along both seams, the two valves often being twisted into a close spiral (Fig. 162, 1).

The **Shepherd's Purse** (Fig. 150) and Honesty are similar to the Stock, but shorter, and their pods are known as

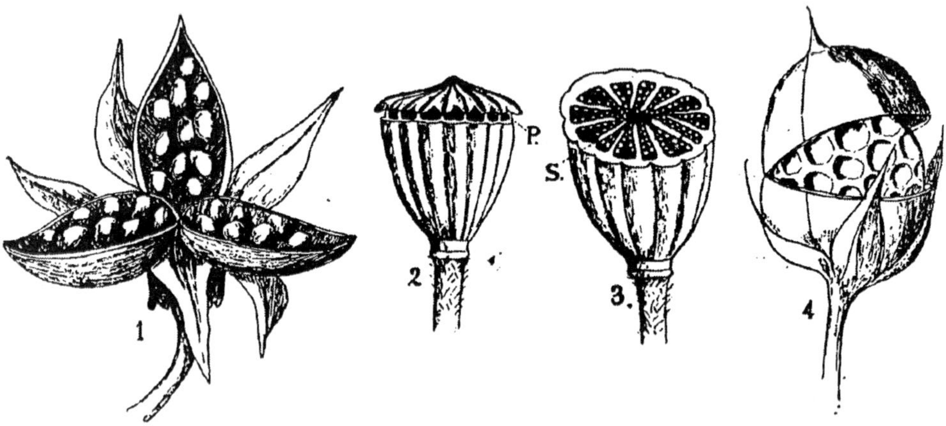

FIG. 151. 1, OPEN CAPSULE OF VIOLET; 2, CAPSULE OF POPPY DEHISCING BY PORES; 3, SAME IN SECTION; 4, CAPSULE OF PIMPERNEL DEHISCING TRANSVERSELY; P, pore; *s*, edges of carpels which project into the ovary but do not meet in the centre.

siliquas. Compare these and note how very different forms of fruit may arise from one type of structure.

A common type of dry splitting fruit which is more or less globular is called a **capsule**, and dehiscence takes place in a variety of ways. Some capsules open above, forming a cup surrounded by teeth, as in the Campion (Fig. 159, 2). In the **Violet** (Fig. 151, 1) it opens by three valves. Other capsules dehisce by pores, as in the **Poppy** (Fig. 151, 2 and 3), where the pores are around the margin; and in the Snapdragon the capsule is oblique and has three pores (Fig. 159, 3). In some Campanulas the pores are at the base (Fig. 159, 5); while in the globular fruits of the Pimpernel (Fig. 151, 4),

Henbane, and Plantain, the capsules split transversely and the upper part comes away as a lid.

Succulent fruits.—We have seen (p. 215) that the pericarp of many dry fruits splits when ripe, and the seeds, surrounded by a thick testa, are dispersed. On the other hand, the more familiar edible fruits have a succulent or juicy coat which does not split when ripe; that is, they are **indehiscent.** Examine the fruit of a **Cherry** (Fig. 152) and note the thin

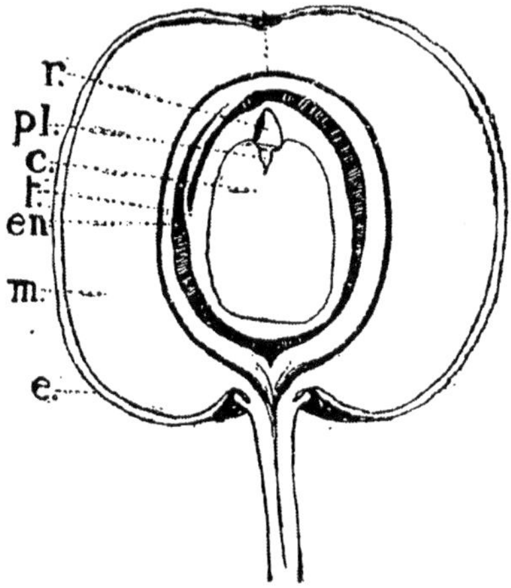

FIG. 152. VERTICAL SECTION OF DRUPE OF CHERRY.—*c*, cotyledon; *e*, epicarp; *en*, endocarp; *m*, mesocarp; *pl*, plumule; *r*, radicle; *t*, testa.

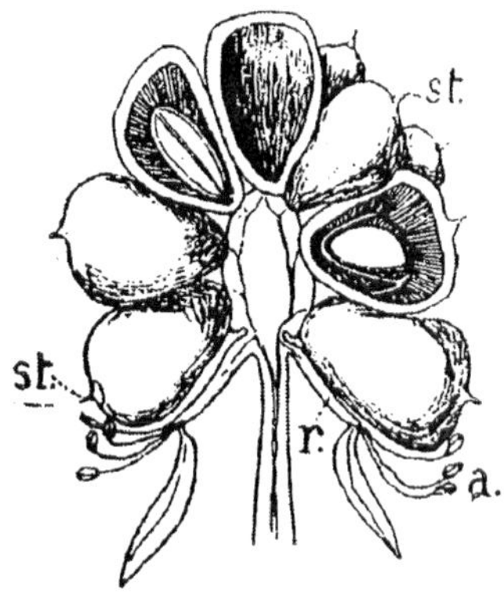

FIG. 153. AETERIO OF DRUPELS OF BLACKBERRY.—*a*, remains of stamens; *r*, receptacle; *st*, stigmas.

outer coat or **epicarp**, the fleshy middle coat or **mesocarp**, and the hard inner coat, the **stone** or **endocarp**, enclosing one seed.

The Plum and the Damson are fruits similar to the Cherry, all three arising from a superior apocarpous pistil. The name **drupe** (Gr. *druppa* = over-ripe olive) is given to fruits of this kind.

Compare the above with the fruits of the **Blackberry** (Fig. 153) and Raspberry. Each of these consists of many small drupes or drupels, borne on an upgrowth from the centre of the receptacle.

The **Gooseberry** (Fig. 154), Grape, Currant, and Tomato are syncarpous fruits, in which the endocarp is succulent as well as the mesocarp, and they have one cavity (**loculus**) which contains several seeds. Such fruits are known as **berries.** The name berry is also given to many common

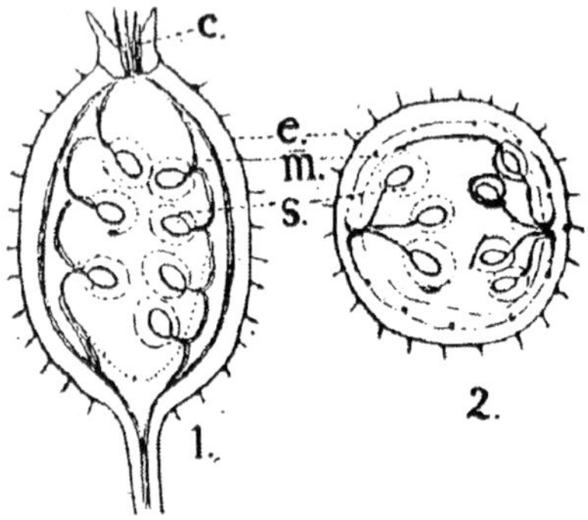

FIG. 154. FRUIT OF GOOSEBERRY.—1, berry in vertical section; 2, in transverse section; *c*, remains of flower; *e*, epicarp; *m*, mesocarp; *s*, seeds embedded in succulent endocarp.

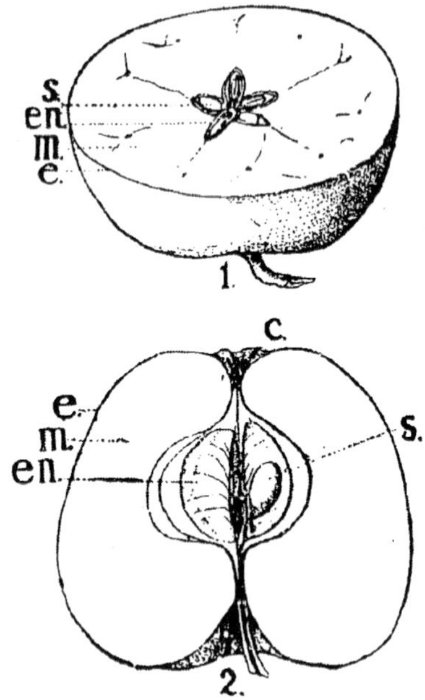

FIG. 155. POME OF APPLE.—1, transverse section, showing the five carpels; 2, vertical section; *c*, remains of flower; *e*, epicarp; *en*, endocarp or core; *m*, mesocarp; *s*, seed.

fruits like the long fruits of Gourd, Cucumber, and Banana. The latter is without seeds, the plants being propagated from rhizomes.

The Date is one-seeded, the seed being hard and stony. The Orange and Lemon have a leathery epicarp, and are divided into several chambers. In berries like the Currant and Gooseberry the fruit is inferior, the calyx being on the top of the fruit; while the Tomato, Grape, Orange, and

Lemon are superior. The Pomegranate is peculiar in that the edible part consists of the succulent testas of the seeds.

The Apple and the Pear, as is the case with the Rose and the Strawberry already examined, are fruits in which the receptacle takes part in their formation; and they are sometimes called **false fruits** or **pseudocarps**. In the **Apple** (Fig. 155) the core is formed from the pistil, and the pips are the seeds, while the fleshy part is formed from the receptacle. Such a fruit is called a **pome**.

The Fig is a compound fruit consisting of a hollow inflorescence, within which are numerous small drupes. The Pine-apple is also a fleshy inflorescence, the axis of which is continued above and bears leaves. The Mulberry is formed from a spike of many flowers; the perianth-leaves of each become united and fleshy, and enclose the ovary, thus resembling the Blackberry in appearance but differing widely from it in origin.

The modifications found in fruits have, in most cases, an obvious connexion with the dispersal of seeds, the various devices for which we will next consider.

CHAPTER XVII

DISPERSAL OF FRUITS AND SEEDS

PRIOR to August 1883, Krakatau, one of the East Indian Islands, was covered with impenetrable forests. On August 26 and 27 of that year, a violent volcanic eruption occurred; the topography of the island was completely changed; and the lava and molten ashes which fell upon the remaining portions completely destroyed the vegetation. This provided a rare opportunity of studying the ways in which a barren island (about twenty-five miles from the nearest mainland) acquired a new flora and fauna.

Colonization of a barren island.—It was found that the first colonists were microscopic Blue-green Algae, Bacteria and Diatoms, which formed slimy patches on the pumice and ash, and provided a suitable medium in which, later, the spores of Ferns and Mosses germinated. All these bodies are so very minute that they float as dust in the air, and are thus capable of being carried long distances by the wind. In this way spores of these plants were carried from the adjacent islands, and formed the first elements of the flora. Then followed flowering plants having light wind-borne fruits, and along the shore appeared seedlings from seeds carried by ocean currents and washed up by the sea, many of them in logs of wood. The rotting logs brought Fungi, and, in the cracks of the bark, small animals. Birds visiting the island brought other seeds, and lastly, man's influence was seen in the introduction of cultivated plants, and in the weeds that followed in his train. In a few years' time, large parts of the island were again covered by rank and luxuriant vegetation. Wind, water, and animals, especially birds, proved to be the three principal agents concerned in carrying the seeds of a new flora to the island.

In a study of the modes of dispersal of the common plants around us, we find the same agents at work; and an interesting collection may be made showing the various devices by which Nature secures this end.

The object to be attained is that seeds should be carried far enough away from the parent plant to prevent overcrowding, and to ensure that they are on suitable ground unoccupied by the same species. We will now study some typical examples of dispersal mechanisms.

A. *Dispersal by wind.*—The essential condition for wind-dispersal is lightness in proportion to bulk, or a floating device which increases the surface without greatly increasing the weight of the seed, e. g.:

(1) Minute spores[1] of Fungi, Mosses, and Ferns, and also microscopic plants. The frequent occurrence of moulds on organic substances is due mainly to the ease with which their abundant spores are carried in the air.

(2) Minute seeds, as of Orchids, which are rendered lighter still in proportion to bulk by a light, loose, outer

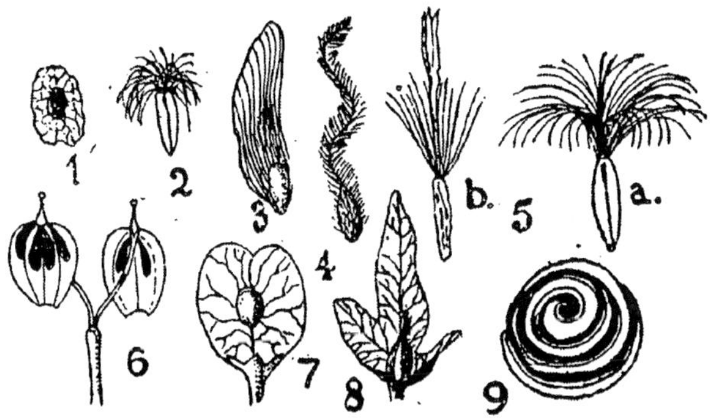

FIG. 156. FRUITS AND SEEDS DISPERSED BY THE WIND.—1, seed of Orchid, much magnified; 2, seed of Willow; 3, seed of Pine; 4, achene of Clematis; 5 *a*, fruit of female flower of Coltsfoot; 5 *b*, barren fruit of male flower of Coltsfoot; 6, mericarp of Hogweed; 7, samara of Elm; 8, fruit of Hornbeam; 9, rolled pod of Medicago.

coat (Fig. 156, 1). Heaths also have extremely small wind-dispersed seeds.

(3) Seed-Parachutes. Small seeds, even though larger than those mentioned above, may bear tufts of hairs serving as a parachute, e. g. Willows (Figs. 156, 2 and 157), Poplars, Willow Herbs (*Epilobium*) (Fig. 158, 2), and Cotton (the cotton fibres of commerce being highly developed seed-hairs).

(4) Fruit-Parachutes. The small fruits of many Compositae and other plants have a pappose calyx, e. g.

[1] Spores are minute reproductive cells capable of growing into new plants. Fungi, Mosses, and Ferns never produce true seeds.

Groundsel, Dandelion (Fig. 147), Coltsfoot (Fig. 156, 5), Thistles, Valerian, Bulrush; and the perianth of the Cotton-grasses (*Eriophorum*) (Fig. 158, 1) is transformed into a tuft of long hairs. The Clematis (Fig. 156, 4) and Mountain Avens have a long, persistent, feathery style. The awn of the Feather-grass is twisted, and ends in a beautiful plume a foot in length. Some Anemones have hairy fruit-coats.

(5) Winged Seeds occur in the Pine (Fig. 156, 3), Larch, and cultivated climbers like *Eccremocarpus* and *Bignonia*.

(6) The Flattened Fruits of the Hogweed (Fig. 156, 6) split into two thin half-fruits and are readily detached from their slender threads during high winds.

(7) Many fruits have a Winged fruit-coat, and the fruit is often flattened, e.g. Birch, Elm (Fig. 156, 7), Common Ash (Fig. 9, 1), Sycamore (Fig. 202, 5), and Maple. The wings of the Hornbeam (Fig. 156, 8) are formed from bracts. Most of these, however, are too heavy to be carried far, except during high gales.

(8) Globular Fruits and Plants may be rolled to a slight extent by the winds, as is the case with the fruits of *Medicago* (Fig. 156, 9) and the whole plants of certain species of *Selaginella*.

(9) Censer Mechanisms. The capsules and follicles of many plants (Fig. 159, 1 to 5) are borne on erect stalks, and the seeds can only escape from the cup-like fruit-case when violently shaken by the wind or by a passing animal; then a few seeds may be jerked out. Commonly the fruit-stalk decays, and the capsules with their seeds fall in a heap to the ground. We have already noticed the more important ways in which capsules open to allow the seeds to escape. Those widely open above are liable to damage by rain, but often the capsule is surrounded by teeth which bend over and close the capsule in wet weather, opening again when the air is drier (Fig. 159, 2). The fruit

FIG. 157. WILLOWS IN FRUIT; ALDERS IN THE BACKGROUND.

FIG. 158. 1, FRUITS OF COTTON GRASS; 2, PAPPOSE SEEDS OF WILLOW HERB.

of the Poppy (Fig. 159, 4) is permanently covered, the small seeds escaping through pores just below the lid. In Campanulas with pendulous capsules the pores are at the base of the capsule (Fig. 159, 5). In either case a strong wind or jerk is necessary to shake out the seeds.

B. *Dispersal by water.*—The dry fruits and seeds of plants growing along the sides of rivers and lakes may be blown into the water and float a short distance before sinking, or washed ashore farther down stream, where they may germinate; generally, however, they sink rapidly. If carried seaward, most of them soon lose their power of germination after entering salt water. Trunks

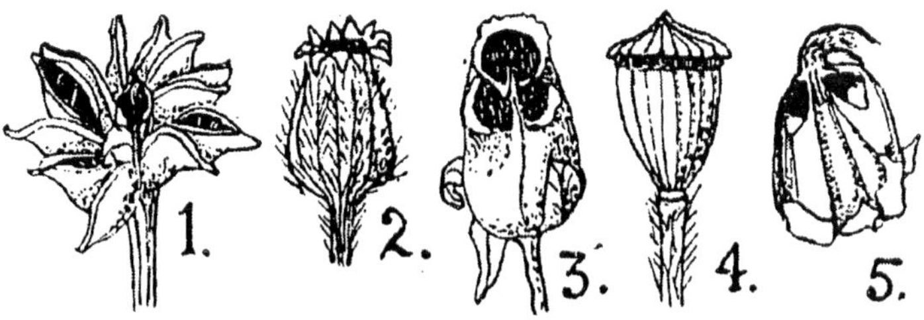

FIG. 159. CENSER FRUITS.—1, follicles of Marsh Marigold; 2, capsule of Red Campion; 3, capsule of Snapdragon dehiscing by pores; 4, capsule of Poppy; 5, capsule of Campanula.

and branches of trees carried down stream often bear seeds and fruits embedded in the mud adhering to them, over long distances. Fragments of plants, especially of water-plants capable of rooting, are often carried considerable distances, and provide an effective means of dispersal. The Canadian Water-weed (*Elodea*), so common in our ponds and canals, has spread extensively in Europe by this means.

The seeds of the White Water-Lily are surrounded by a spongy aril, and between it and the seed-coat is air, which enables the seeds to float until the air escapes, when they sink to the bottom. In the Frog-bit and some Pond-weeds, buds are formed in the autumn which become detached and sink to the bottom; the rest of the plant may die down, the species being renewed the next season by the buds which then renew their growth. Such buds enable the plants to tide over the winter, and are called winter-buds.

C. *Dispersal by animals.*—(1) Fungi growing in a pasture commonly occur on the dung of browsing animals; some of the spores may have been carried thither by the wind, but a common occurrence is that spores adhering to the leaves of plants are eaten by animals and pass uninjured through their food-canal. On the way they are partly digested and this prepares them for germination. Some fungus spores germinate with difficulty until acted upon by a digestive juice. The dispersal of such plants is in the first instance by wind, and in the second by animals which further prepare them for germination.

(2) Birds sometimes carry seeds great distances in mud adhering to their feet.

(3) Fruits and seeds form an important food for many birds. They are attracted by the bright colours of fleshy fruits, of which they eat the edible parts. Such fruits often have either an indigestible, stony endocarp around the seed, as in drupes like the Cherry and Brambles; a hard fruit-coat, as in the achenes of the Strawberry; or, where the endocarp is pulpy, as in berries, a hard seed-coat. While the fleshy parts are digested, the protected seeds often escape; they may pass through the food-canal of the bird, or be ejected in the pellet from the crop. By these means they may be carried some distance from the parent plant and be capable of germination. Thrushes eating such fruits often pick off the fleshy part and by a jerk of the head throw the stones away.

Many examples of fleshy fruits are found in the hedgerows, a fact which is especially interesting when we remember how important hedgerows are as nesting-places for birds. Fruits such as the following are commonly to be found: Hawthorn, Rose, Blackberry, Raspberry, Strawberry, Blackthorn, Plum, Cherry, Crab, Mountain Ash, Barberry, Gooseberry, Guelder Rose, Wayfaring Tree, Elder, Honeysuckle, Ivy, Dogwood, Holly, Spindle Tree, Buckthorn, Woody Nightshade, White Bryony, and Black Bryony.

(4) Animals like the sheep and goat often have large numbers of hooked fruits in their coats. Structures of this kind which have become entangled in the wool are called

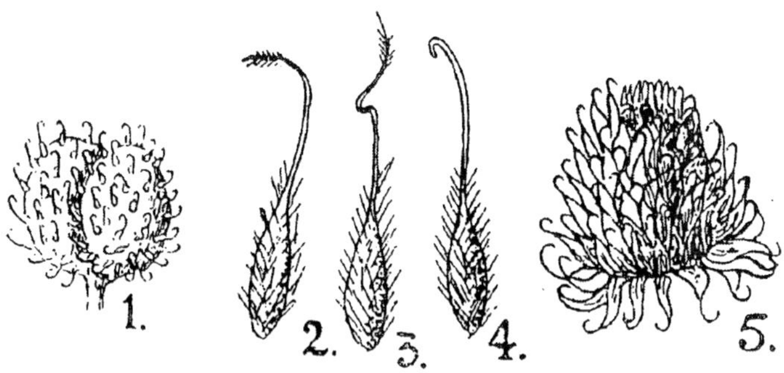

FIG. 160. HOOKED FRUITS.—1, Cleavers; 2, 3, and 4, achenes of Avens showing the bending of the style to form a hook; 5, capitulum of Burdock with hooked bracts.

burrs; some of them are entire fruits, others are hooks only, the rest of the fruit having broken off. When the wool is eventually scoured and these foreign bodies with the dirt are thrown on to the waste-heaps near the factories the seeds often germinate. The plants that spring up indicate the region from which the wool has been obtained.

If several kinds of hooked fruits are examined, it will be seen that the hooks are developed from different organs (Fig. 160, 1–5).

In the Cleavers (Fig. 160, 1), Sanicle, and Enchanter's Nightshade, they are on the fruit-coat. In the Avens (*Geum*) the style is hooked (2, 3, 4). In the Burr Marigold they are barbed pappose bristles; while those of the Burdock (Fig. 160, 5) are hooked bracts.

(5) In some cases Nuts of various kinds may be collected as food by such animals as mice and squirrels; and some of these may be left, which then germinate.

(6) Again, ants are active agents in fruit and seed dispersal, especially in the case of those seeds which have oil-bodies attached to them. This is well seen in the Gorse (Fig. 161, 1), and more clearly still in the Castor Oil seed (2). Similar oil-bodies occur in many plants, e. g. Cow Wheat, Cornflower, and several fruits of Sedges, Rushes,

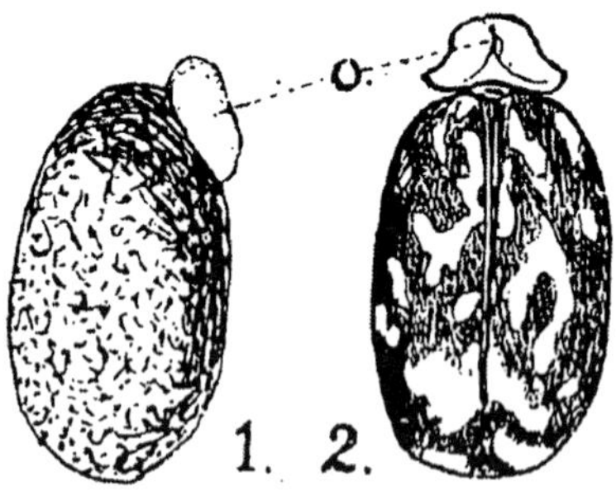

FIG. 161. SEEDS WITH OIL-BODIES.
1, Gorse; 2, Castor Oil; *o*, oil-body.

and Grasses. The ants eat the oil-body, and throw the seeds away; along an ant-run, lines of such plants may be found grown from seeds which have been dropped by ants on their way to the nest. The process may easily be observed if such seeds are laid in their track.

D. *Propulsive mechanisms.*—In addition to the modes of dispersal already mentioned, many plants possess devices which render them independent of wind, water, or animals as carrying agents. (1) Often, as the fruit ripens and dries, tensions are set up in the fruit-coat, which result in a sudden bursting, whereupon the seeds are shot out sometimes several feet.

(*a*) This is well observed in pods of Gorse, Broom, and Sweet-Pea (Fig. 162, 1); (*b*) the two valves of the siliquas of Crucifers, like Bitter Cress (2), split apart suddenly from below upwards; (*c*) in Violets (3) the capsule splits into three valves; then the edge of each closes over and presses on the smooth pear-shaped seeds, which are then forcibly ejected; (*d*) in the Geranium (4) the five carpels separate from below upwards, press against the calyx, and eventually

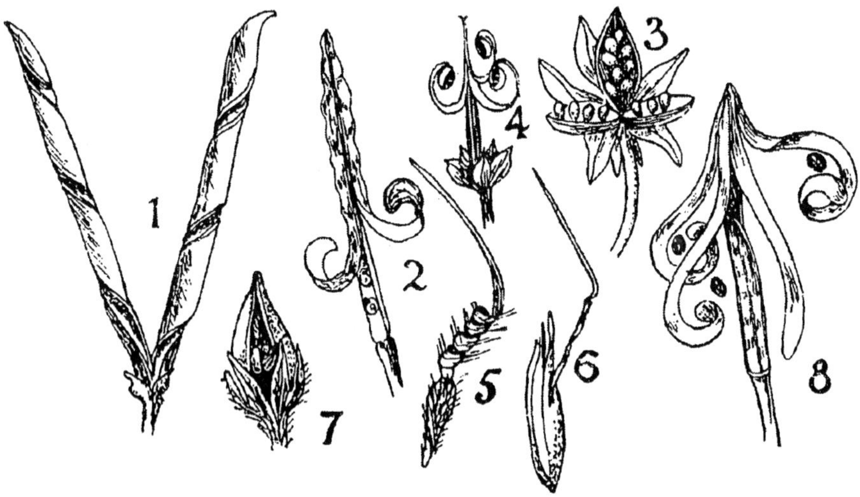

FIG. 162. EXPLOSIVE FRUITS.—1, pod of Sweet-Pea; 2, siliqua of Bitter Cress; 3, capsule of Violet; 4, fruit of Geranium; 5, hygroscopic fruit of Crane's-bill; 6, twisted awn of Oat; 7, capsule of Wood Sorrel; 8, Balsam or Touch-me-not.

gain sufficient force to spring away and throw out the seed as from a sling; (*e*) the Crane's-bills (5) split similarly; but the long awn twists spirally and is jerked off, still retaining the seed. This is buried by the movements of the awn, which untwists when moist and coils up again when dry; (*f*) the awns of some Grasses act in a similar way (6). But awns may straighten suddenly and jerk the grain some little distance. Stiff hairs on the fruit-coat serve as anchors, preventing the fruit from being drawn upwards from the soil. Plants of this kind are able to bury their own seeds.

(2) The turgidity of part of the fruit-coat or seed is the means of expelling many seeds. Interesting examples to observe are the Wood Sorrel (7) and the Balsam (8). In the former each seed has a fleshy aril, the inner layer of which is very turgid. If the ripe fruit is disturbed, the capsule splits, the aril suddenly turns inside out, and the seeds are shot some distance. If the ripe fruits of the Balsam are lightly pressed between the fingers, the fruit-coat splits and the valves roll up inwards with great force, scattering the seeds in all directions.

PART III

SYSTEMATIC BOTANY

CHAPTER XVIII

CLASSIFICATION OF PLANTS

In the preceding chapters we have investigated the structures and functions of the organs which constitute a plant; the development of plants; and their perpetuation. In the course of our study we have met with many individuals, and occasionally analogies and differences have been pointed out. But, so far, no general attempt has been made to classify and arrange the two hundred thousand or so members of the vegetable kingdom. How are we to discover the basis of a satisfactory classification? The question obtains its most interesting solution in a review of the history of botanical science.

History of systematic botany.—There is little doubt that in the early dawn of civilization the culture of plants was studied from the utilitarian point of view: by the agriculturist to provide food for himself and his flocks and herds, and by the physician to prepare useful medicines. Among later nations, and especially among the Greeks and Romans, the subject assumed a new aspect: attempts were made to systematize the vegetable kingdom from the data and facts which had accumulated and been recorded during preceding centuries. But beyond the subdivision by Aristotle and Theophrastus into Trees, Shrubs, and Herbs, this branch of our science was practically dormant until the sixteenth century.

The beginning of a scientific system is indicated in the celebrated *Herbal* of John Gerard, but to better advantage in the *De Plantis* of Andreas Caesalpinus, who divided the vegetable kingdom into fifteen classes, each distinguished by a typical *fruit*. Other distinguishing characters were introduced by subsequent writers. John Ray divided 'flowering' from 'flowerless' plants, and suggested the terms Monocotyledons and Dicotyledons as prime divisions of the former. Robert Morison considered the structures of the flower and of the fruit; and De Tournefort proceeded to more detail by taking the corolla into account.

The illustrious Swedish botanist, Carl Linné, better known by the latinized form of his name, Linnaeus, now enters into our brief review. He introduced a classification based upon the reproductive organs, i. e. the stamens and pistil. The effect of the Linnaean system on systematic botany cannot be over-estimated, and although the system in detail was subsequently replaced by others, the lines of thought and nomenclature which he developed are fundamental.

The Linnaean system was *artificial*; the natural affinities and relations of plants were ignored, although the author was aware that these considerations were essential to a correct classification of the vegetable kingdom. It foreshadowed a *natural* system, a system which would exhibit a continuous sequence of plant-life, from the lowest vegetable organisms to the most elaborated members of the plant world. Such a system could not be derived from a study of the functions and forms of one or more special organs of plants; it could only be deduced from a study of the forms and *development* of these organs.

In the construction of natural systems, the earliest pioneers were the French botanists De Jussieu and De Candolle. Robert Brown and John Lindley in England, and Endlicher in Germany, added much to our knowledge. Later Bentham and Hooker in England, also Eichler and Engler in Germany, materially advanced the natural system. The history and development of botanical classification can be studied in Sachs's *History of Botany*. The foregoing account sufficiently acquaints the student with the following important fact:

A means of classification is to be sought only in the study of the form, function, and *development* of the organs which constitute plants.

Concurrently with the advance in botanical classification, or Systematic Botany, nomenclature received much attention. The binomial system gradually supplanted all others, in which every plant received a compound name, the first representing its *genus* and the second its *species*; and in the case of closely-related forms a third or varietal name was added. Groups of related genera form an *order*; of related orders, a *cohort*; and of related cohorts, a *family* or *class*.[1]

The chief divisions of flowering plants.—A brief summary of the characters of the larger groups will illustrate the use made of the parts of the flower and fruit in classifying plants. All plants which produce seeds, e. g. Pine, Larch, Buttercup, Stock, Primrose, Daisy, Bluebell, and Crocus, belong to one large group, the **Spermaphyta** (Gr. *sperma* = a seed, *phyton* = a plant), and are thus distinguished from such plants as Algae, Fungi, Mosses, and Ferns, which do not produce seeds.

Cone-bearing plants like the Pine and Larch produce naked ovules: that is, the ovules are not enclosed in an

[1] In some systems of classification the term *family* is used instead of *order*, and the latter term takes the place of *cohort*.

ovary before pollination. Such seed-plants are included in a division called **Gymnosperms** (Gr. *gymnos* = naked) and are a very ancient type.

Seed-plants like the Buttercup, Primrose, and Bluebell, produce their ovules in an ovary formed of closed carpels, and belong to a more modern group called **Angiosperms** (Gr. *angeion* = a vessel) ; to this group belong the great majority of the seed-plants of the vegetation of the present day. In the Stock and Primrose the parts of the flower are in fours or fives, and the seeds contain two cotyledons. Such Angiosperms are placed in a class called **Dicotyledons.** On the other hand, the parts of the flowers of the Bluebell and Crocus are in threes, and the embryo of the seed has only one cotyledon. Such Angiosperms form the class **Monocotyledons.**

These classes are further subdivided according to the relationships of the parts of the flower. Dicotyledons with a simple perianth of free petals in one or two whorls are known as **Archichlamydeae** (Gr. *arche* = beginning, *chlamys* = a mantle), while those with a more highly developed perianth, in two whorls, the petals being joined together by their edges to form a gamopetalous corolla, are called **Sympetalae** (Gr. *syn* = together), or **Monopetalae** (Gr. *monos* = one).

Other subdivisions depend on (1) the relations between stamens and pistil, whether the former are hypogynous, perigynous, or epigynous ; and (2) the condition of the ovary, whether apocarpous or syncarpous—one or more celled.

We thus see that the characters of most importance in classification are those to be noticed in a careful examination of the parts of a flower from outside inwards. The classification of the above-mentioned plants may be shown as follows :

SPERMAPHYTA

- Gymnosperms
  - PINE, LARCH
- Angiosperms
  - Dicotyledons
    - Archichlamydeae
      - apocarpous — BUTTERCUP
      - syncarpous — STOCK
    - Sympetalae
      - superior — PRIMROSE
      - inferior — DAISY
  - Monocotyledons
    - superior — BLUEBELL
    - inferior — CROCUS

The Study of a local flora.—A book dealing with the plants of a country or a district, in which the species are arranged and classified in the manner indicated, is called a **Flora**. In the study of the vegetation of your district you will find it more interesting and profitable to devote your attention to the plants of one habitat at a time, than to collect plants indiscriminately. Always have an object in view and follow it with care and intelligence. In each habitat probably one or only a few species, which are best adapted to it, will predominate. These are the plants you should study first and most carefully, neglecting for the time being the rarer ones. Distinguish between social species, i. e. plants of the same kind growing together in large numbers; and those which occur sparingly; also between the large sturdy trees and shrubs and the plants growing under their shade and protection.

Note in detail the form, mode of growth, and the structure of the leaf, and see whether these bear any relation to the plants' environment, such as soil, water-supply, humidity, altitude, and exposure to sun and wind. You will find that, in nature, plants group themselves into **plant-societies** and **associations** according to the conditions of the habitat (Fig. 163).

By means of a Flora determine the species carefully and try to arrange the plants found in the order of importance in the vegetation: first the dominant ones, then the frequent and occasional ones, and so on in descending order. A collection of the characteristic species should be made and classified according to habitat. Each should bear a label giving the following particulars: natural order, genus, species, plant-association or society, position in the association (i. e. dominant or otherwise), locality, and date.

Study in the same way the plants of several different habitats, and contrast the types. In time you will learn to recognize all the more important species and the more interesting facts concerning their distribution.

CHAPTER XIX

CLASS I, DICOTYLEDONS

Archichlamydeae

Dicotyledons are distinguished by having the vascular bundles of the stem arranged in a ring, the leaves are net-veined, the parts of the flower are in whorls of four or five, and the embryo of the seed has two cotyledons.

This class is much larger than that of the Monocotyledons. The broad, net-veined leaf is very characteristic; secondary thickening is general. Dicotyledonous trees form the great deciduous forests of Temperate regions; the evergreen shrubs and peculiar xerophytes of semi-desert and desert regions belong largely to this class, and many of the species forming the rank vegetation of Tropical forests are also Dicotyledons.

Fig. 163. Lake, Marsh, Meadow, Hedge, Wood, Moor, and Mountain.

The Dicotyledons are divided into two great divisions. The first, **Archichlamydeae**, includes plants which have a relatively simple type of flower. The natural orders of this division embrace flowers increasing in complexity from primitive forms like Willows and Buttercups, with many stamens and much pollen, up to the more highly specialized forms like the Sweet-Pea and Violet, or, as in the Hogweed and Chervil, to small epigynous flowers rendered attractive by being massed into compact inflorescences. The second division, the **Sympetalae,** is characterized by the petals being joined into a flower-tube, as in the Heath and Primrose, accompanied by reduction in the number of stamens and greater economy in the production of pollen and honey, culminating in the aggregate flower-heads of the Daisy and Dandelion. In this chapter we shall deal only with the first division.

(A.) **Archichlamydeae**. The distinctive features of this division are: the parts of the perianth are either absent, or in one or two whorls; the inner whorl of petals is free (i. e. the flowers are polypetalous).

Order SALICACEAE. Trees with unisexual flowers in catkins, flowers dioecious, perianth absent. Flowers of the male catkins have two or more stamens in the axil of a bract. Flowers of the female catkins have two carpels in the axil of a bract; syncarpous; ovules indefinite. Fruit a capsule. Seeds with a tuft of hairs at the base (see Fig. 185, p. 278).

The Willows, Sallows, and Osiers are abundant in the North Temperate region, especially in low-lying, wet areas. Many are pollarded (*Salix alba*, &c.); others are coppiced (*S. viminalis*), and used for basket-making (see p. 277). On mountains in Britain, in the Alps, and in Arctic regions, very small creeping forms occur, only one or two inches high, e.g. *Salix herbacea* and *S. reticulata.*

The Poplars (*Populus*) also belong to this order (see Fig. 186).

Order BETULACEAE.[1] Shrubs or trees with simple stipulate leaves. Flowers in terminal catkins. Monoecious, and usually arranged in small three-flowered cymes in the axils of the bracts of the catkin (Fig. 188). Perianth reduced. Ovary inferior; carpels two, syncarpous, two-celled, one ovule in each cell. Fruit a one-seeded indehiscent nut.

This order includes several well-known trees, e.g. Birch (*Betula alba*), Alder (*Alnus glutinosa*), Hazel (*Corylus Avellana*), and Hornbeam (*Carpinus Betulus*) (see Figs. 187, 188, 189, pp. 282–7).

Order FAGACEAE.[1] Trees with simple stipulate leaves. Flowers in axillary catkins, monoecious. Perianth lobes 5–6. Carpels three, syncarpous. Ovary inferior, usually three-celled with two ovules in each. Fruit a one-seeded nut. Nuts enclosed in a cupule (see Fig. 191, p. 290).

The order includes many forest trees, e.g. Oaks (*Quercus spp.*[2]), Beech (*Fagus sylvatica*), and Sweet Chestnut (*Castanea sativa*).

Order RANUNCULACEAE. The perianth is indefinite, often petaloid and polypetalous. Stamens indefinite and hypogynous. Pistil superior. Carpels indefinite and usually apocarpous. Fruit a group of achenes or follicles (Figs. 144 and 149). The flowers may be cyclic: *Aquilegia*; hemicyclic: *Ranunculus*; acyclic or spiral: *Aconitum, Helleborus*.

The plants of this order are chiefly North Temperate,

[1] The orders *Betulaceae* and *Fagaceae* are very closely related, and sometimes placed in one order, *Cupuliferae*.

[2] Spp. after a generic name means there are several species in the genus.

and many occur in Britain, where they are often cultivated for their showy flowers. There is great diversity in the flowers; the perianth is often petaloid without differentiation into calyx and corolla. The following should be examined and their peculiarities noted: Meadow Rue (*Thalictrum*), with small flowers and inconspicuous perianth, *Clematis* (Fig. 109, p. 164), *Anemone* (Fig. 110, p. 164), Marsh Marigold (Fig. 111, p. 164), and Christmas Rose (*Helleborus*),

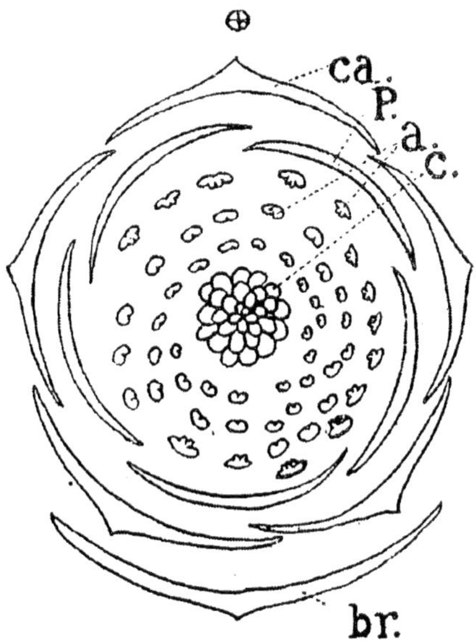

FIG. 164. FLORAL DIAGRAM OF BUTTERCUP.
a, stamens; *br*, bract; *c*, carpels; *ca*, sepals; P, petals.

with a petaloid calyx and no corolla, except in the latter. In Caltha, honey is secreted by the carpels, and in the Christmas Rose by small, tubular 'petals'. In the Buttercups (Fig. 112, p. 166) (*Ranunculus spp.*) there is a calyx and a conspicuous corolla, and the nectary is at the base of the petals. In the Columbine (Fig. 130, 1, p. 186) (*Aquilegia vulgaris*) the five petals form long, honey-secreting spurs. In the Larkspur (Fig. 130, 3–5, p. 188) (*Delphinium*), the spurred posterior sepal contains the two posterior honey-secreting petals. In the Monkshood (Fig. 130, 2, p. 187) (*Aconitum Napellus*) the posterior sepals form a large, blue hood, enclosing two nectaries, which are modified petals. In the two latter the flowers are zygomorphic. Floral formula: K 5 or 3, C 0–5 or ∞, A 5–∞, G 1–∞.

The nectaries of the Christmas Rose, Winter Aconite (*Eranthis*), Love-in-a-Mist (*Nigella*), Monkshood, and Aconite are known as 'honey-leaves', and are probably derived from stamens.

Order CRUCIFERAE. Sepals and petals four each. Stamens six, in two whorls, two outer short, and four inner long ones (tetradynamous). Pistil superior of two carpels (syncarpous), two-celled, divided by a partition (replum). Fruit, a siliqua or silicula (Figs. 1–5).

This order contains upwards of 200 genera and 1,200 species widely distributed throughout the North Temperate and Mediterranean regions. They are mostly herbaceous perennials, but some are annuals, and others are biennials. The leaves are without stipules; the inflorescence is a raceme or corymb, and is usually without bracts.

Very many cultivated plants belong to this order, e.g. the Cabbage (*Brassica oleracea*), from which many useful varieties have been derived by cultivation and selection, such as Red Cabbage, Kale, Savoy, Brussels Sprouts, in which the axillary buds become small cabbages. In the Broccoli and Cauliflower the inflorescence becomes abnormally branched and fleshy, and the Kohlrabi is a tuberous stem or corm.

From other species of *Brassica* are derived the Turnip and Swede; the Radish, Horse-Radish, Watercress, Garden Cress, and Mustard, also belong to this order, as well as some of the commonest weeds of cultivation, e.g. Charlock and Shepherd's Purse; while others are cultivated for their flowers, for example, Wallflower, Stock, Candytuft, Honesty, *Arabis*, and *Aubretia*.

Order CARYOPHYLLACEAE. Herbs with opposite leaves, nodes swollen, inflorescence a dichasium or scorpioid cyme. Flowers regular. Sepals 4–5. Petals 4–5, usually white or pink. Stamens 8–10, the outer whorl usually opposite the petals, hypogynous or sometimes perigynous. Pistil syncarpous, one-celled, placentation free central, ovules indefinite, styles 2–5. The embryo of the seed is curved and surrounds the perisperm.

The order is a large one, and includes many familiar plants, both wild and cultivated. Many species occur in Britain in very varied habitats, as often suggested by their common names, e.g.:

Water Chickweed (*Stellaria aquatica*), Bog Stitchwort (*S. uliginosa*). On the coast we have: Sea Campion (*Silene maritima*), Sea Purslane (*Arenaria peploides*), Sea Spurrey (*Spergularia rupestris*), Sea Pearlwort (*Sagina maritima*). On the mountains: Sea Campion (*Silene maritima*), and alpine species of Chickweed (*Cerastium alpinum*), Moss Campion (*Silene acaulis*), Alpine Campion (*Lychnis alpina*), Pearlwort (*Sagina spp.*). On walls, rocks, dry banks, and fields: Pinks and Carnations (*Dianthus spp.*), Chickweeds, Sandworts and Pearlworts.

In moist woods and hedgebanks: Red Campion (*Lychnis dioica*), Ragged Robin (*L. Flos-cuculi*), Greater Stitchwort (*Stellaria Holostea*), Wood Stitchwort (*S. nemorum*). As weeds of cultivation: Bladder Campion (*Silene inflata*), Corn Cockle (*Lychnis Githago*), Chickweeds (*Stellaria media*, &c.), Corn Spurrey (*Spergula arvensis*).

Many plants of this order are readily distinguished by their opposite, entire leaves and swollen nodes, and also by the inflorescence, which is very characteristic. The main axis ends in a flower (Fig. 165, 2). In the axils of the pair of leaves below, branches arise, each of which bears a pair of leaves. In the axils of the latter leaves, branches arise as before, and so the process may be repeated. The name **dichasium** or **false dichotomy** is given to this type of inflorescence. Sometimes one of the branches at a node outgrows the other, and in the later branches

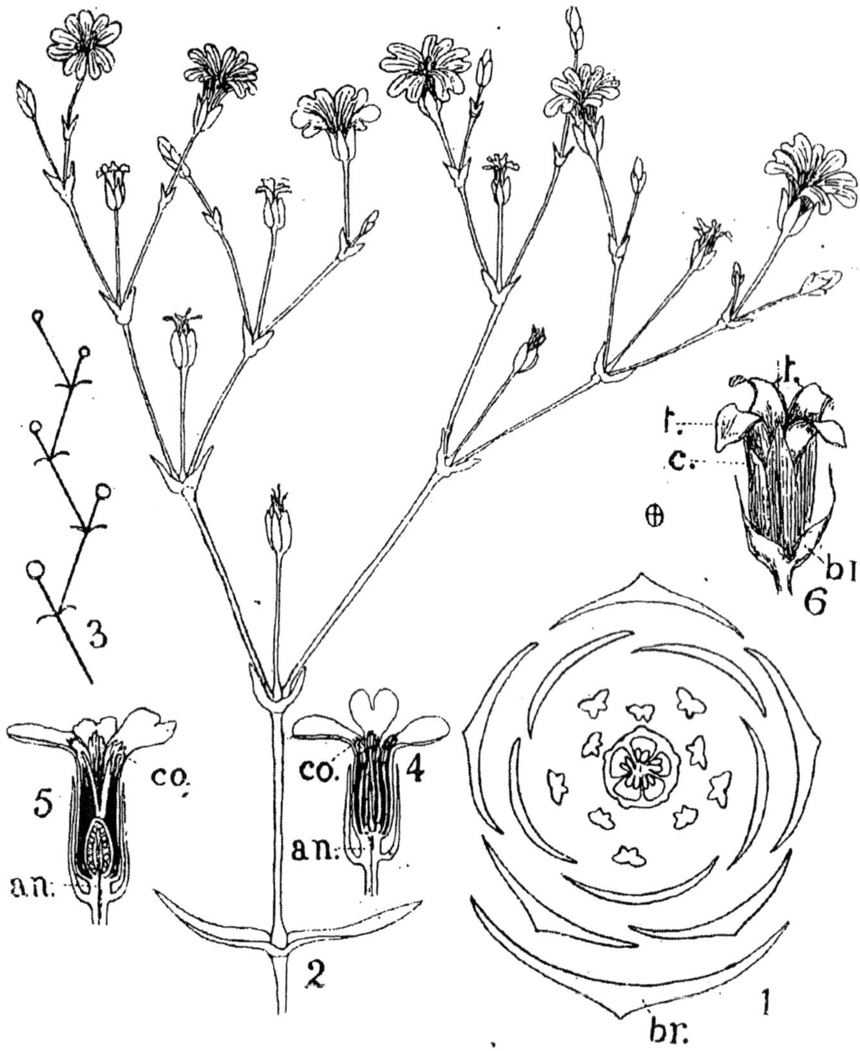

FIG. 165. 1, FLORAL DIAGRAM OF LYCHNIS; 2, INFLORESCENCE OF CERASTIUM; 3, DIAGRAM OF A MONOCHASIAL CYME; 4, VERTICAL SECTION OF MALE FLOWER OF RED CAMPION; 5, VERTICAL SECTION OF FEMALE FLOWER OF RED CAMPION; 6, FRUIT OF DIANTHUS; *an*, androphore; *br*, bract; *c*, calyx; *co*, corona; *t*, teeth of capsule.

one only is developed, the bud in the axil of the other leaf being suppressed, and giving origin to a monochasial cyme or cincinnus (Fig. 165, 3). The flower is typically pentamerous, i. e. its parts are in whorls of five each, and the floral formula is K 5, C 5, A 5 + 5, G $(\underline{5})$.

Many modifications, however, are met with, and the plants of the order may be divided into two groups: (1) A higher group which includes the Pinks, Catchflys, and Campions, and (2) a lower group in which are the Stitchworts, Chickweeds, &c.

The floral diagram of a **Campion** (*Lychnis*) is shown in Fig. 165, 1. Note that the stamens of the innermost whorl are opposite the sepals, and the outermost whorl stands opposite the petals. This arrangement (also found in the Wood Sorrel) is said to be **obdiplostemonous** (L. *ob* = inverse, Gr. *diploos* = double, *stemon* = a filament). The outer stamens are formed first, and ripen before the inner ones.

Cut a flower vertically, and note the arrangements of its parts. The five sepals are united to form a tube (gamosepalous), the internode between the calyx and corolla has elongated and raised the corolla, stamens, and carpels on a stalk (androphore) (Fig. 165, 4 and 5). The corolla has five free petals, the limb often divided. In the Red Campion there is an outgrowth at the junction of limb and claw of each petal, forming together a corona (Fig. 165, 4 and 5, *co*).

Note the arrangement of the ten stamens. The anthers are ripe before the stigmas (proterandrous). Look for the nectary at the bases of the stamens. The pistil has five styles, the ovary consists of five carpels, syncarpous, and one-celled. (In *Silene* there are only three carpels.) Cut transverse and longitudinal sections of the ovary and note the free-central placenta on which are numerous ovules. The gamosepalous calyx and long, narrow flower-tube exclude all but the long-tongued insects, such as bees, moths, and butterflies. The fruit is a capsule opening by four to ten teeth (Fig. 165, 6), and the seeds, ornamented with wart-like outgrowths, are dispersed by wind or animals (see p. 222).

The Red Campion (*L. dioica*) is dioecious; the female plant also differs from the male plant by its larger size and coarser growth. Some species with white flowers, e.g. Night-flowering Campion (*Silene noctiflora*), are closed during the day, but are open and sweet-scented at night, and are visited by night-flying moths.

The flower-stalks of the Catchfly are covered with sticky hairs, to which numerous small insects adhere, hence its name.

The Stitchworts, Chickweeds, and Sandworts differ from the preceding in having a polysepalous calyx, a wider, more open flower, and honey accessible to short-tongued insects. Some, like the Chickweeds, are able to pollinate themselves. Sometimes the petals are so deeply cleft that the corolla appears to have eight or ten petals. In the Stitchworts, Pearlworts, and others, we often meet with reduced flowers.

Order ROSACEAE. Leaves usually stipulate, receptacle more or less hollowed, sepals and petals four or five each, stamens indefinite, perigynous, ovary usually superior and apocarpous (Fig. 166, 1); sometimes (e.g. Apple) it is inferior and syncarpous (2).

There are ninety genera and upwards of two thousand species in this widely distributed order. It contains many familiar and cultivated species, including a large number of our common fruit trees and shrubs. Many spread rapidly by vegetative reproduction, e.g. the Strawberry, Silverweed, and the Blackberry by runners, and the Raspberry by suckers (Fig. 27, p. 61).

The flowers resemble those of Ranunculaceae, but are

distinguished from them by the hollow or concave receptacle, perigynous calyx, corolla, and stamens, and by the fact that these parts of the flower are in whorls and not spirally arranged. The form of the receptacle and the mode of origin and structure of the fruits of this order are interesting; and the following should be studied as showing transitions from perigynous to epigynous, and apocarpous to syncarpous flowers and fruits. The Meadow-sweet (*Spirea Ulmaria*) has a nearly flat receptacle; the fruit

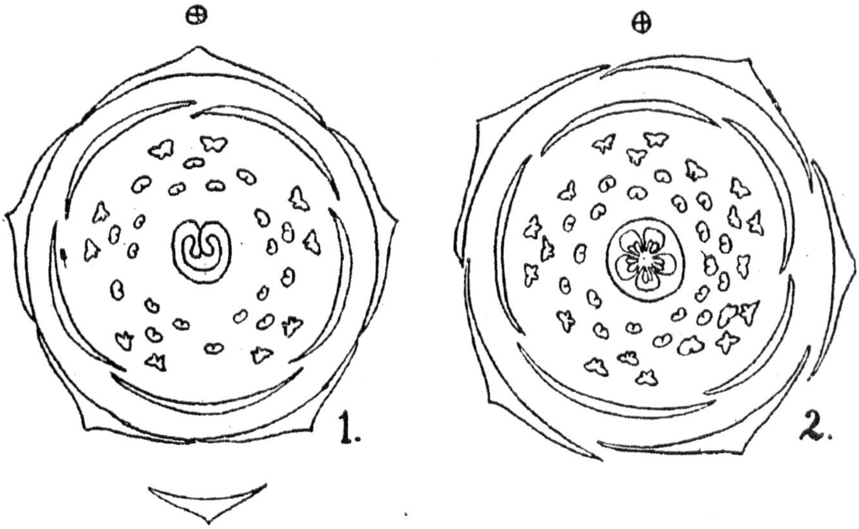

FIG. 166.
1, FLORAL DIAGRAM OF PLUM; 2, FLORAL DIAGRAM OF APPLE.

is a group (aeterio) of two-seeded, twisted follicles. The Tormentil (*Potentilla erecta*) has a persistent calyx and also an epicalyx; the receptacle is convex in the middle and bears many dry, one-seeded achenes. The Strawberry (*Fragaria vesca*) has an epicalyx (Fig. 113, 1), and the central convex part of the receptacle enlarges in the fruit, becomes fleshy, and bears the dry achenes on the outside of it. The Blackberry (*Rubus fruticosus*) and Raspberry (*R. Idaeus*) have no epicalyx; and the column rising from the centre of the flat receptacle bears an aeterio of drupels (see Fig. 153). The achenes of the Mountain Avens (*Dryas octopetala*) have a persistent feathery style for wind-dispersal; and in the Water Avens (*Geum rivale*) the style becomes hooked and the fruit is dispersed by the fur of animals (Fig. 160, 2, 3, 4). The Lady's Mantle (*Alchemilla vulgaris*) has very small, crowded, green, much-reduced flowers; the calyx is four or five lobed, there are no petals, only four stamens, and one or two carpels enclosed in a dry hollow receptacle. The Salad Burnet (*Poterium Sanguisorba*) has no corolla, the flowers are monoecious and crowded into a head, the upper ones are female with feathery stigmas, the lower ones are male with many stamens, the pollen is dry and carried by the wind. The Dog-Rose (*Rosa canina*) has a deep, hollow receptacle contracted above and enclosing several achenes (Fig. 146). In the Apple (*Pyrus Malus*) the five carpels are syncarpous and inferior, and united to the fleshy receptacle (Fig. 155). Other familiar pome fruits are Pear (*Pyrus communis*), Quince (*Cydonia vulgaris*), *Cotoneaster*, Rowan (*Pyrus Aucuparia*), and Hawthorn (*Crataegus Oxyacantha*).

Thus we get development from perigyny to epigyny in Meadowsweet, Tormentil, Blackberry, Rose, Cherry, Apple, and Pear; from apocarpy to syncarpy in Meadowsweet, Tormentil, Blackberry, Cherry, Apple, and Pear; and fruits of special interest in Avens, Rose, Strawberry, Blackberry, Cherry, and Apple (see pp. 217–19 and 225).

Order PAPILIONACEAE (LEGUMINOSAE). Leaves stipulate, flowers in racemes, papilionaceous. Stamens ten, perigynous, united into a tube by their filaments (monadelphous) or nine united and one free (diadelphous). Pistil of one carpel, superior, apocarpous. Fruit a legume (Fig. 167).

Papilionaceae is a sub-order of Leguminosae. In the latter order are included such species as the Acacias, the

Sensitive Plant, Judas-tree, and Divi-divi. It is one of the largest orders of flowering plants and contains 440 genera and upwards of 7,000 species. All the British species have papilionaceous flowers, and belong to the above sub-order. Many species have nodules on their roots, by means of which the plants can utilize atmospheric nitrogen and thrive in soil deficient in nitrates. Many are climbers: the Scarlet-Runner and Kidney Bean have twining stems, but many climb by means of leaf-tendrils. Some are xerophytes

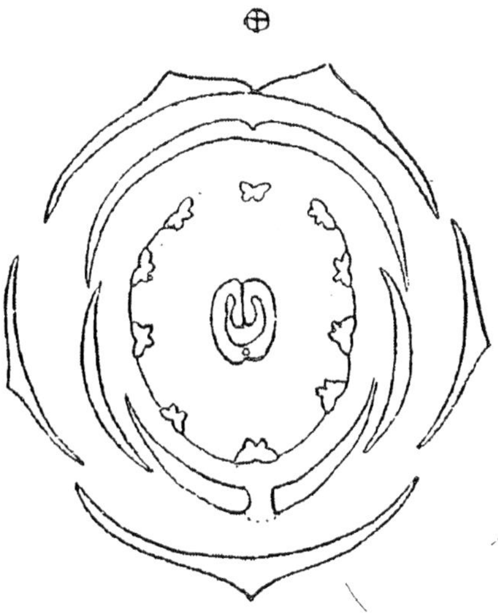

FIG. 167. FLORAL DIAGRAM OF SWEET-PEA.

with leaves reduced to phyllodes, as in some Acacias (Fig. 96, 1), or the leaves are small and the stems angular and green, as in the Gorse and Broom. The stipules are often large, and in some cases—e. g. in the Yellow Vetchling (*Lathyrus Aphaca*)—they perform the functions of leaves, the remainder of the blade being transformed into tendrils.

The leaflets usually perform sleep-movements, and direct the edges of the leaflets to the sky, some assuming the sleep-position immediately when touched, e. g. the leaves of the Sensitive Plant (*Mimosa pudica*). The flowers are adapted to pollination by bees (see p. 188, where a full description of a papilionaceous flower will be found); a few are self-pollinated, e. g. Edible Pea (*Pisum sativum*), and species of Vetch (*Vicia*), some of which have cleistogamous flowers.

In the Gorse (*Ulex europaeus*) (Fig. 133), the Petty Whin (*Genista anglica*), Laburnum (*Cytisus Laburnum*) (Fig. 201), Rest-harrow (*Ononis arvensis*), and Lupin (*Lupinus spp.*), the stamens are monadelphous, and the flowers have no honey, although the Broom has 'honey-guides'. In others the stamens are diadelphous, e. g. Clover (*Trifolium spp.*), Bird's-foot Trefoil (*Lotus corniculatus*), Vetches and Tares (*Vicia spp.*), Pea, Bean (*Vicia Faba*), Scarlet-Runner (*Phaseolus multiflorus*), and Sweet Pea (*Lathyrus odoratus*) (Fig. 131). The seeds are usually rich in proteins and starch stored in the cotyledons, and form important food-stuffs, e. g. Pea, Bean, Pulses. In some, like the Kidney Bean, the pods are eaten. Many are valuable fodder-plants, e. g. species of Vetch, Tare, Clover, Medick, and Sainfoin. The Groundnut or Peanut (*Arachis hypogaea*) develops its pods underground.

Order UMBELLIFERAE. Flowers usually in compound umbels, often zygomorphic. Sepals and petals usually five each. Stamens five, epigynous. Ovary inferior, of two carpels, syncarpous, and on the ovary a honey-secreting disk. Fruit a cremocarp, which splits into two half-fruits (mericarps) (Fig. 148).

This order is remarkable in having small flowers massed into dense, usually compound, umbels, rendering them very conspicuous, and by means of this character most plants of the order can be readily identified (Figs. 119 and 244). Many of the species are poisonous, e. g. Hemlock (*Conium maculatum*); others, like the Carrot (*Daucus Carota*) and Parsnip (*Peucedanum sativum*), are edible, and largely cultivated for their fleshy roots. In the Celery

(*Apium graveolens*) the leaf-stalks are etiolated by banking up with soil and so rendered white and tender.

Other examples are the Common Parsley (*Carum Petroselinum*), Caraway 'seeds' (the fruits of *Carum Carui*), Aniseed (the fruits of *Pimpinella Anisum*), and Coriander 'seeds' (the fruits of *Coriandrum sativum*). Samphire (*Crithmum maritimum*) and Fennel (*Foeniculum vulgare*) occur on sea cliffs, and the Sea Holly (*Eryngium maritimum*) on sandy shores. Some, like the Chervil (*Chaerophyllum*

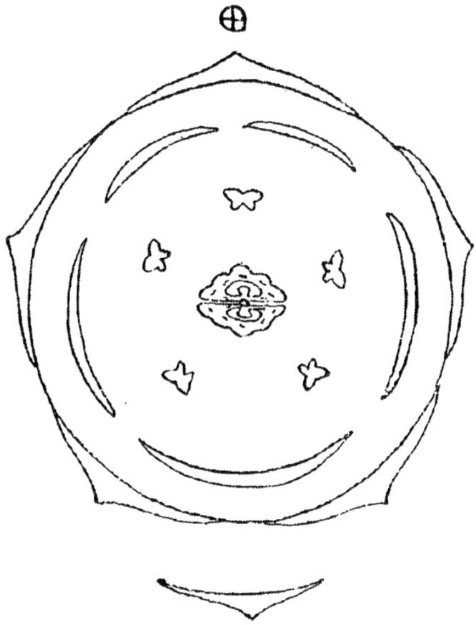

FIG. 168. FLORAL DIAGRAM OF HERACLEUM.

sylvestre), are troublesome weeds in meadows. Several species are marsh plants, e. g. Marsh Pennywort (*Hydrocotyle vulgaris*), Wild Celery (*Apium graveolens*), Dropworts (*Oenanthe spp.*). On stream-sides occurs Sweet Cicely (*Myrrhis odorata*), and several are common in fields and hedgebanks, e. g. Fool's Parsley (*Aethusa Cynapium*), Hogweed (*Heracleum Sphondylium*), Earth-nut (*Conopodium denudatum*), and Hedge Parsley (*Caucalis Anthriscus*).

Usually the fruits consist of two flattened mericarps,

and are dispersed by the wind (see Fig. 148, p. 214). In the Wood Sanicle (*Sanicula europaea*) the fruits are hooked and dispersed by animals.

The Archichlamydeae includes upwards of sixty thousand species. The more primitive forms are distinguished from the higher ones by the parts of the flower being indefinite in number and spirally arranged on the axis. The corolla, when present, is usually polypetalous, and the ovules have two coats. In Ranunculaceae the flower is hypogynous, and the perianth in some species is spiral: in others, cyclic. The stamens and carpels are in general spirally arranged and indefinite. In Cruciferae all the whorls are cyclic. In Rosaceae and Papilionaceae the flowers are perigynous, and in the latter order they are irregular. In the Umbelliferae many small irregular epigynous flowers are massed together in conspicuous umbels. The division is very complex, and the characters, even within a single order, may be very variable.

CHAPTER XX

DICOTYLEDONS

B. Sympetalae

In this division the perianth is in two whorls, and the petals are united (gamopetalous).

Order Primulaceae. Flowers often on scapes, usually regular. Sepals five. Petals five, united. Stamens five, epipetalous and opposite the petals. Ovary superior, syncarpous, one-celled with a free-central placenta. Fruit a capsule, splitting into five valves (Fig. 169).

The plants of this order are mostly perennials with rhizomes or corms. The inflorescence is often a scape, and in the Cowslip (*Primula veris*) the flowers are in a simple umbel (Fig. 121, 2). In the Primrose (*P. vulgaris*) the scape is very short and the flowers appear to arise singly from the short stem (Fig. 121, 1). The flowers are often

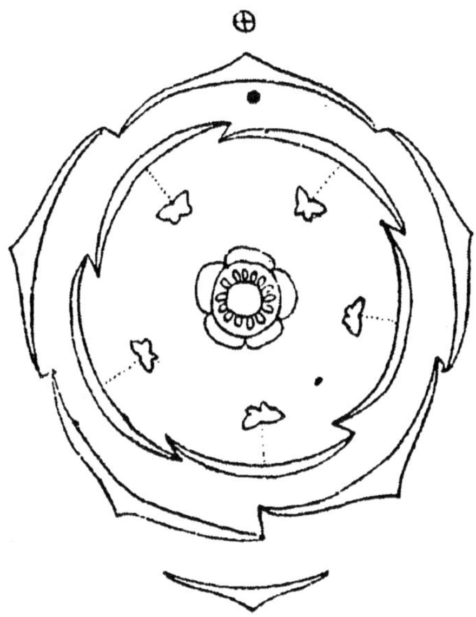

FIG. 169. FLORAL DIAGRAM OF PRIMROSE.

heterostyled (see p. 177) in Primula, Water Violet (*Hottonia palustris*), and Sea Milkwort (*Glaux maritima*). The five stamens are opposite the petals (antipetalous), and the five outer stamens are suppressed. In the Brook-weed (*Samolus Valerandi*) the outer whorl of stamens is represented by five staminodes. The petals of Cyclamen are strongly reflexed. It is not easy to determine the five carpels of the pistil, but the capsule usually splits into five valves, and sometimes abnormal flowers produce five leaves in place of the pistil.

Many species flower in the early spring, and are common on the mountains. Rosette-forming species are frequent, Alpine forms like *Androsace* form compact cushions, and the little Soldanellas send up their flowers through the snow. Many grow in wet places, e. g. Brook-weed, Loosestrife (*Lysimachia*), Creeping Jenny (*L. Nummularia*), and the Bog Pimpernel (*Anagallis tenella*), which occurs in peaty bogs. The Water Violet is an aquatic plant with finely divided leaves which hibernates by means of winter buds. The Scarlet Pimpernel (*A. arvensis*) is a cornfield weed, and the Sea Milkwort occurs in salt-marshes. The Chickweed Wintergreen (*Trientalis*) grows in heaths and upland heathy woods.

Order BORAGINACEAE. Mostly herbs with alternate exstipulate leaves; usually rough with hairs. Flowers often showy in single or double scorpioid cymes which are coiled when in bud; the flowers as they open all face the same way (Fig. 170, 1). Calyx five-lobed. Corolla regular, hypogynous, tubular, lobes five, often spreading; throat more or less closed by projecting scales or hairs. Stamens five, epipetalous. Pistil of two carpels, style gynobasic (Fig. 170, 2). Fruit four one-seeded nutlets (Fig. 171).

The more familiar species are: Forget-me-nots (*Myosotis spp.*), Comfrey (*Symphytum officinale*), Borage (*Borago officinalis*), Evergreen Alkanet (*Anchusa sempervirens*), Hound's-tongue (*Cynoglossum officinale*), Lungwort (*Pulmonaria officinalis*), Viper's Bugloss (*Echium vulgare*).

The flowers of most species show interesting colour-changes during their development, as suggested by the name of one of the Forget-me-nots (*Myosotis versicolor*), which is at first yellow and then blue and violet. Others are white, then change through red to blue, while the Lungwort and Viper's Bugloss are red when young, changing later to violet and blue. The flowers are visited by flies, bees, and moths, and the honey is protected by the narrow tube and overhanging scales or hairs.

Borage and Comfrey are typical bee-flowers. They are

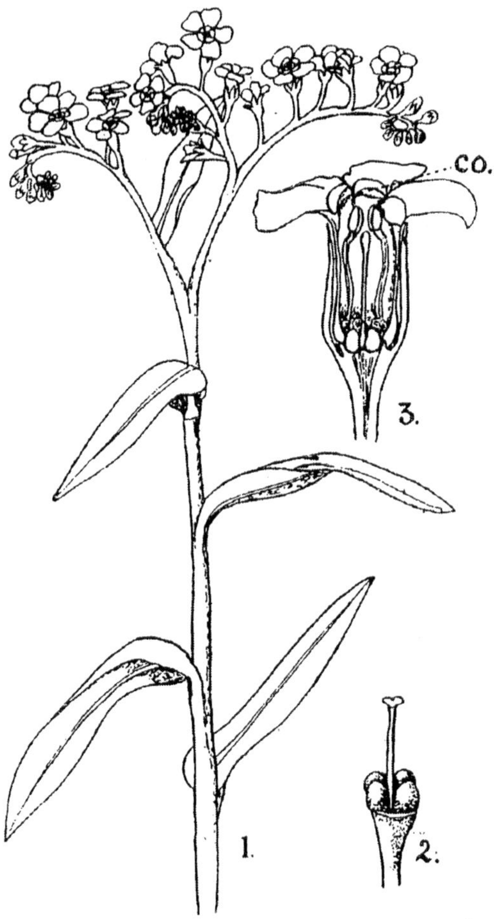

FIG. 170. FORGET-ME-NOT.—1, inflorescence; 2, pistil; 3, vertical section of flower; *co*, corona.

FIG. 171. FLORAL DIAGRAM OF FORGET-ME-NOT.

pendulous, and the cone of stamens showers pollen on the head of the visiting bee in a manner similar to that of the Heaths. The Lungworts have dimorphic flowers, with long and short styles, as in the Primrose, and the corolla-tube is closed by five tufts of hairs between the stamens. The fruit of the Hound's-tongue is provided with recurved hooks, as is also the calyx of the Forget-me-not, which serve as a means of fruit-dispersal by animals.

Order LABIATAE. Stem square; leaves opposite; flower zygomorphic. Sepals five, united. Petals five, united, two-lipped. Stamens usually four, two long and two short (didynamous), epipetalous. Pistil of two carpels, superior, syncarpous, each carpel divided into two cells. Ovary four-lobed, with the style springing from the base (gynobasic). Fruit usually four nutlets (Fig. 172).

There are about 150 genera and 2,800 species in this order. They are very frequent in the Mediterranean region, where many shrubby forms occur which are xerophytes with heath-like habit and back-rolled, hairy leaves. Many are scented, due to volatile oils secreted by epidermal glands, and are often cultivated and used as condiments, or for their oils or perfumes, e. g. Lavender (*Lavendula vera*), Rosemary (*Rosmarinus officinalis*), Thyme (*Thymus vulgaris*), Mint (*Mentha viridis*), Peppermint (*M. piperita*), Marjoram (*Origanum vulgare*), Garden Sage (*Salvia officinalis*).

Vegetative reproduction is common, as in the Garden Mint. The square stem and opposite decussate leaves are very characteristic. The primary inflorescence is usually racemose, but the later branches are cymose. In some, condensed cymes occur at the nodes and, overlapping the leaf-axils, give rise to false whorls of flowers called verticillasters (Fig. 172, 2). The flowers are adapted chiefly to bees (Fig. 172, 3), and some to moths and butterflies. The simpler flowers with shorter tubes, e. g. Thyme and Gipsywort, are visited by miscellaneous insects. The Henbit-Deadnettle (*Lamium amplexicaule*) produces cleistogamous flowers.

FIG. 172. DEADNETTLE.—1, floral diagram; 2, inflorescence; 3, flower in vertical section.

Many species may be found in pastures, meadows, and hedgerows in Britain, e. g.:

Ground Ivy (*Nepeta hederacea*), Self-heal (*Prunella vulgaris*), Hedge Woundwort (*Stachys sylvatica*), Wood Betony (*S. officinalis*), White Deadnettle (*Lamium album*), Red Deadnettle (*L. purpureum*), Yellow Archangel (*L. Galeobdolon*), Wood Sage (*Teucrium Scorodonia*), Bugle (*Ajuga reptans*).

Some are common cornfield weeds, e. g. Hemp Nettle (*Galeopsis Tetrahit*), Corn Mint (*Mentha arvensis*).

Others are marsh plants, e. g. Marsh Woundwort or Water Mint (*M. aquatica*), Greater Skullcap (*Scutellaria galericulata*), Gipsywort (*Lycopus europaeus*).

Order SOLANACEAE. Flowers usually regular. Sepals five, united. Petals five, united. Stamens five, epipetalous; anthers often united. Carpels two, syncarpous, and placed obliquely. Ovary two-celled, ovules indefinite; style terminal. Fruit a capsule (e. g. Henbane), or a berry (e. g. Bittersweet and Potato) (Fig. 173).

The leaves are alternate, but in the inflorescence, as a result of fusion and displacement (adnation), two leaves occur apparently at the same node. The flowers are usually regular, but a few are zygomorphic, and form a transition to the order Scrophulariaceae.

Only four species grow wild in Britain, one of which, the Woody Nightshade (*Solanum Dulcamara*) (Fig. 125), is common in hedgerows, but many are familiar in cultivation. A large number occur in Central and South America, from whence we have obtained such plants as the Potato (*Solanum tuberosum*), *Petunia*, Winter Cherry (*Physalis spp.*), and Tobacco (*Nicotiana Tabacum*). The flower of the Tobacco has a very long corolla-tube, and is pollinated by long-tongued moths.

Many species are poisonous or narcotic, a property which is due to the presence of such alkaloids as atropine, the active principle in belladonna, nicotine, and hyoscyamine, which are derived from the roots, leaves, or seeds of the plants after which they have been named, namely *Atropa Belladonna* (Deadly Nightshade), *Nicotiana spp.* (Tobacco), and *Hyoscyamus niger* (Henbane). The latter is common in waste places in Britain; and occasionally in such places is found the poisonous Thorn-apple (*Datura Stramonium*).

The 'Tea-tree' (*Lycium chinense*) is frequent in hedge-rows but, unlike the Bittersweet, is not a native of Britain.

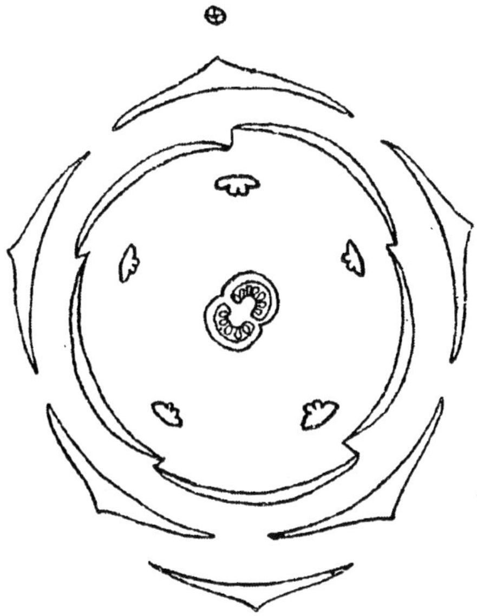

FIG. 173. FLORAL DIAGRAM OF WOODY NIGHTSHADE.

Cayenne pepper is obtained from pods of species of *Capsicum*. Some fruits are edible, e. g. Tomato (*Solanum Lycopersicum*). The Mandrake (*Mandragora officinalis*), connected with which are so many strange superstitions, also belongs to this order.

Order SCROPHULARIACEAE. Flowers zygomorphic, very variable; sepals five, united. Petals five, united, often two-lipped. Stamens four, two long and two short (didynamous), sometimes only two, epipetalous. Pistil of two carpels, superior, syncarpous. Ovary two-celled, style terminal (Fig. 174).

This order contains many poisonous species and is closely related to Solanaceae, but usually the flowers are irregular and the ovary is not oblique. Many species occur in Britain, and they possess interesting peculiarities both in vegetative and reproductive organs. Several are semi-parasites (see pp. 358–9), e. g. species of Eyebright (*Euphrasia spp.*), Bartsia, Lousewort (*Pedicularis spp.*), Yellow Rattle (*Rhinanthus spp.*), and Cow-wheat (*Melampyrum spp.*).

The Speedwells (*Veronica*) occur in varied habitats. The Water Speedwell (*V. Anagallis*), Marsh Speedwell (*V. scutellata*), and Brooklime (*V. Beccabunga*) are marsh or aquatic. Several species are common cornfield weeds; while the Germander Speedwell (*V. Chamaedrys*) is common on hedge-banks; and others occur on heaths and mountains. In gardens many New Zealand shrubby species are cultivated, which are curious xerophytes with small compact leathery leaves resembling species of Cypress, Box, and other evergreens.

Other common species are Toadflax (*Linaria spp.*), Snapdragon (*Antirrhinum spp.*) (Fig. 174, 4, 5), Figwort (*Scrophularia spp.*), Foxglove (*Digitalis purpurea*) (Fig. 174, 6), Musk, and Monkey Flower (*Mimulus spp.*), the latter having sensitive stigma-lobes which quickly close when touched. The flowers of the Foxglove are all brought to one side by the bending of the flower-stalk. The Ivy-leaved Toadflax (*Linaria Cymbalaria*) has flowers which turn to the light, but after pollination, turn away, and the flower-stalk grows and presses the fruits into crannies where they ripen and shed their seeds. The Figworts are pollinated by wasps. The stigma ripens before the anthers, and the stamens and style lie on the lower lip of the flower.

The following flowers should be compared as to number and position of stamens: Mullein (*Verbascum Thapsus*), five stamens (Fig. 174, 1); Figwort and *Pentstemon*, four stamens and a staminode; Toadflax, Snapdragon, and

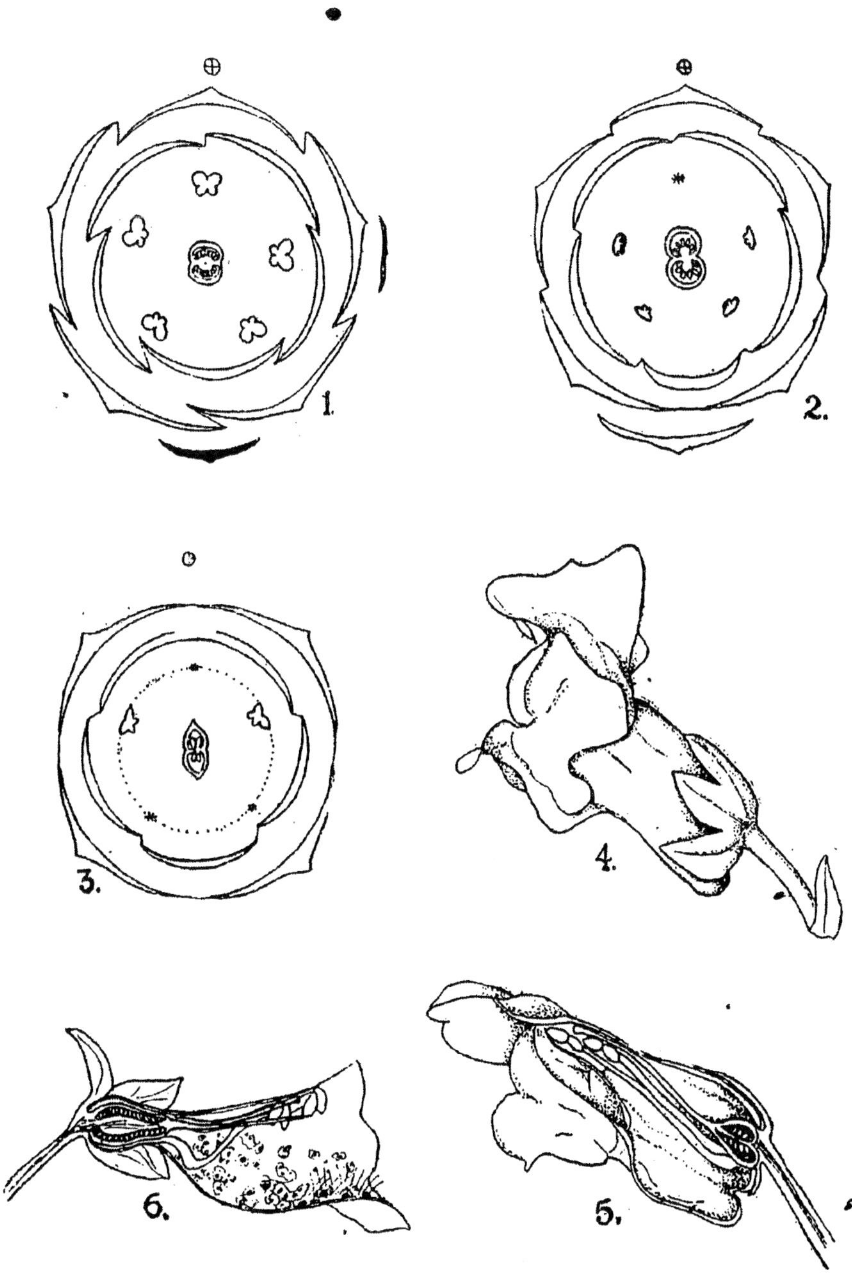

FIG. 174. 1, FLORAL DIAGRAM OF VERBASCUM; 2, FLORAL DIAGRAM OF FOXGLOVE; 3, FLORAL DIAGRAM OF SPEEDWELL; 4, FLOWER OF SNAPDRAGON; 5, VERTICAL SECTION OF SAME; 6, VERTICAL SECTION OF FLOWER OF FOXGLOVE.

Foxglove (Fig. 174, 2), four stamens and sometimes a staminode in the two former. *Veronica* has only two stamens (Fig. 174, 3 and Fig. 126).

Order CAPRIFOLIACEAE. Mostly shrubs or trees, leaves decussate, usually exstipulate. Flowers in cymes, regular, sometimes irregular, usually showy, epigynous. Calyx five-toothed. Corolla five-lobed. Stamens five (or four to ten), epipetalous. Carpels two to five, syncarpous. Ovary inferior, one- to five-celled, with one to many ovules in each cell. Fruit usually a berry or a drupe (Fig. 175, 1).

This order includes a number of shrubs well known in cultivation, several of which are found wild in Britain, e. g. Elder (*Sambucus nigra*), Guelder Rose (*Viburnum Opulus*), Wayfaring Tree (*V. Lantana*), Honeysuckle or Woodbine (*Lonicera Periclymenum*), the Snowberry (*Symphoricarpus racemosus*), Weigelia (*Diervilla florida*). Several others are common in shrubberies.

The Elder has compound, pinnate, and stipulate leaves; small regular hermaphrodite flowers in umbellate cymes; and the fruit is a drupe with one to five stones.

The Guelder Rose has simple leaves with small glandular stipules, and cup-like extra floral nectaries on the leaf-stalk (Fig. 219, 5). The flowers are in corymbose cymes and show an interesting division of labour (Fig. 175, 2). The outer flowers of the inflorescence are neuter, with large, attractive, irregular corollas; but they have neither stamens nor pistil (Fig. 175, 3). The inner flowers are much smaller and perfect (Fig. 175, 4 and 5), producing bright red drupe-like fruits each with one stone. In the cultivated form of Guelder Rose all the flowers are neuter and have large corollas.

The Wayfaring Tree is a characteristic shrub of the woodlands on calcareous soil. The young shoots are

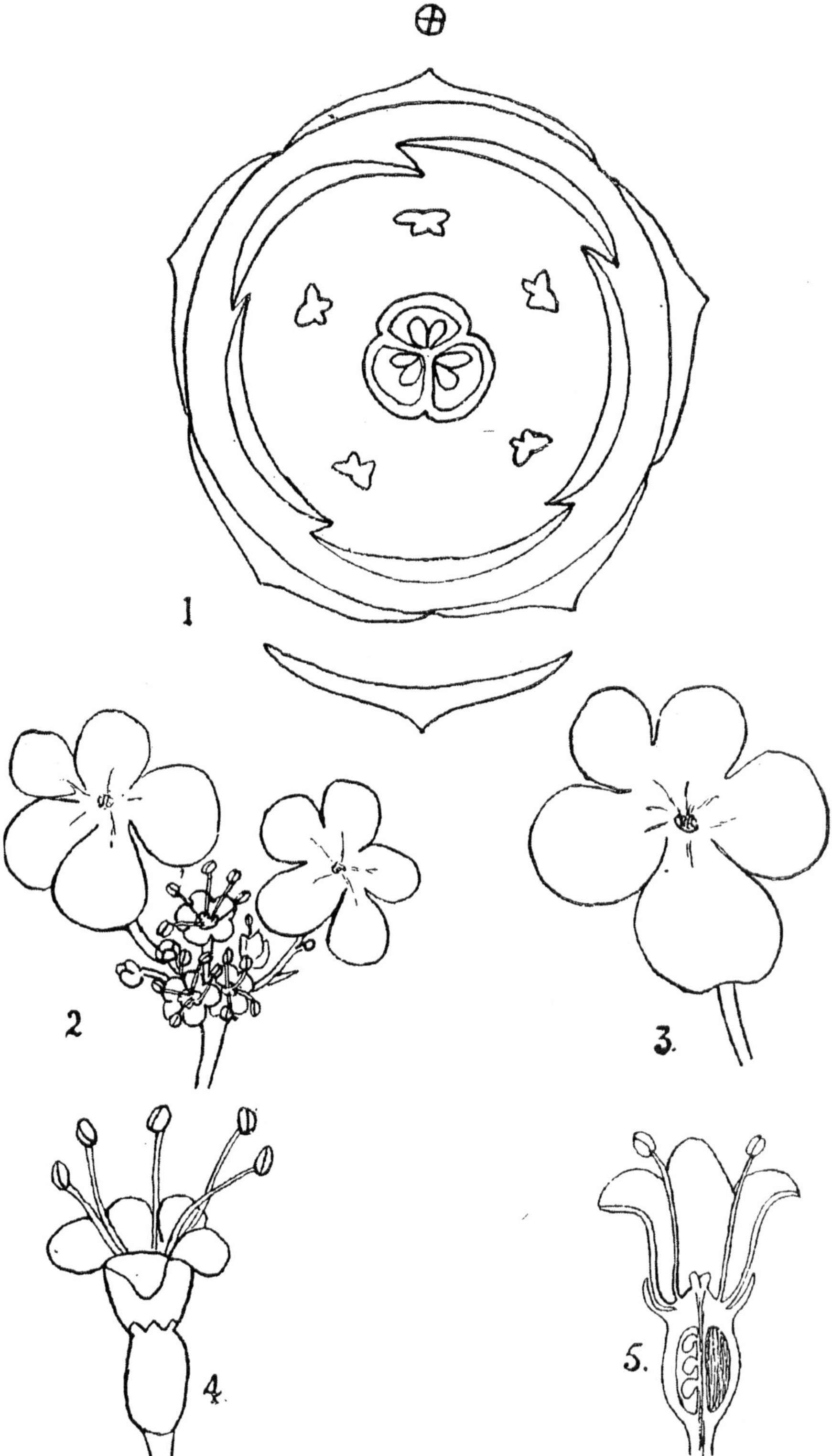

FIG. 175. GUELDER ROSE.—1, floral diagram; 2, portion of inflorescence; 3, neuter flower; 4, perfect flower; 5, vertical section of same.

covered with scurfy hairs, and it is in consequence called the 'Mealy Guelder Rose'.

The Honeysuckle is a twining woodland and hedgerow shrub. The leaves are opposite, entire or sometimes lobed, exstipulate; and the flowers are in capitate three-flowered cymes. The flower is five-lobed, irregular, two-lipped, and tubular. Honey is secreted by the petals and rises in the narrow corolla-tube, so as to be partly accessible to the longer-tongued bees. The flowers are more strongly scented, and contain more honey in the evening, when bees have ceased to collect food. They are well adapted, however, to night-flying hawk-moths, by which they are chiefly visited. The stamens are ripe first and project from the horizontal flower, forming an alighting stage for the moths. The style, at first bent downwards, now becomes horizontal, and the stamens shrivel and bend out of the way. The ripe stigma then receives pollen from another flower when again visited by an insect.

Another species not uncommon in shrubberies is the Perfoliate Honeysuckle (*L. Caprifolium*), where the bases of the opposite leaves are fused (connate) and appear to be pierced by the stem (perfoliate). The Upright Fly Honeysuckle (*L. Xylosteum*) is a common shrub with two-flowered cymes in the axils of the leaves, i. e. the terminal flower of a three-flowered cyme is not developed.

Linnaea borealis is a pretty trailing plant found wild in Scotland and common in the Pine-woods of the Alps. The small fruit is enclosed by two bracts covered with sticky hairs, which, adhering to the coats of animals, serve as a means of dispersal.

Order COMPOSITAE. Flowers in a compact head (capitulum), surrounded by an involucre of bracts. Sepals five, small or pappose. Petals five, regular in disk-florets, and irregular in ligulate and ray-florets.

Stamens five, epipetalous; anthers united in a tube around the style (syngenesious). Pistil of two carpels; ovary inferior, one-celled with one ovule (Fig. 176). The fruit is a pseudo-nut often bearing a pappus for wind dispersal (Fig. 147, p. 213).

No other order contains so many species as Compositae. Upwards of 800 genera and 11,000 species are known,

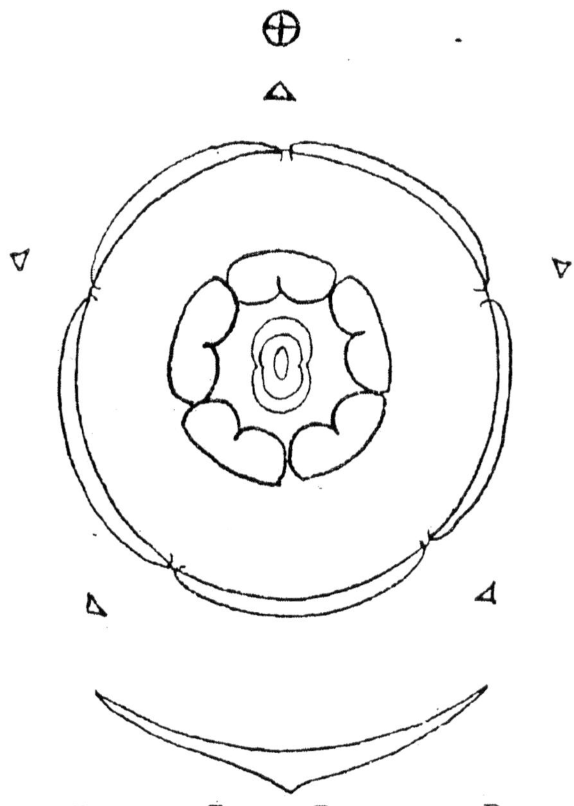

FIG. 176. FLORAL DIAGRAM OF DAISY.

and of these 45 genera and more than 250 species occur in Britain. They are widely distributed, occur in very varied habitats, and show much variation in form. Nevertheless they are easily distinguished from all other plants except species of Scabious, Teasel, and a few plants allied to the Campanulas, which resemble them in the form of their inflorescence.

The characteristic feature of the order is the highly specialized inflorescence; many small flowers being condensed into a conspicuous head or capitulum. Division of labour is so well developed that a capitulum resembles a single flower (see p. 155, Fig. 104), and the arrangements for pollination by insects and dispersal of the pappose fruits by the wind, are so perfect that the flowers represent the highest stage of development yet reached by flowering plants.

The order is divided into two main groups:

(1) *Tubuliflorae*.—Plants without milky latex and the florets of the disk tubular, not strap-shaped, e. g. Daisy, Coltsfoot (see pp. 178–80, Figs. 122 and 123), and Thistle.

(2) *Liguliflorae*.—Plants with a milky latex and the florets all strap-shaped, e. g. Dandelion (see p. 181, Fig. 124), Goat's-beard, and Hawkweed.

Interesting modifications are met with in the florets of the same capitulum as regards distribution of stamens and pistils, and the following should be studied:

Usually the ray florets are female and the disk-florets hermaphrodite as in the Daisy (*Bellis perennis*) (see p. 178), Dog Daisy (*Chrysanthemum Leucanthemum*), and Corn Marigold (*C. segetum*). In the Sunflower (*Helianthus annuus*) the ray-florets are ligulate and neuter; the disk-florets tubular and hermaphrodite. The Cornflower (*Centaurea Cyanus*) has tubular and neuter ray-florets and tubular and hermaphrodite disk-florets. The Butter-bur (*Petasites vulgaris*) is dioecious; the male heads are few-flowered (about thirty), produce honey and pollen, and have a barren ovary and style. The female heads are larger (about 150 florets); two or three of the outer ones are male, the rest being female and producing no honey or pollen. The Groundsel (*Senecio vulgaris*) has no ray-florets, is inconspicuous and self-pollinated. The Ragwort

(*S. Jacobaea*), belonging to the same genus, has conspicuous flowers with ligulate ray-florets and is pollinated by insects.

In a few cases the fruits are dispersed by animals; e. g. the Bur Marigold (*Bidens tripartita*) has hooked fruits and in the Burdock (*Arctium spp.*) the bracts end in recurved hooks and the whole head may be dispersed.

Many Composites are cultivated for their flowers, e. g. Sunflower, Aster, Dahlia, Chrysanthemum, and Cornflower.

Some are interesting Alpine plants, like the Cudweeds (*Gnaphalium spp.*) and the Edelweiss (*Leontopodium alpinum*).

The young flower-heads of the true Artichoke (*Cynara Scolymus*) are eaten. The Jerusalem Artichoke (*Helianthus tuberosus*) has underground tuberous stems with 'eyes' like the potato (see p. 130).

The Sympetalae (or Monopetalae), which contains more than 42,000 species, shows much greater uniformity of flower-structure than does the Archichlamydeae. The parts of the flower are definite in number and cyclic; the corolla is usually gamopetalous and the stamens epipetalous. The ovules have a single integument. The more simple forms, including the Heath and Primrose, have hypogynous flowers, the parts are in five cycles or whorls, two of which are stamens, and the number of the carpels is the same as in the other whorls. In the orders to which belong the Lilac, Forget-me-not, Deadnettle, Nightshade, and Speedwell, the flowers are hypogynous, the parts are in four cycles, the carpels fewer than in the other whorls. In the higher types the flower is zygomorphic. The highest stage of development is reached in the order Compositae, the flowers of which possess the following combination of characters: the corolla is gamopetalous and epigynous, and the anthers are syngenesious. The fruit is small, seed-like, and often provided with a pappus. The inflorescence consists of a large number of small dimorphic flowers condensed into a compact head.

CHAPTER XXI

CLASS II, MONOCOTYLEDONS

In the second division of Angiosperms, namely **Monocotyledons**, the embryo has only one cotyledon and is generally surrounded by endosperm. The plants are usually herbaceous; the stem has many scattered and closed vascular bundles (i. e. there is no cambium between the wood and the bast), and secondary thickening is rare. The leaves are usually parallel-veined and linear. The parts of the flower are in threes, often in five whorls with three parts in each whorl. Most Monocotyledons are perennial herbs and many hibernate by means of rhizomes, corms, or bulbs. Many are characteristic of regions exposed to long, dry periods, e. g. steppes, prairies, and semi-desert areas. The linear grass type of leaf is dominant, and plants of this class cover enormous areas, as in the grassy vegetation of temperate regions, which is a characteristic feature in the scenery. We are familiar with it in our pastures, meadows, and cornfields; in grass moors and cotton-grass mosses; in reed swamps and the marginal vegetation of our ponds, lakes, and rivers.

Tree-like forms are exceptional, and are chiefly confined to tropical regions, where they form a conspicuous feature in the vegetation; the most striking examples are the Bamboos, the Dracaenas (e. g. the Dragon Tree), Agaves, Aloes, and Palms. Many of these have a peculiar mode of secondary thickening.

The orders in this class include many species well known as important food plants, and also many garden favourites, e. g. Gramineae (Grasses, cereals), Liliaceae (Lilies), Amaryllidaceae (Daffodils), Iridaceae (Irises), Orchidaceae (Orchids).

Order GRAMINEAE. True Grasses, plants usually herbs. Internodes of stem hollow. Leaves in two rows; the base forms a long split sheath, and at the junction of this with the linear blade is a membraneous ligule. Flowers in spikelets (p. 201, Fig. 138) enclosed by bracts or pales. Perianth absent, but sometimes the two lodicules are regarded as a perianth. Stamens usually three with slender filaments and versatile anthers. Pistil of one carpel, and generally two feathery stigmas. Ovary superior, one-celled, and contains one ovule.

Floral formula P 0, A 3+0, G $\underline{1}$.

This is one of the largest orders of flowering plants and contains upwards of 300 genera and 3,600 species; they occur in all regions of the globe and often form dominant features in the vegetation, especially in temperate zones. They are of great economic importance, and are a valuable source of food for many domestic animals, as the order contains not only the chief fodder plants, like the meadow and pasture grasses, but such cereals as Rice, Wheat, Maize, Oat, Barley, Rye, &c. In tropical countries, and in China, the Bamboos are used by the natives for innumerable purposes, e. g. for food, shelter, clothing, furniture, weapons, and implements.

Order LILIACEAE. Plants mostly perennial herbs, hibernating by means of rhizomes or bulbs. Inflorescence usually a raceme, more rarely a cyme. Flowers regular, hermaphrodite; perianth of six free or united lobes, often petaloid in two whorls of three each. Stamens usually six, hypogynous. Carpels three, syncarpous; ovary superior, three-celled with many anatropous ovules. Fruit a capsule or berry. Floral formula P 3+3, A 3+3, G $\underline{(3)}$ (Fig. 177).

This large order contains many species familiar either as

wild plants or cultivated in gardens for their showy flowers, e. g. the Bluebell (*Scilla nutans*) (Figs. 87 and 133), Hyacinths (*Hyacinthus spp.*), Tulips (*Tulipa*), Star of Bethlehem (*Ornithogalum umbellatum*), Lily of the Valley (*Convallaria majalis*), Herb Paris (*Paris quadrifolia*), Bog Asphodel (*Narthecium ossifragum*), Onion and Garlic (*Allium spp.*), Solomon's Seal (*Polygonatum officinale*), Fritillary, Funkia, and the Autumn Crocus (*Colchicum autumnale*). Some are climbers, like the beautiful Gloriosa with leaf-tip tendrils, and Lapageria with twining stems. Some are very large: e. g. Yuccas, Dracaenas, Aloes, and the New Zealand Flax (*Phormium tenax*). Species of Asparagus, Smilax, and Butcher's Broom (*Ruscus spp.*), many of which are climbers, develop peculiar, leaf-like stems (**phylloclades**) (Fig. 96, 2).

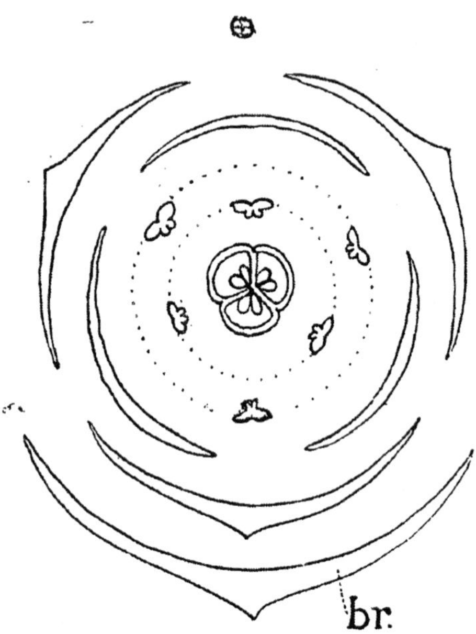

FIG. 177. FLORAL DIAGRAM OF HYACINTH.—*br*, bract.

Order AMARYLLIDACEAE. Plants similar to Liliaceae but with ovary inferior. Mostly herbaceous perennials with bulbs or rhizomes. Inflorescence cymose, and in bud enclosed in a spathe consisting of two fused bracts. Flowers regular, sometimes zygomorphic, hermaphrodite; perianth of six united petaloid lobes, in two whorls of three each. Stamens six, epipetalous; anthers introrse. Carpels three, syncarpous; ovary inferior, three-celled, ovules numerous, placentation axile. Fruit a capsule or berry (Fig. 134). Floral formula P 3+3, A 3+3, $G_{(\overline{3})}$.

The order is a large one, and most of the species occur in dry climates, tropical or sub-tropical. They hibernate during the unfavourable season by means of their bulbs or rhizomes. Many are cultivated for their large, showy flowers: e.g. the Daffodil (*Narcissus Pseudo-narcissus*) with a single flower, Jonquil (*Narcissus Jonquilla*) with a cymose umbel of flowers, Snowdrop (*Galanthus nivalis*), Snowflake (*Leucojum*), *Agave*, *Alstroemeria*, *Amaryllis*, *Crinum*, and *Eucharis*.

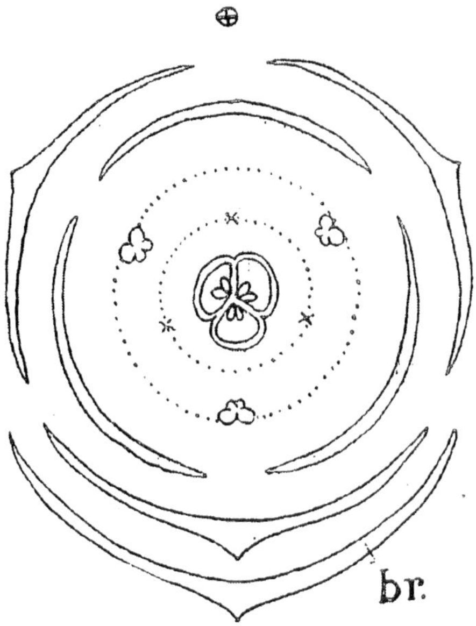

FIG. 178. FLORAL DIAGRAM OF IRIS.—*br*, bract.

Order IRIDACEAE. Perennial herbs, hibernating by means of rhizomes, corms or bulbs. Flowers regular or zygomorphic; perianth petaloid of six lobes in two whorls united to form a tube. Stamens three. Carpels three, syncarpous; ovary inferior, three-celled with indefinite ovules; style branched, often petaloid; fruit a capsule. Floral formula P 3+3, A 3+0, $G_{(\bar{3})}$, (Fig. 178).

Many of the species are adapted to a life in countries subject to considerable dry periods; and many of them occur in South Africa, Tropical America, and the Mediterranean region. They include a number of garden favourites, e. g. *Crocus* (Figs. 85 and 135), *Iris* (Fig. 136), *Ixia*, *Gladiolus*, *Freesia*, and *Tritoma*. Very few occur in Britain; the most familiar are the Yellow Flag (*Iris Pseud-*

acorus), Foetid Iris or Gladdon (*Iris foetidissima*), *Crocus vernus* and *C. nudiflorus*.

Order ORCHIDACEAE, **Orchids.** Perennial herbs. Flowers hermaphrodite and irregular. Perianth usually petaloid, of six lobes, the inner median one generally forming a lip or labellum. Stamens reduced to one, rarely two. Carpels three, syncarpous; ovary inferior, one-celled; placentation parietal and bearing

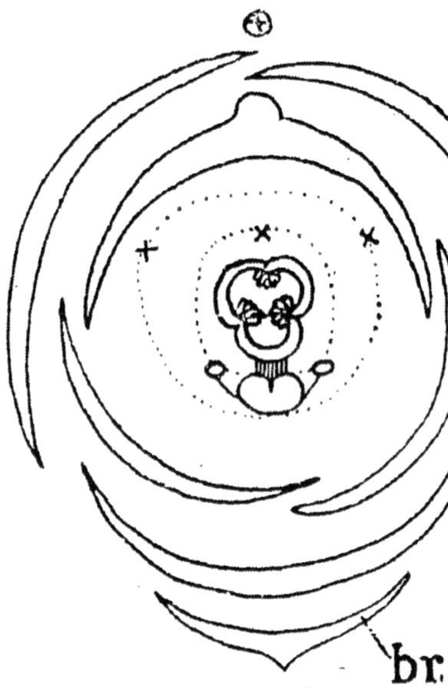

FIG. 179. FLORAL DIAGRAM OF ORCHIS.—*br*, bract.

numerous minute ovules. Axis of flower prolonged as a column above the ovary, bearing the stamens and stigmas. Fruit a capsule and contains very many minute seeds. Floral formula P 3+3, A 1+0 or 0+2, $G_{(3)}$ (Fig. 179).

This order is a very large one, with 400 genera and 6,000 species; eighteen genera and fifty species occur in Britain, many of which are rare. They are most abundant in the tropics, and differ widely in structure according to their mode of life and habitat; many occur as epiphytes on the trunks and branches of trees; and are dispersed by means of their very minute and light seeds. Some are saprophytes (pp. 355–7), and grow on humus, but most of the species in temperate regions are terrestrial. They are largely cultivated for the brilliant colours and often curious and extraordinary forms of their flowers, which are the most highly specialized of the Monocotyledons (Fig. 137, p. 199). To some of them characteristic names have been given.

The more common and interesting British species are:

Early Purple Orchis (*Orchis mascula*), Spotted Orchis (*O. maculata*), Butterfly Orchis (*Habenaria spp.*), Bee Orchis (*Ophrys apifera*), Spider Orchids (*O. arachnites* and *aranifera*), Fly Orchis (*O. muscifera*), Lady's Tresses (*Spiranthes spp.*), Coral-root (*Corallorrhiza innata*), Lady's Slipper (*Cypripedium Calceolus*), Helleborines (*Epipactis spp.*), Tway Blade (*Listera ovata*).

The class Monocotyledons contains about 24,000 species. The simpler and more primitive forms have no perianth and the parts of the flowers are spirally arranged and indefinite, e. g. Pond-weeds (*Potamogeton*). The flowers of Grasses are protected by bracts, have few stamens and carpels, and are pollinated by the wind. The Water Plantain and Flowering Rush have a double perianth, the parts being in two whorls of three each. The flowers of the higher forms are cyclic and have often five trimerous whorls, e. g. Lily and Bluebell. The most highly developed forms have epigynous and irregular flowers, with one or two stamens, e. g. Orchids.

PART IV

COMMON TREES AND SHRUBS

CHAPTER XXII

CONE-BEARING TREES

In the vegetation of the earth, trees occupy the first place. By virtue of their size, wide-spreading branches, and dense foliage, they exert a dominating influence on more lowly plants growing beneath them, and when growing together in large numbers, as in a forest, not only give a characteristic aspect to the scenery, but affect in no small degree the climate of the country in which they grow. They yield many products of great value to man, provide shelter for his home and for his domestic animals, and add much to the beauty of his surroundings. A study of plants, therefore, is incomplete without a knowledge of trees, and in this section a number of the more common kinds have been selected for study.

Scots Pine

Scots Pine (*Pinus sylvestris*) is commonly planted in Britain, sometimes forming large plantations, and frequently scattered amongst other trees in woods. It is sometimes known as the Scotch Fir. In Scotland and Norway it forms extensive forests. The smaller trunks are

FIG. 180. A YOUNG PINE, SHOWING BRANCHES IN FALSE WHORLS.—Each ends in a terminal bud surrounded by a few lateral buds.

FIG. 181. THE NARROW-LEAVED WILLOW IN WINTER.

used for pit-props, and the larger ones as deals. It is an evergreen tree growing to a height of eighty to one hundred feet. The bark is thick, rough, and dark below, but a deep orange above, where the bark peels off in thin flakes.

The branches arise in false whorls (Fig. 180), three or four at nearly the same level, and spread out horizontally. The oldest whorl is the lowest, and they are gradually younger towards the top; hence the conical form of the tree. Each branch in turn bears whorls similar to those of the main axis. The lower, older branches become broken and die, and often the leader also dies; then the upper branches grow considerably and form the broad crown so common in old pines. The first whorl is formed in the third year, and one in each following year; so the age of the tree may easily be determined.

Examine the end of a branch and note the arrangement and structure of the buds (see Fig. 67). The parts are best seen when the buds open, about the middle of May. The terminal bud is surrounded by three or four side buds at nearly the same level. Growing on the bud-axis are many spirally arranged scale-leaves with brown membraneous tips. In the axils of all but the lowest two of these are buds.

Trace the growth of these as the bud opens. Each small bud becomes a short or dwarf shoot (Fig. 182, 1), bearing several scales round its base, and at the end arise two narrow leathery needles. Note their shape and how they are packed together in the bud. They are semicircular in section with their flat, upper surfaces applied to each other. A bud thus gives rise to two kinds of shoots and two kinds of leaves: (1) long shoots, which bear scale-leaves only (Fig. 182, 2), and (2) dwarf shoots, which bear scale-leaves and a pair of green needle-leaves (Fig. 182, 1). Pine needles remain on the tree three or four years, some falling each season.

Examine the old fallen needles, and determine the structures which are thrown off. It is the dwarf shoots that fall and not merely the needle-leaves.

Examine an old branch and note that the bases of the scale-leaves of the long shoots persist and harden, and so produce the roughness of the branch.

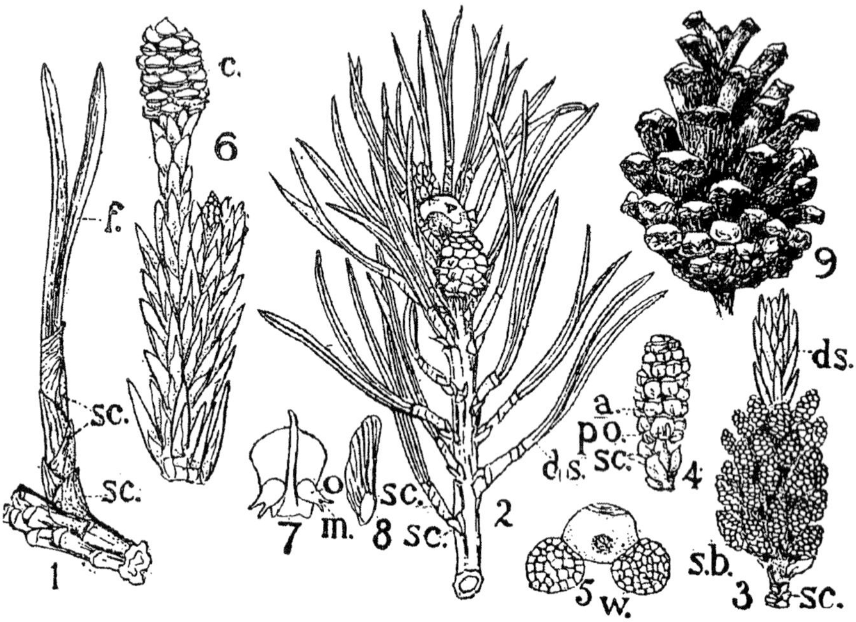

FIG. 182. SCOTS PINE.—1, dwarf shoot; 2, elongated shoot bearing scale-leaves with dwarf shoots in their axils: female cone near end of branch; 3, male cone; 4, a single staminate branch; 5, pollen-grain; 6, branch bearing young female cone; 7, ovule-bearing scale; 8, winged Pine-seed; 9, old female cone; *a*, stamens; *ds*, dwarf shoots; *f*, foliage-leaf; *m*, micropyle; *o*, ovule; *po*, pollen-sac; *s.b*, staminate branch; *sc*, scale-leaves; *w*, wing.

The 'flowers' of the pine are in cones and differ in several important respects from typical flowers. The seeds are not developed in the same cones as the stamens, but arise on different branches of the same tree (monoecious). The male cone (Fig. 182, 3) arises at the end of a branch, and consists of a central axis which bears a tuft of dwarf shoots at the tip. Below, and in the axils of the scale-leaves, short branches arise which bear a few scales below, and numerous spirally-arranged stamens above (4). Each stamen has a very short filament; and the anther bears two pollen-sacs on the under surface (*po*). When ripe, they split longitudinally and the pollen-grains escape in immense numbers. Each pollen-grain (5) is provided with two air-bladders which serve as floats, and it may be carried a great distance by the wind. When the pollen is shed, the staminate shoots fall off, and the dwarf shoots at the tip develop their pairs of needles.

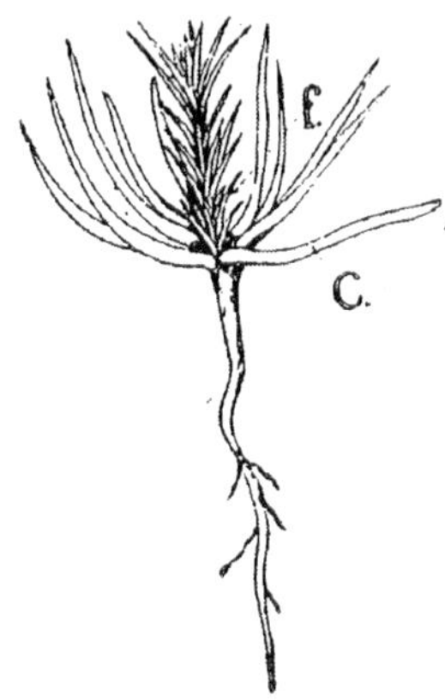

FIG. 183. SEEDLING PINE.—*c*, cotyledons; *f*, first green needle-leaves.

The female cone (2, 6, 9) arises near the end of a branch and in the position of a lateral bud. At first it is about a quarter of an inch long, and appears to be terminal, but later it is seen to be lateral (2). It remains on the tree three years, growing larger each season. The cone consists of a central axis on which are scale-leaves, and on the upper surface of each scale-leaf grows a much larger, thick, flat, woody scale, i. e. the carpel. On the upper surface of each carpel are two straight ovules (orthotropous) (7), with their micropyles (*m*) pointing towards the axis. They are not enclosed in an ovary; there is no style and no stigma; hence there is no pistil as in typical flowers.

When ready for pollination the axis elongates, lifting the carpels apart, and between them the pollen-grains pass and are carried directly on to the micropyle of the ovule. Pollination occurs in May of the first year, but fertilization does not take place until June of the second year. When ripe the carpels become woody (9), gape open from above downwards, and allow the seeds to escape, each carrying with it a thin shaving from the carpel, which serves as a wing for seed-dispersal (Fig. 182, 8).

The ovule of the Pine, unlike that of other flowering plants, becomes filled with endosperm before fertilization. When ripe, part of this persists around the embryo, which has a radicle, plumule, and eight needle-shaped cotyledons. On germination the tips of the cotyledons remain in the seed and absorb the endosperm. The plumule elongates and bears, for the first two years, not scale-leaves, but green needle-leaves. Then as new ones form, they become more scale-like, and buds arising in their axils give rise to dwarf shoots, each with two needle-leaves (Fig. 183).

Larch

Though commonly planted in woods, the Larch (*Larix europaea*) is not a native tree in Britain. It grows rapidly to eighty or one hundred feet, and as in the Pines, the terminal bud continues growth, and for a time the tree is conical. The leader is eventually lost, and the tree then develops an open crown of sparse and delicate foliage. Its bark is fissured, scaly, and grey, tinged with pink, and is early developed on the young shoots.

The tree is easily recognized by its knotted, slender, furrowed branches, which arise alternately and not in whorls. Long and short shoots are formed; on the long ones the leaves arise singly, while on the thick, slow-growing short shoots they are numerous and in tufts (Fig. 184). In wet seasons the dwarf shoots may elongate and form long flexuous drooping twigs. Buds are relatively few and are scattered on the shoot. Notice the large number of scale-leaves and foliage-leaves which have no buds in their axils. The buds stand off at right angles and are covered with very many brown resinous scales. Most of these fall off as the bud opens, but the lower ones persist and harden. The leaves are about an inch long and needle-like, but thin, flat, and soft. In early spring they are bright green in colour and very conspicuous; and they transpire freely. In the autumn they darken, turn brown, and fall off.

The Larch, being the only European Conifer with deciduous leaves, is enabled to grow in situations fatal to other Conifers, and it extends farther northward and attains a greater height than any other tree. The small leaves offer little protection to the slender shoots, and dead twigs are common on the Larch. The 'flowers' are in cones, similar to those of the Pine, and both male and female occur on the same tree. The male cone (Fig. 184, 1 *m.c*), however, is simpler than in the Pine; it consists merely of a central axis bearing numerous stamens and no needle-leaves. Each stamen (2) has a green limb at the tip. The pollen-grains are numerous, dry, and carried by the wind. The female cone (1 *f.c*) has a tuft of green leaves at the base; it is bright red when young, and the scales are lax and flexible. The barren scales (3 *c.s*) are longer than in the Pine, and may be seen projecting beyond the tips of the ovule-bearing scales (*o.s*), the

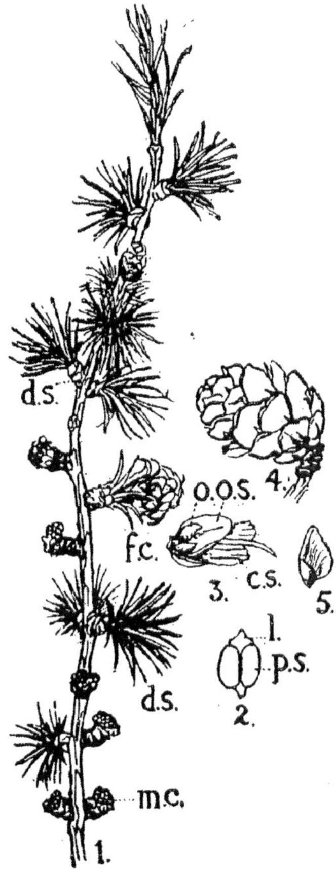

FIG. 184. LARCH.—1, branch bearing dwarf shoots with fascicled leaves; 2, stamen; 3, cone-scale bearing on its upper surface an ovuliferous scale; 4, mature female cone; 5, winged seed; *c.s*, cone-scale; *d.s*, dwarf shoots; *f.c*, female cone; *l*, limb; *m.c*, male cone; *o*, ovule; *o.s*, ovule-bearing scale; *p.s*, pollen-sac.

midrib of the latter being prolonged beyond the scale as a narrow, curved process. The cone (Fig. 184, 4) is mature in the following spring. It is smaller, more lax, and has thinner and more flexible scales than the Pine, and the cones remain on the old twigs many years before breaking off. The seed is winged (5), and is dispersed by the wind. For two or three years, the seedling, unlike the parent, is evergreen.

The Pine and Larch belong to a very ancient group of plants, and differ in many important respects from Angiosperms. The pollen-grains are more complex and deposited direct on the micropyle of the ovule; there is no ovary, style, or stigma; the embryo-sac of the ovule becomes filled with endosperm *before* fertilization; the egg-cell is enclosed in a flask-shaped structure known as the archegonium, an organ characteristic of simpler plants such as ferns and mosses. The seed is naked, i. e. not enclosed in an ovary, the latter character suggesting the name Gymnosperms (Gr. *gymnos* = naked) for the group to which the Pine, Larch, and other cone-bearing trees, belong.

CHAPTER XXIII

CATKIN-BEARING TREES

Willow

Two kinds of Willow are very generally recognized: the 'Palm' and the Osier. The 'Palm' or Goat-Willow (*Salix capraea*) (Fig. 185) grows on dry banks and in woods and hedges, is of shrub-like habit, from fifteen to thirty feet high, and has short, knotted branches loaded in early spring with bright yellow catkins. The male flowers each have two stamens. The leaves, which appear later, are broad and oval and have somewhat kidney-shaped stipules.

The Osier (*S. viminalis*) is common in wet hollows, by stream- and river-sides, and especially in low-lying, marshy districts, where it is frequently coppiced, i. e. cut close to the ground. From both adventitious and dormant buds on the stool, very long, flexible, switch-like branches grow, which are used for basket-making. The catkins are long and slender, the male flowers have two stamens, and the capsules are hairy. The leaves (Fig. 78) have narrow stipules; the blades are from four to eight inches long, lanceolate, pointed, and silky beneath.

Another species, the White Willow (*S. alba*) (Fig. 181), is common in similar situations, and attains a height of from eighty to ninety feet. It has narrow leaves, silky white on both sides, and the male flowers have three stamens.

Other species and varieties with quick-growing shoots and narrow leaves, besides the Osier, are coppiced, and pollarding is common with the larger species. In pollarding, the large branches are cut off several feet above the ground; and new branches, springing from dormant and adventitious buds around the cut surfaces, form a dense crown.

Some Willows growing on sand-dunes and moors are much smaller, being only one to three feet high (Fig. 157), while some alpine species are not more than one or two inches high, and form a flat carpet on the ground. No other genus of British trees has such a bewildering number of species, varieties, and hybrids (i. e. crosses between the different forms), as the Willows.

The buds, often pressed against the stem, are covered by one scale, composed of two fused leaves. The larger flower-buds give rise to short shoots ending in a catkin (Fig. 185, 1 and 2), and the smaller leaf-buds grow into long, leafy shoots. The end bud, and sometimes more, of the branch dies (Fig. 78), and growth is then continued by the next bud below. The flowers (Fig. 185, 3 and 4) have been described on p. 160, and are in catkins, male and female on separate trees, i. e. they are dioecious. In the Goat-Willow the catkins appear before the leaves, but in the Osiers leaves and

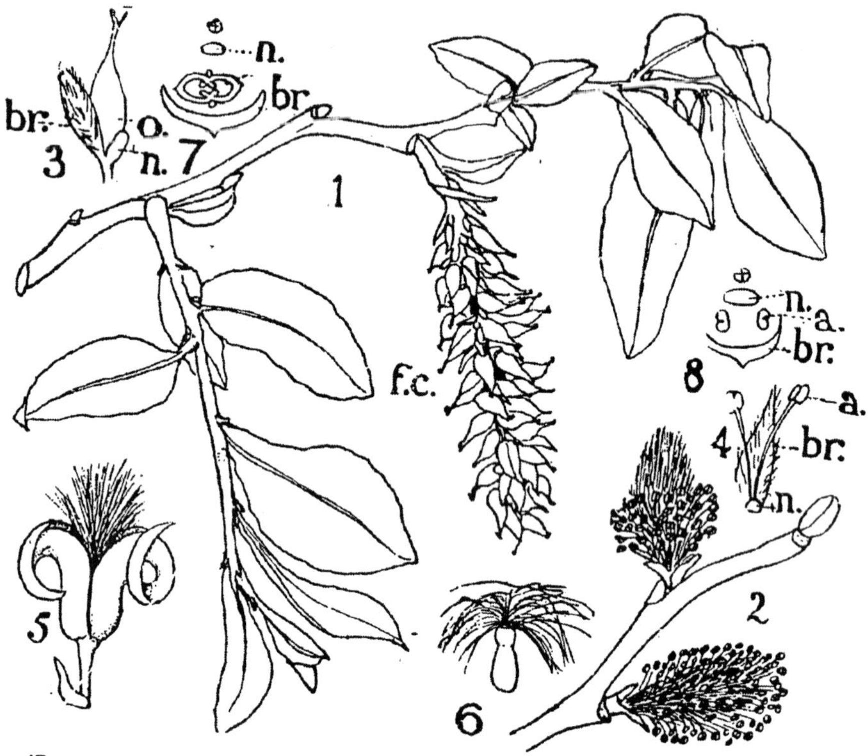

Fig. 185. The Goat-Willow.—1, leafy shoot bearing female catkin; 2, branch bearing two male catkins; 3, female flower; 4, male flower; 5, capsule dehiscing; 6, pappose seed; 7, floral diagram of female flower; 8, floral diagram of male flower; *a*, stamens; *br*, bract; *f.c*, female catkin; *n*, nectary; *o*, ovary.

catkins are out together. The fruit is a capsule (Fig. 185, 5) opening by two recurved valves. The seeds are numerous, and each is provided with a tuft of hairs as an aid to wind-dispersal. Fig. 157 is a photograph of fruiting Willows on a sand-dune; they have the appearance of being coated with cotton wool.

Poplar

Poplars, like most Willows, are trees of damp places, and they grow best in a deep moist soil. In such situations several species are commonly planted, e.g. the White Poplar (*Populus alba*), the young branches and leaves of which are covered with white cottony hairs; the Aspen (*P. tremula*), whose orbicular toothed leaves are green and not cottony; and the Black Poplar, of which the tall Lombardy Poplar is a conspicuous and easily-recognized variety.

The Balsam Poplar is often planted, the opening buds of which are very sticky. The Black Poplar (*P. nigra*, Fig. 186) grows quickly and attains a height of ninety to a hundred feet. Its long, slender, smooth branches curve upwards and form a loose, somewhat pyramidal, crown. The long and pointed buds are covered by four scales, which are modified stipules like the bud-scales of many trees.

As in Willows, &c. (Fig. 78), the end bud dies, and growth is continued by the next bud below; therefore, branching is sympodial, not, as in the Pine and Larch, monopodial.

The leaves are uprolled in the bud, and stipulate, but when the bud opens, the stipules fall off along with the bud-scales. The leaf-stalk is tough but very flexible, flattened laterally, and the rhomboid or somewhat triangular and toothed blade readily quivers in the wind.

This modification, found in some other Poplars as well, may be useful in two ways: (1) By moving readily with the wind, the leaves will produce less strain on the branches; and (2) the movement of the blade will favour transpiration, cause an increased upflow of sap, and therefore increased food-supply, and, in trees growing in a deep moist soil, with a good water-supply, will favour rapid growth.

In the vicinity of the tree we often find young shoots springing from the ground and resembling seedling Poplars.

If these are traced they will be found to spring from long roots of the parent tree which grow horizontally just below the surface, the main roots being deep in the soil. Such shoots are called suckers, and they afford a means of vegetative propagation. In the late summer or early autumn the ground near Poplar trees is often strewn with leafy

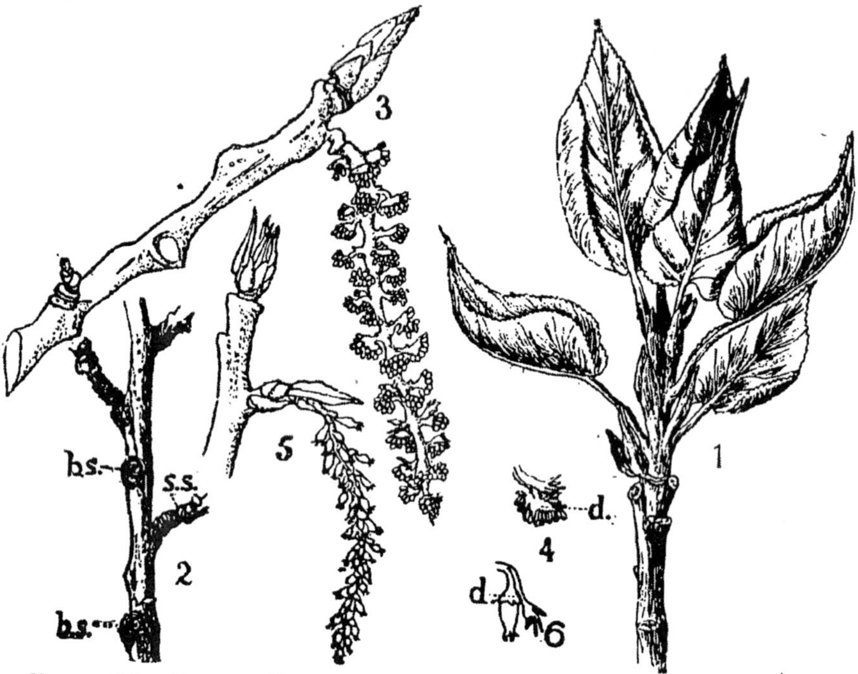

FIG. 186. BLACK POPLAR.—1, leafy shoot; 2, twig with two branch-scars; 3, male catkin; 4, male flower; 5, female catkin; 6, female flower; *b.s*, branch-scars; *d*, cup-like disk; *s.s*, bud-scale scars.

shoots, varying in length from one to six feet. These are deciduous shoots cut off by a separation-layer, as in the leaves (Fig. 186, 2 *b.s*). Compare this with what occurs in the Pine.

The flowers are in catkins and, as in the Willows, are dioecious, but have no nectaries and secrete no honey. They appear before the leaves. The male catkin (3) is lax, pendulous, and about two inches long. The bracts are fringed, and in the axil of each is a flower (4), consisting of a cup-like disk (*d*), and bearing thirty to forty stamens with dull red anthers. Much pollen is produced and dispersed by the wind. The flowers of the female catkin (5 and 6) are also axillary. A cup-like disk surrounds the ovary, and the pistil consists of two united carpels. The ovary is one-celled, and above it are two large branched stigmas (6). The capsules, when ripe, split by two valves; and the seeds, each bearing a tuft of hairs at the base, are dispersed by the wind.

Hazel

The Hazel (*Corylus Avellana*) (Fig. 187) is a shrub or small tree from ten to fifteen feet in height, often forming a conspicuous, shrubby undergrowth in Oak and Ash woods. It is frequently coppiced, and from the old stools which remain, shoots grow out freely, numerous branches thus arising close to the ground. As in the Poplar, suckers spring from adventitious buds on the roots. The cork arises immediately beneath the epidermis (as in Fig. 38), and for several years forms a smooth shining bark, on which are prominent transverse lenticels (Fig. 187, 1 *l*), but, later, the bark peels off in ring-like scales.

Branches of two kinds occur: first the main stem and the old branches, on which the lopsided leaves are arranged in two rows; and secondly the quick-growing stool-shoots and suckers, on which they are in three rows and have larger and more uniform blades. The buds are oval and covered with bud-scales, the nature of which may be easily made out by examining an opening bud. Note the transition from the outer brown scales to scales consisting of pairs of stipules covered with silky hairs, while the innermost pairs have each a small blade between them. The stipules only remain for a short time after the bud opens, but last longer on the leaves of the stool-shoots and suckers. The young shoots are zigzag and hairy, and produce a leaf at each angle.

The leaves (Fig. 187, 2) are short-stalked; the blade is large and somewhat orbicular, with a doubly serrate margin and a pointed apex, and the surfaces are rough and hairy. In the bud the leaves are pleated, i. e. folded between the lateral veins and then upwards along the midrib.

Buds are formed in the leaf-axils, but the terminal bud dies. In the following year the highest lateral bud grows

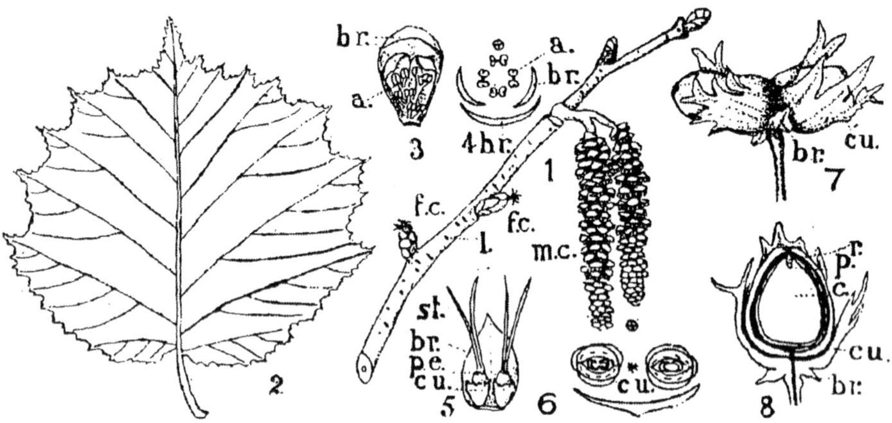

FIG. 187. HAZEL.—1, flowering branch; 2, foliage-leaf; 3, male flower; 4, floral diagram of the male cyme; 5, two female flowers in axil of bract; 6, floral diagram of female cyme; 7, fruiting branch; 8, fruit in longitudinal section; *a*, anthers of branched stamens; *br*, bracts; *c*, cotyledon; *cu*, cupule; *f.c*, female catkin; *l*, lenticel; *m.c*, male catkin; *p*, plumule; *pe*, perianth; *r*, radicle; *st*, stigma.

into a long, zigzag shoot; but those below form dwarf shoots, some of which produce a tuft of leaves, while others become flower-buds. Those which will form male catkins do not rest during the winter, but elongate the same year; hence we find tightly-packed male catkins hanging on the trees in winter (Fig. 187, 1). In mild weather they may open in December, but in severe weather they may remain closed until the end of February or the beginning of March.

The Hazel is monoecious. The male catkins (Fig. 187, 1

m.c) are pendulous, and one and a half to two inches long when open. Each bears a number of bracts, and within each bract are two smaller bracts (3 and 4 *br.*). The flower consists of four halved or split stamens. The filament is divided into two, and each bears at its end half an anther. Thus the flower appears to have eight stamens.

The female catkin (1 *f.c*) resembles a leaf-bud, but is rather larger, and, when mature, bright red stigmas project from it. Its outer structure is like a leaf-bud, and consists of a covering of brown scales followed by stipules and small leaves. In the centre are four or five bracts, each with two flowers in their axils (5). The ovary is inferior and bears a minute perianth (*pe*); on the top are two long red stigmas. At the base is a cupule composed of three scales (*cu*).

After fertilization, the wall of the ovary hardens into a woody shell, the perianth and stigmas shrivel up, and the three scales at the base enlarge and form a leafy cupule enclosing the nut (7 and 8). The seed is attached by a long stalk, has a thin brown testa, a small radicle and plumule, and two large cotyledons stored with oil (8 *c*).

Birch

The Birch (Fig. 188) is characteristic of dry upland woods and heaths, and occurs frequently in the wet fen woods. It is at once recognized by its white, papery bark, and slender switch-like branches. It is a small graceful tree, the trunk being from eight inches to a foot in diameter and attaining a height of forty to fifty feet. The base is covered with a rugged black bark; above, it is white and shining, and peels off in thin flakes; it is marked transversely by long dark brown lenticels. The younger branches are brown at first, changing later to white.

Two species (and several varieties) occur, and are most readily distinguished by their young shoots. The common

Birch (*Betula tomentosa*) is the most abundant. Its branches are greyish-brown, slender, but seldom droop, and the fresh young twigs are hairy. The Silver Birch (*Betula alba*) is a more elegant tree and is not so common. The branches are long and slender and often droop gracefully. Its young twigs are covered with resinous warts.

The buds are ovoid and covered with stipular scales, and the same gradation is met with as in the Hazel. The leaves are alternate, scattered and small; the base is small and leaves a small scar; the stalk is slender and the blade variable. Usually it is broadly ovate to cordate, with a doubly serrate margin, and the surface is glabrous with prominent veins beneath.

The Birch is a tree of fresh air and sunshine. It has a very open canopy, its small scattered leaves (Fig. 188, 2) do not form mosaics, and neither shade each other nor cast much shade on the ground. It grows badly under the shadow of other trees, and is therefore called a light-demanding tree. Usually the undergrowth is equally light-demanding.

The male and female flowers are in separate catkins on the same tree (monoecious). The male catkins are developed in the autumn, and are seen on the trees throughout the winter, two or three together at the ends of the twigs (Fig. 188, 1 *m.c*). In the spring, as the leaves come out, the catkins elongate, droop, and shed an abundance of pollen. The flowers are arranged on the catkin in three-flowered cymes (3 and 4). On the upper side of each bract are two smaller bracts (3 *br*); then three flowers, each with two split stamens, which thus resemble four stamens.

The female catkins are enclosed in buds during the winter, but in February they begin to open (1 *f.c*). At the base three or four leaves form on dwarf shoots, and each shoot ends in a slender catkin. As in the male catkin the flowers are in threes (4). Each bract has two small scales above it

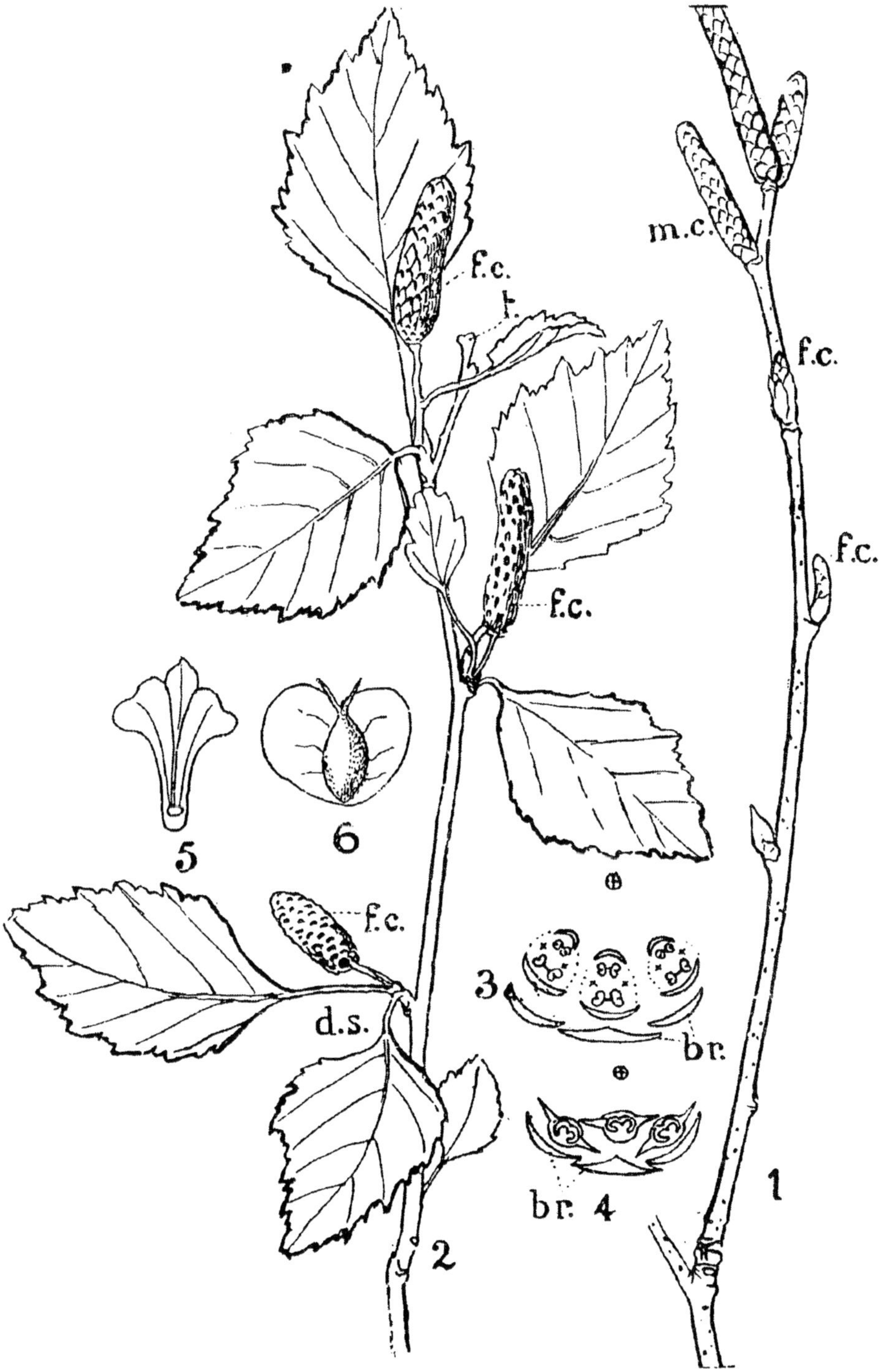

FIG. 188. BIRCH.—1, winter shoot; 2, leafy shoot; 3, floral diagram of male cyme; 4, floral diagram of female cyme; 5, bracts of female flower; 6, winged nutlet; *br*, bracts; *d.s*, dwarf shoot with foliage-leaves and fruiting catkins; *f.c*, female catkins· *m.c*, male catkins; *t*, dead terminal branch.

(Fig. 188, 5) and three small flowers. Each flower consists of a pistil of two carpels; the ovary is two-celled, flattened, and bears two stigmas. There is no cup, and only one ovule develops. The ovary when ripe becomes a winged nutlet (6). The catkin elongates as it ripens (2 *f.c*), and the bracts with the two scales attached, as well as the winged fruits, are scattered by the wind.

Alder

The Alder (*Alnus glutinosa*) (Fig. 189) is a characteristic tree by stream-sides and in low-lying marshy districts, where, along with Willows, it often forms a characteristic thicket. It is usually a small tree, rarely more than fifty feet high, and, growing from the base of the trunk, are often many stool-shoots, which give it a shrub-like appearance. On the roots large clusters of nodules grow, similar in function to those found in leguminous plants. Like the Birch, it is a light-demanding tree, and, when young, grows rapidly and soon frees itself from the shade of its neighbours.

The bark is brownish-black and fissured, with wide scaly ridges. The young branches and buds are greenish-brown to red or violet, and when seen from a distance an Alder thicket is often a rich purple. The longer quick-growing shoots are smooth, but with conspicuous reddish lenticels, and somewhat triangular in section on account of the prominent decurrent leaf-bases (Fig. 189, 1 and 2 *l.b*).

The buds are rather large, triangular, and distinctly stalked (*b*) by a slight elongation of the axis beneath the lowest bud-scale. The leaf-scars are ovate to rhomboid, with five leaf-traces, often reduced to three by the fusion of the three lower ones. The bud-scales are stipules (Fig. 189, 3), coated with a waxy secretion, and are not easy to separate. Note the relationship between leaves and

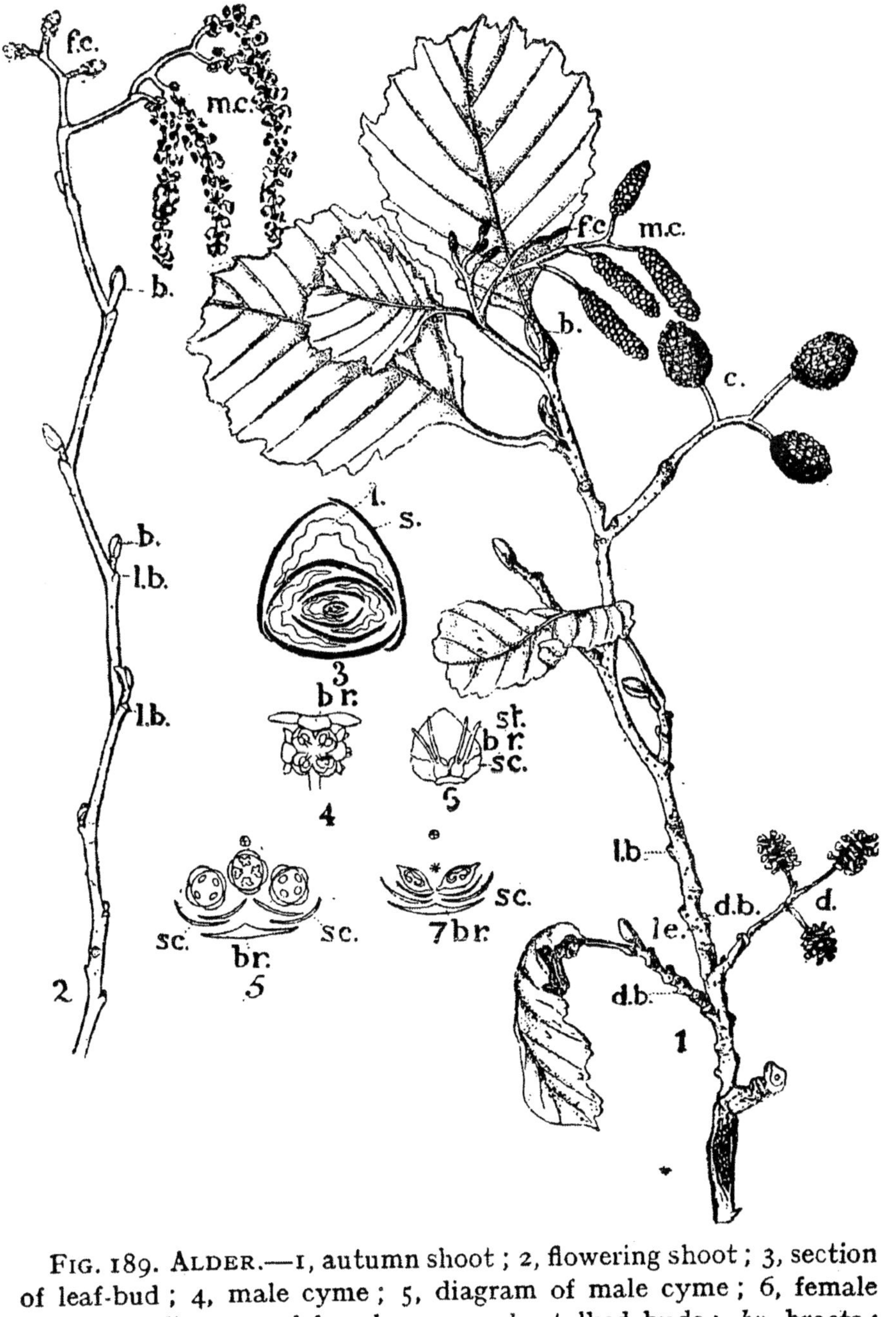

FIG. 189. ALDER.—1, autumn shoot; 2, flowering shoot; 3, section of leaf-bud; 4, male cyme; 5, diagram of male cyme; 6, female cyme; 7, diagram of female cyme; *b*, stalked buds; *br*, bracts; *c*, 'cone' with fruits; *d*, dead 'cones'; *d.b*, dormant buds; *f.c*, female catkins; *l*, leaf; *l.b*, decurrent leaf-bases; *le*, lenticels; *m.c*, male catkins; *s*, stipules; *sc*, scale-leaves; *st*, stigmas.

stipules as shown in the diagram. On the lower part of the shoot are numerous small dormant buds.

The leaves are stalked, and the base decurrent and stipulate. The stipules fall off when the leaf is mature; the blade is obovate, doubly serrate, and has a rounded or often notched apex. The leaves are folded fan-wise in the bud.

The flowers are in catkins, and both male and female occur on the same tree. Both are developed in the summer and may be seen on the tree before the leaves fall (Fig. 189, 1). They are thus exposed throughout the winter and open about February in the following year, before the leaves appear (2). The male catkins (1 and 2 *m.c*) are long and pendulous. On the upper surface of each bract, and united to it, are four scales (5 *sc*), and three flowers, each flower having a four-lobed perianth and four stamens. The female catkins (1 and 2 *f.c.*) are small and erect; the bracts and scales are the same as in the male catkins, but there are only two flowers, the central one not being developed (6 and 7).

When fertilized the ovary becomes a dry, flattened, one-seeded nutlet. Each bract with its four scales grows and becomes a green five-lobed woody scale, the whole resembling a small Pine cone (1 *c*). The nutlets ripen in the autumn but are retained until the following spring; the scales then dry and separate and allow the nutlets to fall out, when they may be dispersed by the wind or fall into the stream and be carried some distance before being washed ashore. The old dead blackened cones remain several years on the trees before they are broken off (1 *d*).

Note the different years' flowering shoots represented on the branch in Fig. 189, 1. At the growing end are the catkins (*f.c* and *m.c*), which will remain on the tree all the winter, and open in the following spring. Below are the ripe female cones (*c*) of the present season, and lower still are the old cones (*d*), which shed their fruits the previous year.

Fig. 190. Beech Wood in Winter.

Beech

The Beech (*Fagus sylvatica*) (Fig. 190) is one of the largest of British trees and occurs most extensively in the chalk districts of the south of England. In the north it is commonly planted, but doubtfully native. It attains a height of one hundred to one hundred and twenty feet; and in woods it develops a tall, straight trunk, extending to the crown and having few branches below. In the open it forms large branches low down on the trunk, and in consequence its wide-spreading crown comes nearly to the ground. From its base several massive buttresses are given off, which extend into the shallow main roots. Its bark is thin, very smooth and olive grey, and on the exposed side of the trunk numerous shoots may develop, which serve to protect it.

Two kinds of branches are formed, and give a characteristic aspect to the tree: (1) the quick-growing, slender, zigzag shoots with a bud standing off at each bend (Figs. 74 and 191), and (2) the slow-growing dwarf shoots, the ages of which may be determined by counting the sets of scale-scars on them (see p. 115).

The end bud of the long shoot sometimes dies, then growth is continued by the next bud below, which gives rise to a long shoot. The lower, lateral buds are displaced to the upper side of the leaf-scar and form dwarf shoots. Each produces three or four crowded leaves, which vary in size. By the bending of the leaf-bases all the blades are brought into a horizontal plane and form an excellent mosaic (Fig. 191, 1).

The lowest leaves of a shoot produce only small buds in their axils, and these remain dormant; no buds are formed in the axils of the leaves of dwarf shoots, growth being continued by the end bud. The buds are long, thin, oval, and pointed, and covered by about twenty light-brown membraneous scales, which are stipules. The leaves are folded fan-wise and covered with silky hairs (Figs. 72, 75, 76). As the bud opens and the leaves, which are in two rows, mature, the fringed stipules fall off. The leaf-base is small, and leaves a small, oval scar with three leaf-traces; the

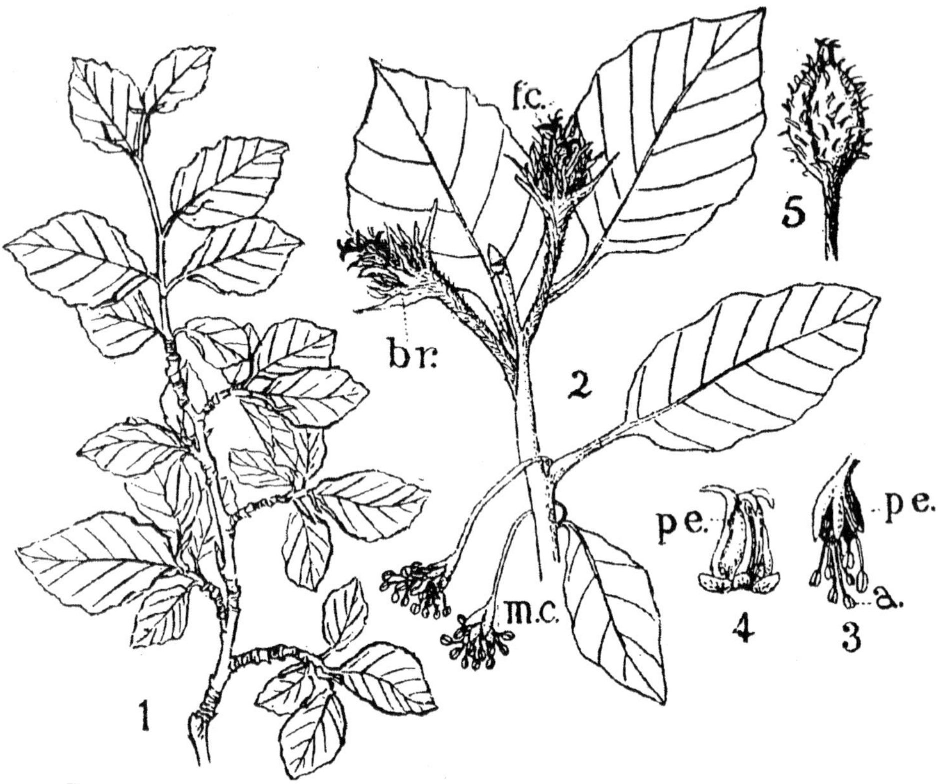

FIG. 191. BEECH.—1, leafy shoot showing leaf-mosaic; 2, flowering shoot; 3, male flower; 4, female flower; 5, fruit enclosed in a spiny cupule; *a*, stamens; *br*, bracts; *f.c*, female catkins; *m.c*, male catkin; *pe*, perianth.

stalk is short and hairy; the blade oval, thin, and tough, smooth above and silky beneath; the margin is wavy and, when young, fringed with hairs.

In young trees and in cut Beech-hedges, the leaves turn a light brown in the autumn and remain on the twigs all the winter. The spreading, plate-like branches of the

Beech, forming tiers of mosaics facing the sky, produce a closer canopy and cast a deeper shade than any other British tree, and the vegetation beneath is very scanty. As regards light, it stands at the opposite extreme to the Birch : it is a shade-enduring tree.

The flowers are in heads, which arise in the axils of the leaves of the current year, and therefore come out after the leaves. The stamens and pistils are in separate flowers on the same tree (Fig. 191, 2). The male flowers are in a globular cluster at the end of a long, pendulous stalk. Each flower has a perianth of four to seven hairy lobes (3), and from eight to twelve stamens. The female flowers (4) arise higher on the branch ; the stalks are short, thick, and erect ; and each inflorescence contains only two flowers. A single flower consists of a three-celled ovary with two ovules in each cell, and three red stigmas. On the top of the ovary is a perianth with about six lobes. Surrounding the two flowers is a hairy cupule, which, after fertilization, becomes thick, woody, and spiny (5), and when ripe splits into four valves. One nut is formed in each flower ; it is triangular, with a smooth brown coat, and contains only one seed.

Oak

The Oak (Fig. 192) is the largest and most characteristic of British trees, and formerly Oak forests covered a large part of England. Two species are common, and they sometimes characterize distinct habitats. The Sessile Oak (*Quercus sessiliflora*) is the dominant tree on shallow, poor siliceous soils, and is typical of the woods on the Pennine slopes, and other similar hilly regions. The Peduncled or Stalked Oak (*Q. Robur*) is often the prevailing tree in lowland woods, with a deep rich siliceous soil over clays and loams. On soils containing much lime, both species tend to occupy a very subordinate place in the vegetation.

Often the two species grow together, then hybrids between them are frequent. The Oak grows to a great age and size, and weather-beaten specimens, like the Cowthorpe Oak in Yorkshire, may have a trunk seventy feet in girth. The trunk attains a height of one hundred to one hundred and fifty feet, and is much branched. A characteristic feature of the tree is its gnarled and contorted branches, which end in clustered twigs. The bark is rugged, with deep vertical furrows; the ridges break transversely and form oblong scales (Fig. 193).

The buds are crowded around the ends of the twigs (Fig. 194). If the end bud persists, it grows into a long shoot, but frequently it dies and the lower buds grow out as a cluster of short, leafy twigs. Dwarf shoots, though common, are not regular, as in the Beech; and the buds in the lower leaf-axils and in the axils of the bud-scales are small and remain dormant (Fig. 194, 1 *d.b*). The buds are stout, blunt, and oval, and are covered with about twenty stipular scales. The twigs, with numerous small oval lenticels, are somewhat angular on account of the prominent leaf-cushions, and the leaf-scars have three or more groups of leaf-traces.

The leaves are alternate (Fig. 194, 2), and the small brown stipules soon fall off. The base is curved, swollen, and continued as two lines down the stem, and on either side of it is a stipule-scar. In August and September the separation-layer is seen distinctly across the leaf-base. The leaf-stalk is grooved above and longer in the Sessile than in the Stalked Oak; the blade is obovate, deeply and irregularly lobed (sinuate), and somewhat leathery. In the Sessile Oak, the base of the blade is more or less tapering, and has many branched hairs on the under surface. In the Stalked Oak, the base of the blade is produced into two recurved, ear-like lobes (auricled), and the hairs on the under surface are few and simple.

FIG. 192. OAK IN WINTER.

FIG. 193. OAK BARK.

1 2

FIG. 195. ELM BARK.—1, Mountain Elm; 2, English Elm.

FIG. 196. BARK OF COMMON ASH.

The flowers are in male and female catkins, both on the same tree, and appear with the leaves in April or May. The male catkins (Fig. 194, 3) arise either in the axils of the

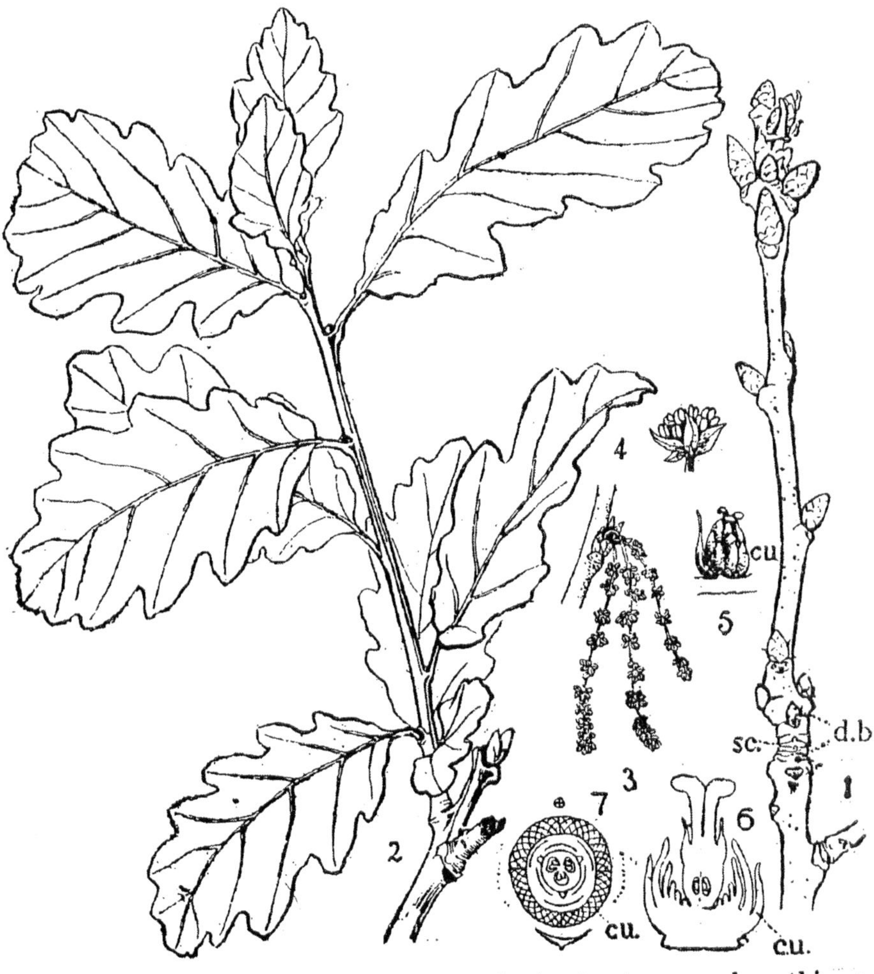

FIG. 194. OAK.—1, winter twig; 2, leafy shoot; 3, male catkins; 4, male flower; 5, female flower; 6, vertical section of female flower; 7, diagram of female flower; *cu*, cupule; *d.b*, dormant buds; *sc*, scale-scars.

lower leaves or lower on the shoot from buds of the previous year. They are long, pendulous, and lax, with the flowers arranged in groups on the slender axis. Each flower has a perianth with six fringed lobes and from four to twelve stamens (Fig. 194, 4). The female catkins arise near the end of the twigs and bear only one to five flowers. The stalk of the catkin is long in the Stalked Oak and much shorter in the Sessile Oak. Each flower (5 and 6) arises in the axil of a bract and is surrounded by a shallow cupule. The pistil consists of three united carpels; the ovary is inferior and three-celled (7), with two ovules in each cell, but the ovules are not formed until after pollination. The stigmas are broad, red, and three-lobed. Usually only one ovule develops into a seed, though very small ones may also be found within the polished shell of the acorn. For the structure of the acorn see Fig. 143.

A comparison of the flowers of such catkin-bearing trees as Hazel, Birch, Alder, Beech, and Oak shows that they differ in several important respects from those of Willows and Poplars. In the Willows and Poplars (Salicaceae) the flowers are dioecious, naked, i. e. they have no perianth, and hypogynous. The pistil consists of two united carpels; the ovary is one-celled with many ovules; the fruit is a dehiscent capsule, and the seeds have a tuft of hairs. In the Hazel, Birch, and others, the flowers are monoecious, and have an epigynous perianth. The pistil has two or three united carpels; the ovary is two-celled with one or two ovules in each cell, and the fruit is an indehiscent, one-seeded nut.

In each case the flowers are small and inconspicuous. The stamens produce a large quantity of dry pollen carried by the wind to the large branched stigmas of the female flowers.

CHAPTER XXIV

TREES WITH MORE HIGHLY DEVELOPED FLOWERS

THE flowers of some of our forest trees and shrubs are so conspicuous, that they have earned the popular name of 'flowering trees' and 'flowering shrubs'. They differ from the previous types in usually having both stamens and pistils in the same flower (i.e. they are hermaphrodite), and often have a well-developed perianth. The less conspicuous of these are the Elm, Sycamore, and Common Ash; on the other hand, the Rowan, Laburnum, Horse-Chestnut, and Lilac, have very showy flowers, attractive to insects, which pollinate them.

Elm

The Wych Elm (*Ulmus montana*), also known as the Scots or Mountain Elm, is a native of Britain and is northern and upland in its distribution, in which districts it commonly occurs in the damp woods and hedgerows. Another species, the English Elm (*Ulmus campestris*), is not native, and occurs most commonly in the lowlands and southern half of England (Fig. 195, 1 and 2). There are also many varieties.

The Elms are tall trees, eighty to one hundred and twenty feet high, with a deep and coarsely-fissured bark (Fig. 195, 1), resembling that of the Oak. The crown is large and spreading, and the lower part of the trunk is often thickly clothed with stool-shoots.

The buds and twigs are more hairy, and the latter thicker, in the Wych Elm than in the English Elm. The terminal bud often dies, and growth is continued by the next lower bud (Fig. 77). This grows into a long shoot, while the lower lateral buds form dwarf shoots as in the Beech, with a similar type of leaf-mosaic (Fig. 197). In flowering

FIG. 197. LEAF-MOSAIC OF ELM.

branches the small, pointed, upper buds produce leafy shoots, but below these are larger globular buds which produce clusters of flowers (Fig. 198, 1).

The leaves are ovate to obovate and often lopsided ; the margin is doubly serrate with a pointed apex, rough above and velvety beneath. The leaves of the English Elm are

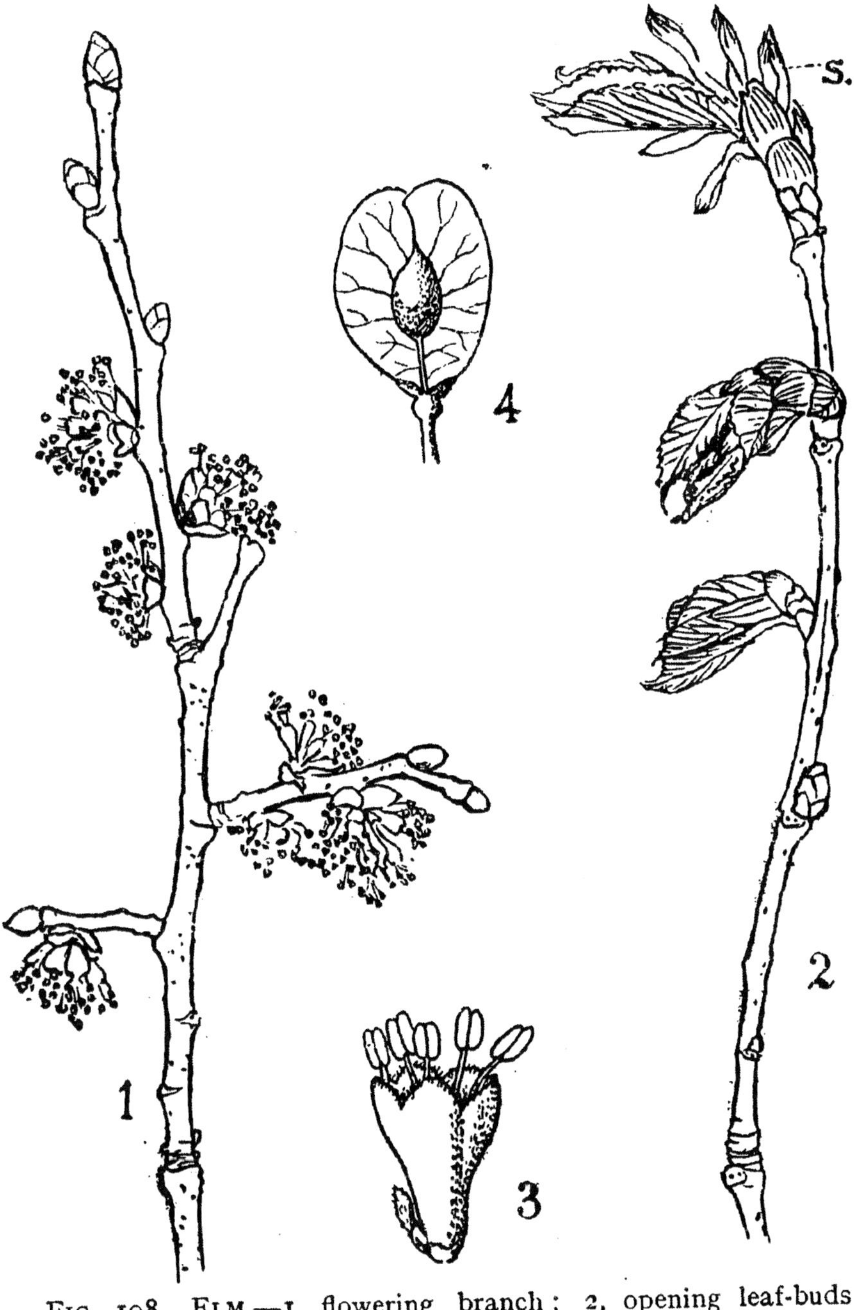

FIG. 198. ELM.—1, flowering branch; 2, opening leaf-buds; 3, Elm flower; 4, winged fruit; *s*, stipules.

smaller, ovate to cordate, and nearly smooth above. The stipules (Fig. 198, 2) fall off as the bud opens, but those on the stool-shoots remain for some time.

The flowers are developed in globular buds in the position of dwarf shoots, each containing a cluster of sixteen to eighteen flowers in small cymes. They open in March or April, before the leaves appear. Each flower, unlike the previous types, is hermaphrodite, i.e. stamens and pistil are in the same flower (Fig. 198, 3). The perianth is inferior and bell-shaped, with five or six fringed lobes; the stamens are five or six and opposite the lobes. The pistil is superior, of two united carpels. The ovary is two-celled and flat, with two stigmas.

After fertilization the ovary-wall expands into a thin, flat, veined wing surrounding the single seed. The end is often deeply notched and the two edges overlap (Fig. 198, 4).

The fruits (samaras) are developed in dense clusters, the wings are green, and a tree in full fruit appears at a distance to be in leaf. When the fruit is mature, the wing dries, becomes grey-brown in colour, and is dispersed by the wind. In this country the fruits of the English Elm often do not ripen their seeds.

Rowan

The Rowan (*Pyrus Aucuparia*), as is usual with well-known plants, bears several other popular or local names, such as Roan Tree, Wickens or Quicken Tree, and Mountain Ash, and is very conspicuous in the autumn, when covered with its bright scarlet berries. In distribution it follows pretty closely the Sessile Oak, but, although abundant in places, it never becomes the dominant tree of the wood. It occurs mainly on siliceous soils, but especially in heath woods in hilly districts, where it ascends far up the moorland valleys, becoming little more than a shrub. It is a small tree, fifteen to thirty feet high. The older part of the trunk is covered with a thick, furrowed bark, while it is thin and smooth above. Suckers are readily formed from the roots, and the branches are arranged spirally and come off from the tree at an acute angle.

The twigs are stout and covered with a smooth, shining, grey to deep red-brown bark; the lenticels are few but distinct, transverse, and yellowish. The dwarf shoots (Fig. 199, *d.s*) are stout, prominent, and densely ringed; and their ruggedness is intensified by the rather prominent crescent-shaped leaf-bases, each with five leaf-traces. The cortex, when bruised, and also the flowers, have an unpleasant smell of decaying fish.

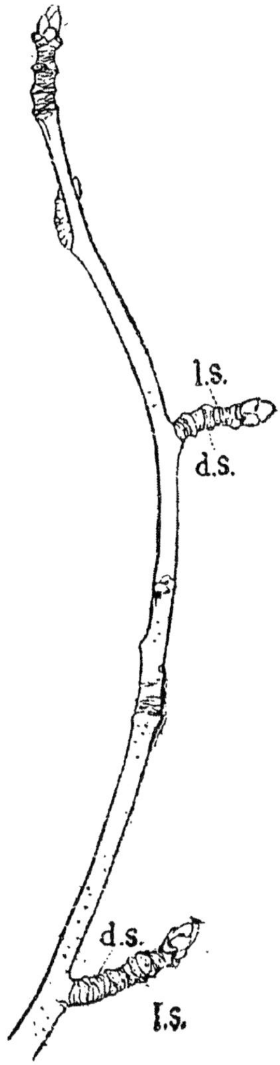

FIG. 199. WINTER SHOOT OF ROWAN.—*d.s*, dwarf shoots with numerous scale-scars; *l.s*, foliage leaf-scars.

The terminal buds are ovoid, large, and shining, and covered with five or six velvety and fringed bud-scales; the lateral buds are smaller and pressed against the stem; the end buds of the dwarf shoots elongate very little and produce each season from three to five leaves crowded together.

The leaves are alternate and stalked, and have prominent leaf-bases and deciduous stipules. The blade is pinnate and divided into eleven or more sessile, oblong leaflets, always with an odd terminal stalked leaflet. The leaflets are one to two inches long, serrate, smooth above, and slightly hairy below, and have pinnate veins.

The inflorescences are developed on the dwarf shoots, and each is a much-branched flat-topped cyme, rendered conspicuous by the massing of a large number of small flowers. The flowers (Fig. 200) open in May or June. The receptacle forms a deep cup with the five sepals, five petals, and about twenty stamens attached to the rim. The pistil consists of two or three united carpels which are in turn united to the receptacle-cup (2). The styles are

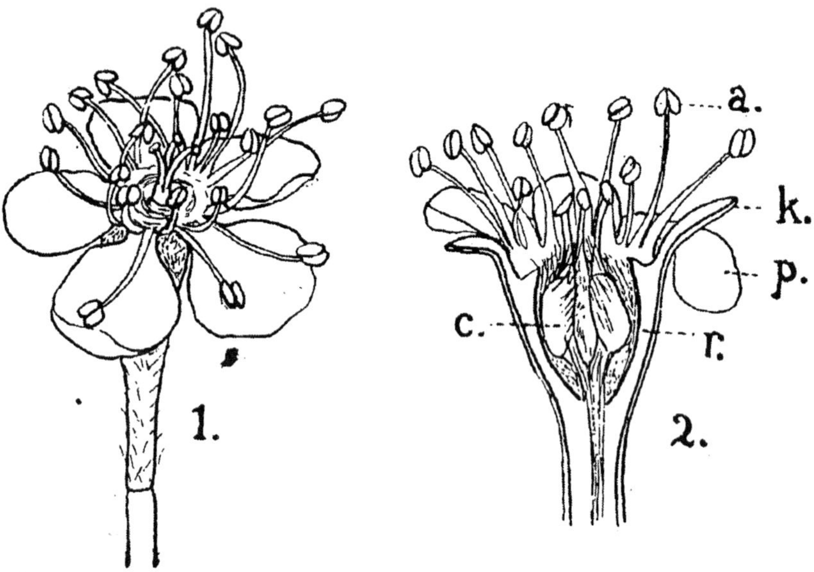

FIG. 200. ROWAN.—1, flower; 2, vertical section of flower; *a*, anther; *c*, carpels; *k*, sepal; *p*, petal; *r*, receptacle.

free, the same in number as the carpels, and the stigmas are ripe before the stamens. The stamens are of three different lengths; the outer and longer ones stand above the stigmas, while the innermost and shortest ones are incurved (1). Round the base of the styles is a honey-secreting ring, and the honey is partly protected by hairs projecting from the styles.

Numerous insects visit the flowers, such as beetles, flies, and bees, but, if their visits are ineffective, self-pollination takes place. After fertilization the receptacle becomes fleshy ; the carpels form the core enclosing the seeds, and the receptacle, at first green and small, enlarges and becomes fleshy and bright scarlet.

The fruits are small pomes and are formed in the same way as the apple. They are dispersed by birds--thrushes, redwings, and fieldfares being fond of them. The specific name of the tree (*Aucuparia*) is derived from the fact that its berries were used to entice redwings and fieldfares into nooses of hair suspended in the woods.

Laburnum

The Laburnum (*Cytisus Laburnum*) (Fig. 201), so often grown as an ornamental tree, is not a native of Britain, but of Central Europe ; where in spring it covers the lower mountain slopes in 'sheets of gold', as does the Gorse on the mountains of Wales and in the west of Ireland. It is a small tree, fifteen to twenty feet in height, with lax, often drooping, branches. The bark remains smooth for many years, then becomes fissured longitudinally ; the branches are greenish-brown to olive, and the young shoots are grey-green and covered with silky hairs.

The end buds are white, silky, and surrounded by several prominent leaf-bases (1), on which are narrow persistent stipules, giving the buds a fringed appearance. The lateral buds are rather smaller and flattened, and rest on prominent leaf-bases. Many of these buds are suppressed or dormant, especially on the concave side of the branch ; hence the lax branching. Some form prominent densely-ringed dwarf shoots, resembling those of the Rowan (1 and 3 *d.s*).

The leaf-scars are small and semi-lunar, and have three leaf-traces. The buds are covered by two or three rather loose scales which show transitions to foliage-leaves, thus proving that they are reduced leaf-bases (2). The leaves on the long shoots are alternate and separated by

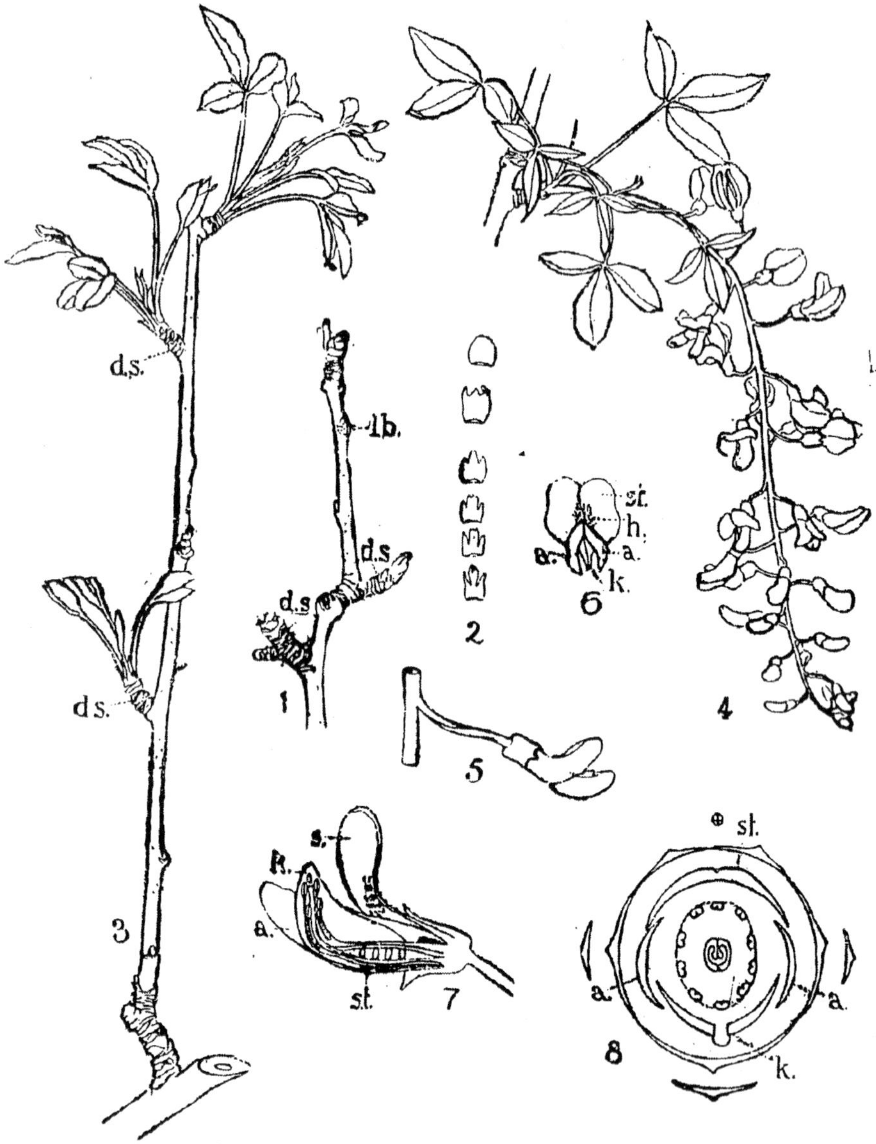

FIG. 201. LABURNUM.—1, winter shoot; 2, bud-scales and young leaves; 3, opening buds of leafy shoot; 4, flowering shoot; 5, flower in side view, flower-stalk twisted; 6, flower in front view; 7, vertical section of flower; 8, floral diagram; *a*, alae; *d.s*, dwarf shoots; *h*, honey-guides; *k*, keel: *l.b*, lateral bud; *st*, standard; *s.t*, stamen trough.

distinct internodes, while those on the dwarf shoots arise in tufts (Fig. 201, 3). The leaves are alternate and compound; the base is small with narrow persistent stipules; the petioles are long, and the blade is trifoliate. Each leaflet is attached by a short stalk, and is entire, ovate, and pointed; the upper surface is smooth; the under surface silky white, especially when young.

The inflorescence is a lax drooping raceme (4). The flowers are irregular, pea-like (papilionaceous), and open in May. The calyx is two-lipped and consists of five united but unequal sepals. The corolla has five petals, as in the Sweet-Pea. The drooping habit of the raceme inverts the flowers. This, however, is righted by the twisting of the young flower-stalk (5), and thus the standard is brought into its usual conspicuous and erect position (6). The honey is quite concealed and secreted in a swelling at the base of the standard, which has two dark honey-guides directed towards the nectary. The receptacle is slightly hollowed, and the petals and also the stamens are joined to the side of it, being therefore perigynous (7). The ten stamens are all joined (monadelphous) (8), and are ripe before the stigma. The pistil has only one carpel; and the style, together with the stamens, is upturned in the keel.

The flowers are visited by bees, and after each visit the keel springs back into its place ready to be revisited. The fruit is a pod, constricted between the seeds and covered with dark brown hairs. The seeds and other parts of the tree are poisonous.

Sycamore

The Sycamore (*Acer Pseudo-platanus*) (Fig. 202) is a native of Middle Europe, but not of Britain. It is, however, a very familiar tree in woods, parks, and hedgerows. Often, when the Oak is cut down in the woods, the Sycamore is planted in its place and so becomes a common woodland tree. It also tends to spread by self-sown seedlings. In Scotland it is known as the Plane Tree, but it must not be confused with the true Plane so commonly planted in London, which is a form of *Platanus acerifolia.*

It is a large tree, fifty to sixty feet high, with a wide-spreading, somewhat pyramidal, crown. The ash-coloured bark is smooth, and in the old tree scaly, but not fissured. The terminal bud continues growth ; therefore branching is monopodial. The buds are in crossed pairs (Fig. 70), the terminal and also the flowering buds being larger than the lateral buds. The branches are slaty-grey to reddish-brown, and are dotted with numerous lenticels. The details of a shoot, the structure of a bud, and the formation of the leaf-mosaic have already been described (pp. 113–14, Figs. 70, 71, and 73).

Beneath a Sycamore on the side of a road the pavement is often covered with shining rain-like drops, due to a sticky, sugary excretion called 'honey-dew', from aphides which infest the leaves. They suck the sap and exude drops of honey-dew, which spread over the leaves like a varnish. When the aphides are abundant, the drops fall from the tree like fine rain.

Two opposite buds are formed immediately below the terminal one, and if the latter produces a flowering shoot the axis ceases to grow in length. The flowering shoot is eventually thrown off and leaves a scar between the two lateral buds, which in time give rise to a forked branch (false dichotomy).

The flowers arise in large end buds. The inflorescence (Fig. 202, 1) is a pendant raceme of umbel-like cymes, each with three or four flowers opening in May or June. The flowers vary in the raceme. Usually the terminal one of a cyme (2) is complete, and consists of five sepals and five petals all similar, greenish-yellow, and free. There

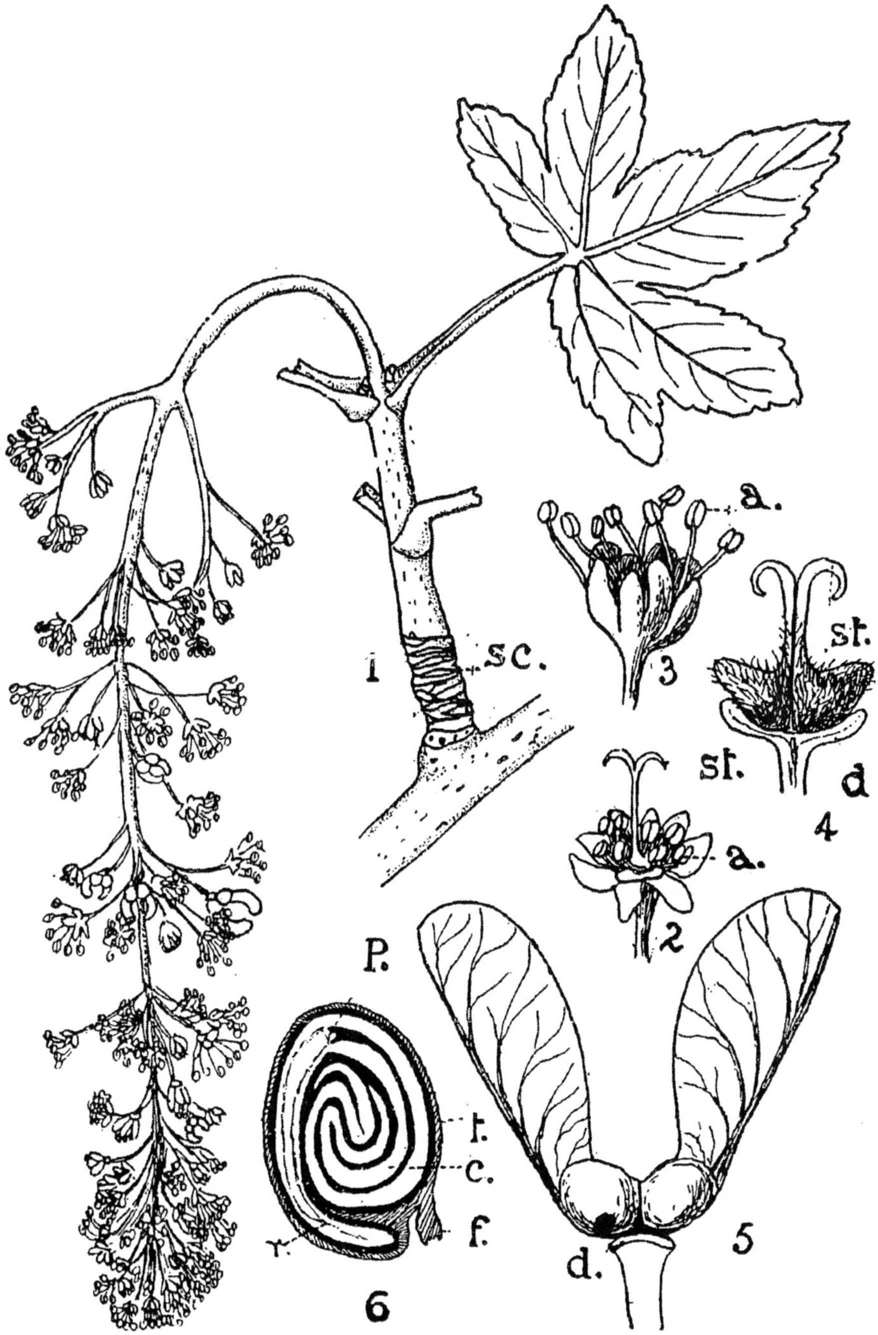

FIG. 202. SYCAMORE.—1, flowering branch; 2, hermaphrodite flower; 3, male flower; 4, pistil; 5, double samara; 6, vertical section of seed showing the folded cotyledons; *a*, anther; *c*, cotyledon; *d*, disk; *f*, funicle; *p*, plumule; *r*, radicle; *sc*, scale-scars; *st*, stigmas; *t*, testa.

are eight stamens together with a pistil which is superior, and consists of two, sometimes three, united carpels.

The ovary is two-celled, each cell having two ovules, but only one develops into a seed. The lateral flowers are often staminate, and the pistil, when present, is small and abortive (Fig. 202, 3). In the complete flowers the stamens ripen before the stigmas.

Honey is secreted on the prominent disk at the base of the pistil (4 *d*), and, being exposed, is accessible to short-tongued insects like flies, which freely visit the flowers.

The fruit or ' key ' (see p. 214) is a double samara (5), and when ripe, splits into two half-fruits. They fall spirally and the wing aids in wind-dispersal, but, except in high winds, they are not carried far from the parent tree.

Horse-Chestnut

The Horse-Chestnut (*Aesculus Hippocastanum*) (Figs. 66, 68, 69, and 103) is a native of Greece and Asia, where its seeds are ground and mixed as a medicine with horses' food ; hence its specific name, and the English equivalent ' Horse-Chestnut '. Another explanation of the name is that ' horse ' or ' coarse ' is applied to it to distinguish it from the edible Chestnut (*Castanea*). It is commonly planted in Britain as an ornamental tree. It grows seventy to eighty feet high, with an erect trunk, three to four feet thick at the base, and with a broad pyramidal crown. The bark is smooth for many years, and then becomes grooved and scaly. As in the Sycamore, branching is monopodial, and in the young trees very regular, with a tendency for the inner branches to be smaller. The branches curve downwards and outwards, and in open situations the end twigs are markedly upturned and end in very large, sticky, red-brown buds, the structure of which has already been studied (pp. 108–13).

As is commonly the case with buds, the last leaves formed in the season are greatly reduced and become the oldest and lowest scales of the winter bud, and sometimes traces of blades are found on their tips. Note the number and the different lengths of the internodes in a year's growth, and compare the shoot with one-year shoots of other trees. Often the lowest nodes are short, then follow longer ones, and finally shorter ones again at the end of the season. The leaves appear early, and this often leads to irregular growth. For if the leaves are killed by late frosts, as is frequently the

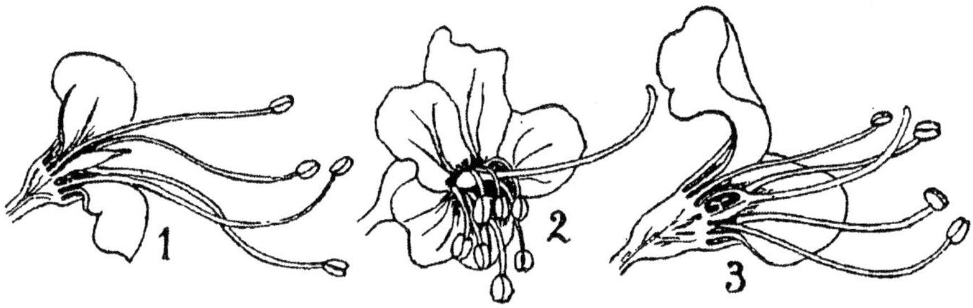

FIG. 203. FLOWERS OF HORSE-CHESTNUT.—1, male flower; 2, hermaphrodite flower: female stage; 3, later stage: stamens raised to level of stigma.

case, new leaves are produced from buds which otherwise would not have opened until the following season.

The flowering buds are very large (Fig. 103); the inflorescence is a big erect panicle; the main axis is racemose and the branches are cymose. After flowering, the inflorescence is cut off by a cork layer forming a large scar, which has a bud on both sides (Fig. 68, 5); with the growth of these, false dichotomy results. Three kinds of flowers often occur in the same inflorescence through abortion: (1) upper male flowers, which open first (Fig. 203, 1); (2) perfect (hermaphrodite) flowers, in which the stigma ripens before the stamens (Fig. 203, 2); (3) abortive flower-buds, which fall off without opening.

The flower is irregular, has five sepals, five petals (though sometimes there are only four), seven stamens, and a superior pistil of three united carpels. The ovary is three-celled, with two ovules in each, and above is a single long style. At the bases of the young petals are yellow spots which later turn red. The anthers and pollen are also red.

The flowers are visited by bees and are well adapted to the size and habits of the humble-bee. The upper flowers with an abortive ovary, open first (Fig. 203, 1). Later, the perfect flowers open, the style projecting horizontally, and the stigma is ripe before the stamens, which, at this stage, hang downwards out of the way (2). Thus pollen may be carried by bees from the male flowers to the stigmas of the perfect flowers. Finally, the stamens turn upwards, parallel to the style, and shed their pollen, and so may effect self-pollination (3). The bee presses its legs between the petals, and pushes its proboscis into the flower to obtain honey from the disk on the outside of the stamens. In doing so, the hinder part of its body touches the stigma and also the ripe anthers, and at the same time it carries away pollen on the bases of its middle and hind legs. The fruit is formed from the ovary and becomes a large and spiny capsule. When ripe it splits into three valves, each containing two large brown seeds.

Common Ash

The Common Ash (*Fraxinus excelsior*) (Fig. 204) is a native tree, widely distributed in Britain in very different habitats. It is especially characteristic of the woods of the rocky and scree-covered limestone hills in the north and west of England, in which it is usually the dominant tree. In non-calcareous areas it is common in the wet soils along stream-sides and is frequent in the wet carr woods of lowland and fen districts. It is sometimes coppiced. The tree attains a height of eighty to a hundred feet, and the trunk extends almost to the top of the oval pyramidal, but loose, crown. The bark (Fig. 196) has the form of a meshwork of oval longitudinal fissures. Shoots often spring from dormant buds on the trunk. The branches arise in crossed pairs and are greenish-grey to olive-green.

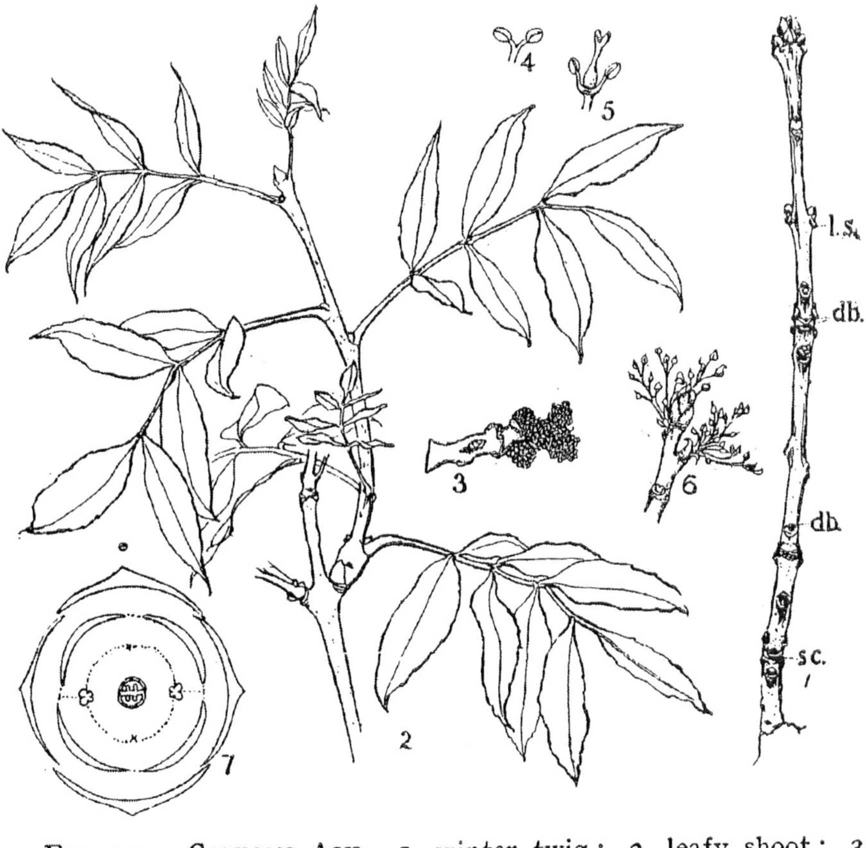

FIG. 204. COMMON ASH.—1, winter twig; 2, leafy shoot; 3, flowering shoot; 4, male flower; 5, hermaphrodite flower; 6, young fruiting branch; 7, floral diagram of hermaphrodite flower; *db*, dormant buds; *l.s*, leaf-scar; *sc*, scale-scar.

They remain smooth for a long time; then become finely fissured and have a few scattered, longitudinal lenticels. The twigs (Fig. 204, 1) are thick, smooth, often upturned at the ends, and sometimes form false whorls. Dwarf shoots are common and very knotted. The terminal buds, which are the largest, are short, and covered with velvety black hairs. Though sometimes separated by a short internode, the lateral buds are generally opposite and in crossed pairs. A flattened appearance is given to the nodes by the large leaf-bases in which the buds are partly embedded. The leaf-scars are large and shield-shaped, with many somewhat fused leaf-traces. The bud-scales are in crossed pairs; two to four may be seen on the outside, and, as in the Horse-Chestnut and Sycamore, they are leaf-bases.

The leaves, which appear late, are in crossed pairs, but sometimes they are alternate (Fig. 204, 2). The base is large and has no stipules; the petiole and midrib are grooved above, especially opposite to the leaflets; hairs occur in the groove, and sometimes small insects inhabit it. The blade is large, compound pinnate, and with from seven to thirteen leaflets, which are ovate, lanceolate, and irregularly serrate; the apex is long and pointed (acuminate). In young trees, during damp weather, drops of water exude from water-pores at the leaf-tips. When the leaves fall, a separation-layer forms across the bases of the leaflets as well as across the leaf-base. This occurs also in the Horse-Chestnut.

The Ash, like the Birch, is a light-demanding tree and endures shade badly. When planted along with, and under the shade of, quicker-growing Pines, the young main shoot grows rapidly towards the light and so forms a tall, slender trunk; hence Pines, in such circumstances, are called 'Nurses' by foresters.

The flowers appear in April or May and before the leaves. The inflorescences (Fig. 204, 3) are dense, racemose cymes of a dark purple colour, due to the purple-brown anthers and stigmas. The flowers (4 and 5) are polygamous, i.e. staminate, pistillate, and hermaphrodite flowers may occur on the same tree, and sometimes the trees are dioecious. The male flowers have no perianth, and consist merely of two stamens joined at the base (4). The female flowers have a much-reduced calyx and a superior pistil of two united carpels, and there are two large stigmas. The hermaphrodite flowers have no perianth, but two stamens below the pistil (5).

The flowers are wind-pollinated, and only one seed matures. As the fruit ripens, the free end of the ovary enlarges into a flat, leathery wing, and forms a samara carried by the wind. A young fruiting branch is shown in 6, and the floral diagram in 7. The structure of the fruit has already been studied (p. 24, Fig. 9, 1 and 2).

Lilac

The Lilac (*Syringa vulgaris*) (Fig. 205) is a native of the wooded slopes of Persia and Central Europe, and was introduced into Britain at the beginning of the seventeenth century, when many of our common ornamental trees were brought to this country. It is a shrub ten to fifteen feet high, and is often surrounded by numerous suckers from the roots. The suckers grow rapidly, forming long, straight, switch-like shoots. This increase in vegetative growth tends to reduce its flowering activity, hence the removal of the suckers in cultivation.

The bark is greenish-brown, fissured and scaly; and the small, slender branches are grey to olive, with conspicuous oval lenticels.

The end bud of the ordinary branches often dies, and since the lateral buds are in crossed pairs, this leads to the forked branching (false dichotomy) which is such a striking feature of the shrub. The large inflorescence-bud is also terminal, and forked branching occurs here after flowering, as in the Sycamore and Horse-Chestnut.

Many variations in bud-suppression will be found by careful examination of a Lilac shrub (Fig. 205, 1 to 5): (1) The lowest two or three pairs of buds on a shoot are very small, and usually remain dormant (*db*); (2) the

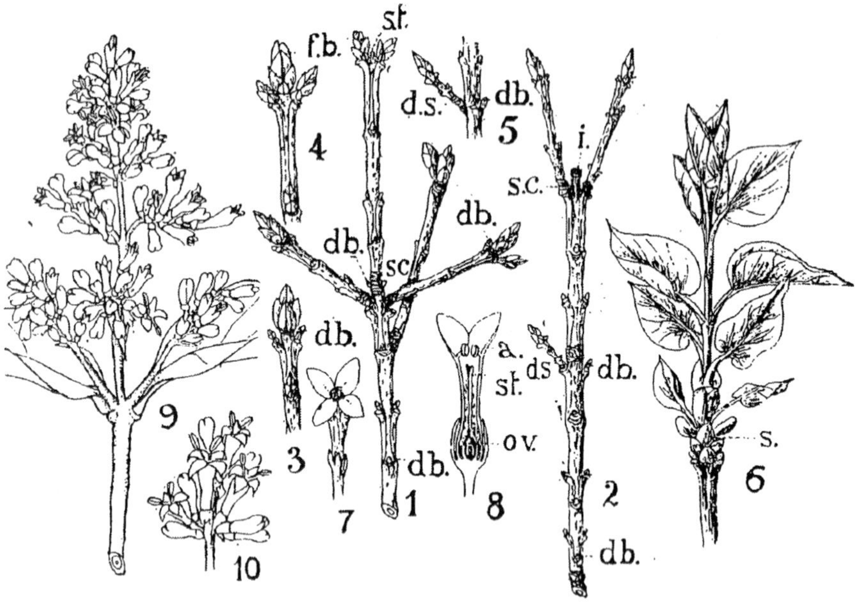

FIG. 205. LILAC.—1, winter shoot: terminal buds suppressed, lateral buds vigorous; 2, forked branch formed from lateral buds, between which are the remains of an inflorescence; 3, branch with persistent terminal buds, the two lateral buds dormant; 4, terminal flower-bud; 5, dwarf shoot; 6, young leafy shoot, showing transition from scale-leaves to foliage-leaves; 7, flower; 8, vertical section of flower; 9, inflorescence of Privet; 10, part of inflorescence showing flowers in small cymes; *a*, anther; *db*, dormant bud; *ds*, dwarf shoot; *f.b*, flower-bud; *i*, remains of inflorescence; *ov*, ovary; *s*, bud-scales; *sc*, scale-scars; *st*, stigma; *s.t*, suppressed terminal bud.

terminal bud often dies and growth is continued by the next pair below (1 *s.t*); (3) the terminal bud may persist, and suppression occur in one or both of the next buds below (3 *db*); (4) often one of a pair lower down is dormant (5 *db*); (5) occasionally buds grow very slowly and form dwarf shoots (2 and 5 *db*).

A comparison should be made of shrubs showing different degrees of bud-suppression, and it should be determined what effect this has on their form; also how cutting a shrub induces dormant buds to become active.

The buds are large and slightly sunk in the prominent leaf-bases. They are covered by four or five pairs of green scales which are strongly keeled and, being in crossed pairs, render the bud of a square shape in cross-section. The scales are reduced leaves (leaf-bases); the foliage-leaves within are in ten to twelve crossed pairs, and their blades are not folded, but lie edge to edge. Below each bud is a small, crescent-shaped leaf-scar, with one long leaf-trace representing several fused veins. The leaf-cushion is prominent and is continued as two ridges through the length of the internode, the position of the ridges changing with each node. The leaves have no stipules; the stalk is long, and the blade cordate to ovate; the margin is entire, ending in a long point (acuminate), and the surface is smooth (Fig. 205, 6).

The inflorescence is a large, erect, loose panicle of small, but showy, flowers, similar to those of the Privet (Fig. 205, 9 and 10). The small calyx has four united sepals; the four lilac to purple petals are united into a long narrow tube with four limbs spreading crosswise, and on the corolla-tube are two short-stalked stamens (7 and 8). The pistil is superior and consists of two united carpels, and the style is divided above into two stigmas which stand just below the anthers. The last named nearly fill the entrance to the tube and protect the honey, which is secreted by the ovary and rises somewhat in the tube.

Honey, scent, and colour attract numerous insects, the long tube favouring the long-tongued species. If their visits are not effective, pollen may fall on to the stigma, and bring about self-pollination.

The fruit is a two-valved capsule; each capsule containing about four seeds, with a slight membraneous wing.

Both Privet and Common Ash belong to the same natural order (Oleaceae) as the Lilac. The Common Ash has no perianth, though other species of Ash have, and so approach more closely to the Lilac. The Privet flower (Fig. 205, 10) is very similar to the Lilac, and they should be compared.

FIG. 206. VEGETATION OF LAKE, WOOD, MOOR, AND MOUNTAIN.

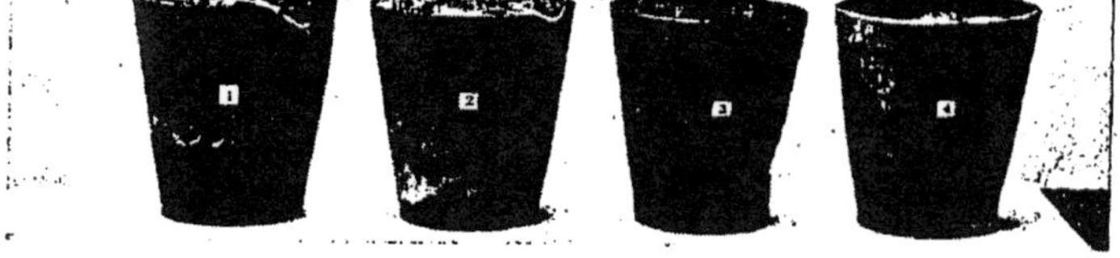

FIG. 207. MUSTARD SEEDLINGS IN DIFFERENT KINDS OF SOIL.—
1, surface soil; 2, sand; 3, clayey subsoil; 4, clay.

PART V

ECOLOGY

CHAPTER XXV

PLANT HABITATS AND COMMUNITIES

Vegetation of the valley and mountain.—In a walk from the bottom of a valley to the top of a mountain many striking changes in the vegetation are met with (Fig. 206). Below is the river, with, perhaps, here and there, ponds or lakes bordered with reeds, rushes, and other moisture-loving plants, and trees such as Alders and Willows. The flat ground beyond, composed of alluvium laid down in the past by the river, is highly cultivated and occupied by cornfields and meadows, bounded by hedgerows, with occasional undrained patches of marsh. On the rising ground these meadows gradually give place to pasture and woodland or uncultivated heath; stone walls often replace the hedgerows; and, as we ascend, the plants vary in character according to soils, drainage, water-supply, aspect, altitude, and the like. Higher still, the trees disappear and give place to wild, bleak moorland, perhaps covered with deep, wet, acid peat (Fig. 209); while the rocky peaks forming the summits are covered with a vegetation very unlike that met with at lower levels. On such high peaks the plants are exposed to great extremes of climate, heat and cold, wet and drought, bright sunshine, and dense, wet mist, driving and often drying winds, and

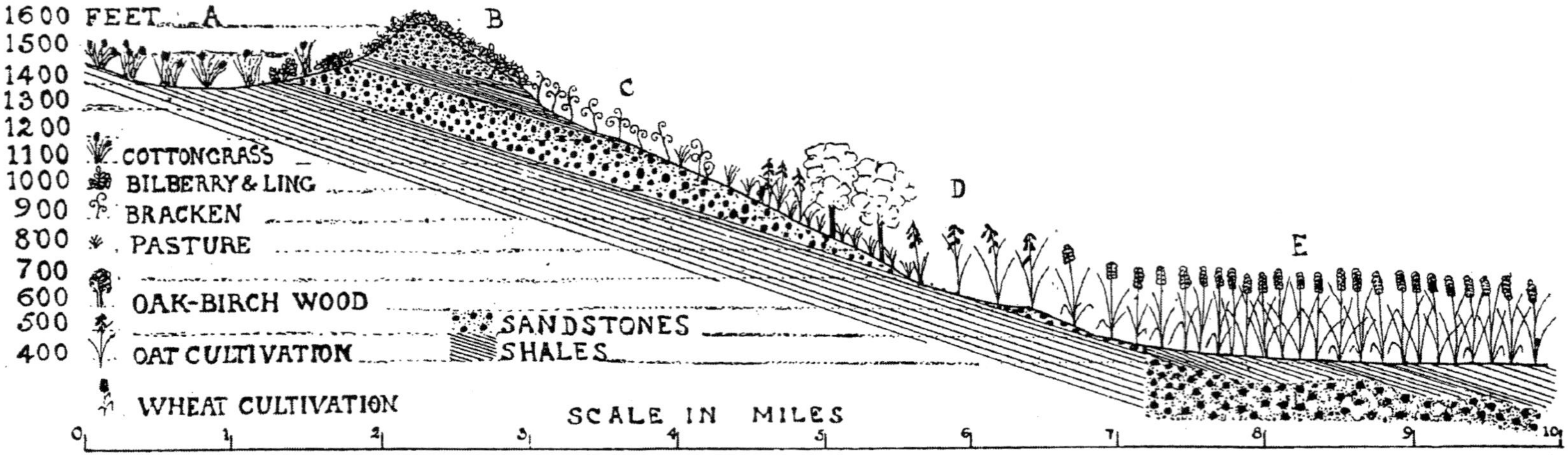

FIG. 208. TRANSECT SHOWING THE VEGETATION OF A MOUNTAIN-SLOPE AND VALLEY.

also to a rarefied atmosphere. These changes of altitude, slope, climate, aspect, soils, and water-supply sort out, as it were, groups of plants, each with its own characteristics, and so completely is this done that it imprints itself on the landscape.

The diagram, Fig. 208, is a transect extending from the summit of a hill to the valley below, and shows the striking changes in the types of vegetation. The high ground is treeless. To the west is a wet, peat-covered Cotton-grass moor (A). Rising above is a rocky summit with stony slopes, dominated by Bilberry (*Vaccinium Myrtillus*) (B), and along with it are Ling (*Calluna vulgaris*) and rolled-leaved Grasses. These moorland species extend down the slope, but here the most conspicuous species is the Bracken (C). The zone of Oat cultivation (D) now begins, but the area is largely given up to pasture. In the higher part of this zone the Oak reaches its upward limit. At the lower level is the zone of Wheat cultivation (E), but, as in the Oat zone, pasture-land predominates.

Factors affecting the distribution of vegetation.—In Britain the factor which has the greatest influence on plants is the soil, and some of its constituents are much more effective in determining the distribution of plants than others, e. g. water-content, humus, acidity, and lime, which are therefore called determining factors. A given habitat will, with the varying seasons, be subject to a more or less definite cycle of climatic conditions, e. g. temperature, rainfall, atmospheric humidity, and wind.

The conditions affecting plant growth due to topography, e. g. large mountain masses, altitude, exposure, and slope, are known as topographic or physiographic factors; those due to soil-conditions are edaphic factors; temperature, precipitation (i. e. rainfall and humidity), and winds are climatic factors.

The vegetation of a given area will thus be under the

influence of a series of factors, topography, soil, and climate, which will permit or favour the growth of certain species to the exclusion of others, and such vegetation will have a definite character. In other words, the nature of the habitat must determine largely, not only the form, but the kind of plant growing in it. The study of plants in relation to their habitats is called **ecology** (from Gr. *oikos*=house or habitat).

Influence of water-supply on plant form.—So important is the water-supply to plants that, in proportion as the amount available is large or small, they often develop forms and structures suited to the conditions of such habitats. In extreme cases this is so marked that special names have been used to designate them. For example, plants whose structural peculiarities enable them to grow in water are known as aquatic plants or *hydrophytes* (Gr. *hydor* = water, *phyte* = plant). Plants growing in marshy ground, as on the sides of ponds, ditches, and rivers, or in wet hollows in woods, are called *hygrophytes* (Gr. *hygros* = moist). At the other extreme are plants adapted to life in habitats with an uncertain water-supply and under conditions favouring strong transpiration, e. g. sand-dunes (Fig. 210), moors, and deserts. Such plants are called *xerophytes* (Gr. *xeros* = dry), while plants growing in a salt-marsh are known as *halophytes* (Gr. *hals* = salt). Every gradation, however, is found between hydrophytes and xerophytes, and it is impossible to draw a sharp line between them, but it has been found convenient to speak of plants adapted to habitats intermediate between the two extremes as *mesophytes* (Gr. *mesos* = intermediate). Included under the name mesophytes are plants very varied in habit, form, and structure. Those which show marked seasonal differences, e. g. plants with deciduous leaves, which are mesophytes in summer and xerophytes in winter, are called *tropophytes* (Gr. *tropos* = change).

FIG. 209. WASTING PEAT ON A COTTON-GRASS MOOR.

FIG. 210. A SAND-DUNE.

FIG. 211. A SALT-MARSH.

Plant-communities.—The whole of the vegetation of a habitat composed of plant-communities which are determined by the generally constant soil and climatic conditions in that habitat is called a **plant-formation.** The vegetation of sand-dunes, also that of salt-marshes (Fig. 211), furnishes good examples of plant-formations; other examples are aquatic, marsh, fen, and moor formations. There are also extensive and complex formations like those on siliceous soils and those on calcareous soils.

Minor variations of the habitat within the formation give rise to well-marked plant-communities, such as the Heather moor, the Cotton-grass moor, the Grass heath, the Sessile Oak wood, the Pine wood, and the Limestone Ash wood, which are easily recognizable during a country walk as characteristic features in the landscape.

These plant-communities within a plant-formation are called **plant-associations**, and are usually dominated by one or a few species of plants, often with a characteristic form and habit, e.g. the small, rolled-leaved evergreen shrubs of the Heather moor, the tussocks of Cotton-grass on the Cotton-grass moor, or the dominance of true grasses on the Grass heath; and, in the case of woodlands, the prevailing tree, Oak, Pine, Ash, or Beech, influencing and being accompanied by a peculiar undergrowth.

Compounds of the names of the dominant species are used in naming the several associations, e.g. Heath association, Ash wood association, Oak-Birch-Heath association, Alder-Willow association, &c.

Plant-associations contain within them a number of smaller communities, the **plant-societies.** These consist of species more or less related to each other as regards periods of active growth, shoot-systems of varying heights and shade-requirements, underground parts of different kinds and at different depths, drawing upon different constituents or tapping different layers in the soil. A good illustration is the society of Soft Grass, Bracken, and Bluebell so common in the Sessile Oak wood associations.

The more important plant-associations indicated should, wherever possible, be studied in the field; but much may be learnt from a detailed study of those types of vegetation which lie close at hand, e.g. the plants of a hedge and ditch, a meadow, a wooded escarpment, or a bit of moorland. As a preliminary to the study of plant-associations we will make a few observations on soils.

CHAPTER XXVI

THE SOIL

Origin of soils. Sedentary and transported soils.—If we examine a section of soil in a quarry, as in Fig. 212, we can form some idea of its origin. At the surface is a dark layer containing the roots of the plants forming the surface vegetation. Below this is a lighter layer, the subsoil, which grades off into the hard rock beneath. Acted upon by the atmosphere, rain, and frost, the upper parts of the rock have been broken into fragments, the smallest particles being nearest the surface and forming the soil. Such a soil has been derived from the rocks below, and its surface layer has been darkened by organic matter, chiefly the decaying remains of plants, which have grown in it. A soil of this kind is said to be **sedentary**.

A section along a river bank is very different. Often down to a considerable depth we find no trace of rock from which the soil could have been formed. The soil is made up of particles varying in size from fine grains of sand to pebbles and even boulders, all more or less rounded and water-

Fig. 212. Section of Soil in a Quarry.

Fig. 213. Soil on Glacial Drift.

worn. Everything is suggestive of an old river bed, and such alluvium has been carried there by water. In many places large areas are covered with similarly mixed materials, but more angular, and whose rocks are of different kinds. These have been deposited by ice-sheets and glaciers, and are known as glacial drift (Fig. 213). The soil in these areas is not derived from the underlying rock, but from material which has been carried from a distance. Such soils are therefore called **transported soils**, and are often very complex in character and liable to vary much even in short distances.

Effect on growth of different soils.—In all cases we find that the roots of plants occur mainly in the dark soil. Why is this ? Do they find more available food there than in the subsoil ? Test this by sowing a few seeds in pots, one filled with dark surface soil, a second with sand, a third with subsoil, and a fourth with clay, and compare the results.

Fig. 207 is from a photograph of such an experiment and shows how differently they have fared, though all the other conditions are the same. Those in the surface soil are sturdy and healthy, in sand they have not grown so well, while those grown in subsoil are very poor and starved, and have evidently been unable to extract from it a suitable amount of food ; the seeds in the wet clay failed to germinate.

Composition of the soil.—As is shown by water-culture experiments, the compounds which form plant food must be soluble, and contain at least the elements oxygen, hydrogen, nitrogen, sulphur, phosphorus, potassium, magnesium, calcium, and iron. But we have seen that a plant obtains carbon from the carbon dioxide of the air in sunlight, and the other elements are absorbed by the roots in the form of compounds such as are used in a culture-solution. Other elements also occur, e. g. sodium, silicon, and chlorine, but some of these are not essential to all green plants. A soil, therefore, for vigorous plant-growth, must contain these necessary constituents, and it is from plant-remains in the soil that much of this food is derived; hence the greater fertility of the surface soil. But surface soils vary considerably both in physical properties and chemical constitution, according to the proportions of sand, clay, lime, humus, or organic matter they contain.

Samples of different soils should be obtained and the properties of the different constituents studied. By means of a few simple experiments many important facts may be discovered.

Take a little garden-soil and weigh out 10 grammes. Spread it out to dry for a few days at the temperature of the room, and weigh again.[1] How much has been lost in drying? Now place the soil on a tin lid, heat it for a short time at 100° C., and weigh again. Has more been lost? Finally, burn the soil and weigh again. How much has been lost by burning?

The part that burns away is organic matter, chiefly decaying remains of plants. Note the change in colour after burning; the dark soil has become 'terra cotta'. Ten grammes of garden soil when air-dried lost 2·56 grammes of water, and a further 0·33 when heated to 100° C., and lost 0·88 of organic matter on burning. What is the proportion of water to the humus in the soil? Determine this proportion in different soils.

Put some garden-soil into a jar of water and stir thoroughly. Note the floating fragments of humus. Pour off the muddy water into a large vessel and repeat the washing until the water clears quickly. Allow the muddy water in the large vessel to stand for a few days and notice how long it takes to settle. Pour off the clear water and

[1] The water-content will vary even in the same soil, being greater on a wet than on a fine day.

compare the two. The material left after washing consists of stones, grit, and sand.

Or put a little soil into a tube, add water, and shake. Allow it to settle and note how the layers arrange themselves. The coarsest are at the bottom, succeeded by layers of finer and finer materials, the water above holding the finest particles for days in suspension, while floating on the surface will be numerous fragments of decaying leaves and stems. The part which settles slowly from the muddy water is very fine and sticky like clay, but dark-coloured with the organic matter. Spread a layer of this on a tin lid and bake it: see what happens. It shrinks and cracks just as clayey ground does in hot dry weather. Now add water to it. Does it regain its original properties? Baking has destroyed its adhesiveness.

Take three glass slips, place on each a drop of water, then a little sand, loam, and clay respectively, and cover with cover-glasses. Examine them under the microscope and note the sizes, forms, and appearances of the particles. Examine humus and peat in the same way, and look for fragments of tissues, e. g. fibres and vessels. Can you find any threads of mould on the decaying parts of plants?

Dark garden-soil consists of small stones, grit, sand, clay, organic matter, and water. It differs strikingly from subsoil in containing much organic matter, and when this is present plants grow well in it. It seems likely, therefore, that from this plants get much of their food.

But before the organic matter is of use it has to be decomposed and converted into soluble inorganic salts, as plants can only take up their food-materials in solution. Hence a fertile soil must contain organic matter, must have the means of decomposing it, and must possess a suitable water-supply to dissolve the salts when formed.

Organisms in the soil and their work.—What are the agents in the soil which act upon the organic matter? Can we isolate them from the soil, or induce them to grow on an artificial food-substance, so that we may examine them? The following experiment enables us to do this:

Sterilize some water by boiling, cover it to keep out dust, and allow it to cool. Thoroughly clean and sterilize, by boiling, seven test tubes and three small dishes (Petri dishes are the most convenient), invert them to drain off the water, and close each tube with a plug of clean cotton wool. Label the tubes A to G and the dishes 1, 2, and 3. Boil a leaf of gelatine in a little water, and with this half-fill tubes A, B, and C. Pour A into dish No. 1 and cover. Take a little garden soil, sterilize part of it by baking, leaving the rest untreated. Half-fill tubes D and E with sterilized water and to D add a little baked soil and shake up. With a sterilized pipette transfer a few drops of the muddy liquid from tube D to E to dilute it. Pour several drops of this into the gelatine in tube B, mix them, pour into dish No. 2, and cover as before.

Using the remaining tubes, repeat this with the ordinary unbaked soil and pour the mixture of gelatine and diluted soil-water into dish No. 3. Cover and set all three aside in the dark for a day. Look at them occasionally and compare the results. How do you account for the differences?

If the experiment has been carried out carefully, Nos. 1 and 2 remain unchanged, but in No. 3 a number of specks appear on the gelatine, which soon increase in size, and form distinct patches or colonies. If a little of one of these be examined under the microscope, myriads of tiny rods or bacteria will be seen moving in the liquid. These, together with the moulds, are the living beings of the soil which decompose the organic matter, and prepare an important part of the mineral food which green plants take up in solution. Culture No. 1 shows they were not present in the boiled gelatine, and No. 2 shows that the bacteria in this part of the soil were killed by baking.

The organisms in the soil are very various and do different kinds of work. Some take up the work where others leave off, and carry it a stage farther. Much oxygen is used up in the process, and in consequence the air in the soil is poorer in oxygen and richer in carbon dioxide than the atmosphere above ; however, the supply of oxygen is kept up by diffusion.

In addition to bacteria and moulds, other and bigger organisms are at work, such as earthworms, millipedes, centipedes, beetles, and the larvae of many insects, all acting on the soil in complex ways. For example, the burrows of earthworms serve as air- and water-channels and make it easier for roots to penetrate ; while the worm-castings contribute much to the fine dark soil on the surface, and render it more soluble.

Nitro-bacteria and the origin of nitrates.—The process called **nitrification** is an excellent illustration of the work of soil-organisms. During the changes which take place in the organic matter in the soil, owing to the action of one set of bacteria, nitrogen and ammonia are formed. The nitrogen escapes into the air and the ammonia is acted upon by another bacterium and converted into nitrous acid. This acid readily reacts with certain mineral substances in the soil (e. g. carbonates) to form salts called nitrites. A third group of organisms converts the nitrous acid into nitric acid ; and the action of the free nitric acid on salts of calcium, potassium, and sodium gives rise to compounds like potassium nitrate (saltpetre), calcium nitrate (lime-saltpetre), and sodium nitrate (Chili saltpetre). These compounds are of great value to plants ; they are very soluble in water, and therefore can be absorbed by the root-hairs ; they are the only source from which most flowering plants obtain their nitrogen.

The organisms which bring about these important changes are called *nitro-bacteria*, and they differ from other organisms in being able to assimilate carbon dioxide in darkness. To carry on their work, they not only need suitable mineral food, but also free oxygen and moisture. Their action is stopped by too dry a soil and by strong sunlight. From this we see why aeration of the soil is essential, and why over-watering, which drives out the air, is injurious.

An exception to the rule as to the source of nitrogen for green plants is found in members of the Pea family (Leguminosae) and a few others such as the Alder and Sea Buckthorn. On the roots of these plants, tubercles are formed (Fig. 256, 3 *tu*) as a result of attacks by bacteria-like organisms in the soil, which enter by the root-hairs and cause the swellings. These organisms fill the tissues of the nodules and are able to take up the free nitrogen of the air and convert this into nitrogenous compounds, which in turn are passed on to the 'host'-plant as food, or given up to the soil when the plants decay. By virtue of this alliance, the nodule-bearing plants are able to thrive in soils deficient in nitrates. Such a union of organisms is called symbiosis (see p. 356). If soil, deprived of nitrates by previous crops, is sown with Clover or other leguminous plants, and the latter ploughed in as green manure, the loss is made good and the land becomes richer. Another exception is found in insectivorous plants, which are able to supplement their supply of nitrogen from animals which they capture and digest in peculiarly modified leaves (see p. 361).

Test for Nitrates.—Dissolve a little potassium nitrate in water, add three or four drops of diphenylamine sulphate, and then a little strong sulphuric acid. The deep blue colour produced indicates the presence of nitrates. Test a sample of garden soil for nitrates. First dry the soil, plug the tube of a funnel with cotton wool, filter paper, or asbestos, and fill up with the dry soil. Carefully pour water on it, and collect a small quantity of the water which drops from the end of the funnel, and test for nitrates as before.

Soils vary greatly in different localities; some are siliceous, and contain varying proportions of grit, sand, and clay. Others are calcareous, and contain varying proportions of

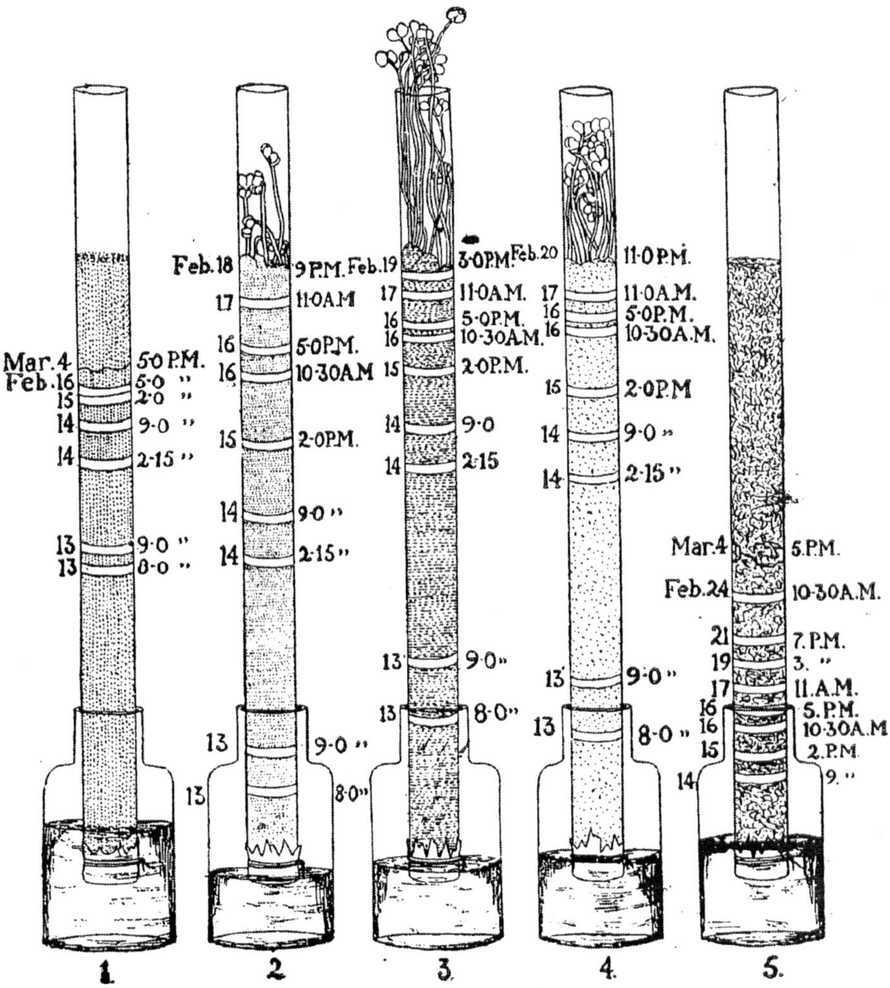

FIG. 214. ABSORPTION OF WATER BY DIFFERENT SOILS.
1, sand; 2, clay; 3, surface soil; 4, subsoil; 5, bracken peat.

lime. There are also great differences in the proportion of organic matter present, and the amount of water they hold. These differences are more or less reflected on the vegetation they support. The plants growing on siliceous soils are often different from those on calcareous soils, and give a distinct aspect to the country. Still there is much overlapping, and very few plants occur which are unable to exist in either kind of soil. The vegetation on deep, wet

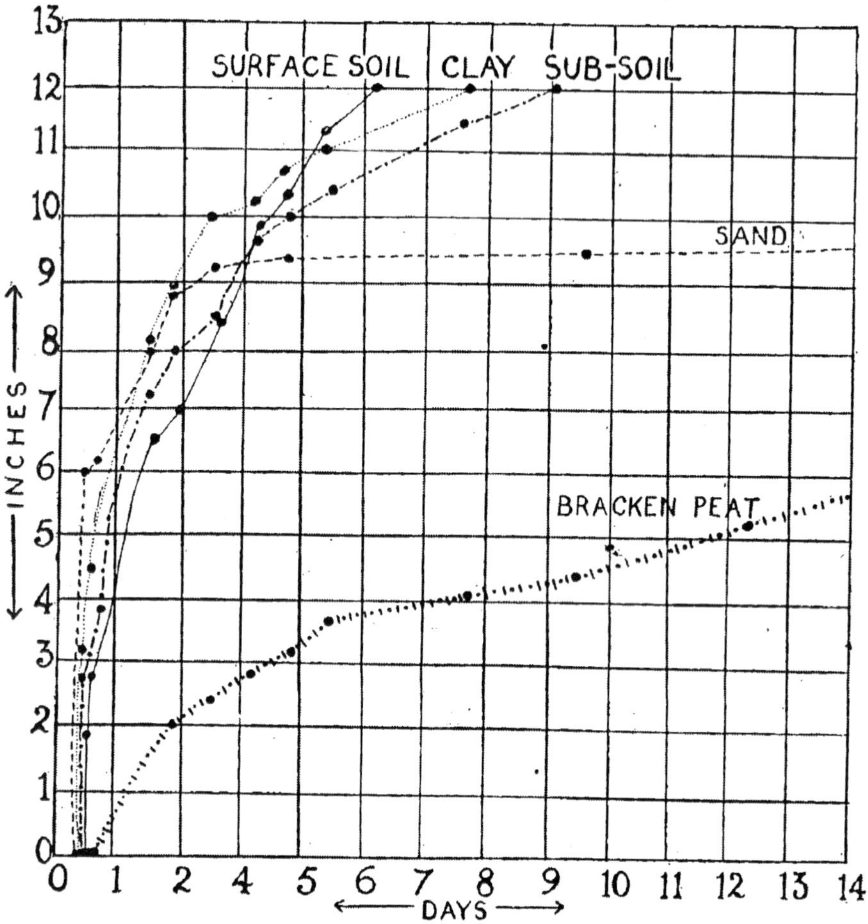

FIG. 215. CURVES SHOWING THE DIFFERENT RATES AT WHICH WATER ASCENDS IN DIFFERENT SOILS.

peat differs from both; this soil is more exclusive, many species being unable to subsist on it.

Properties of soils.—Different kinds of soil should be collected and their properties compared, e.g. sand, loam, clay, lime or chalk soil, humus from a wood, and peat. Take samples of each, dry them, and put them into separate tubes, closing one end with muslin, as in Fig. 214. Tap the tubes several times until the soil has settled down; then put the closed end into water, and note the rate at which the water rises, marking each period with a strip of gummed

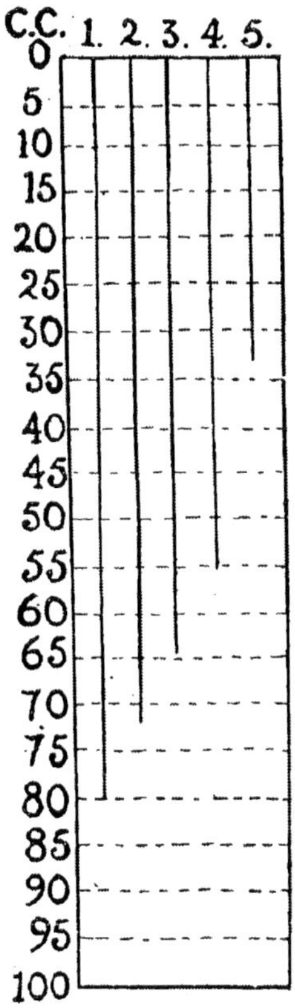

FIG. 216. DIAGRAM SHOWING AMOUNT OF WATER ABSORBED BY THE SOILS IN EXPERIMENT FIG. 214.—1, surface soil; 2, clay; 3, subsoil; 4, bracken peat; 5, sand.

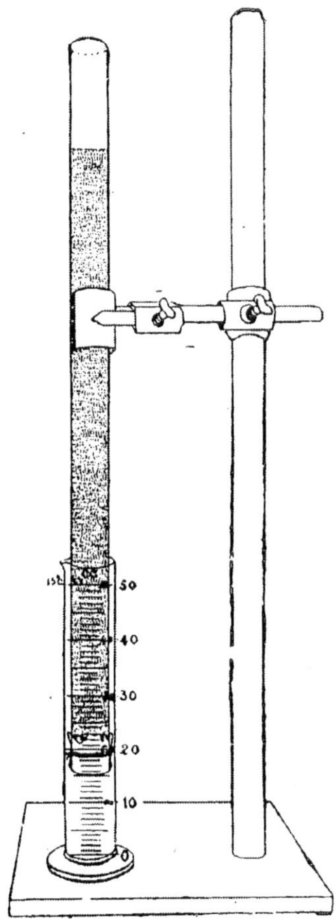

FIG. 217. EXPERIMENT TO DETERMINE THE RATE OF PERCOLATION IN THE SOIL.

paper. Sow ten mustard seeds at the top of each tube of soil and note when they germinate and how they grow in the different soils.

Does the water rise at an equal rate in each tube? In which case does it rise highest? How does the proportion of humus affect (1) the height to which the water rises and (2) the water-holding capacity of the soil? Continue the records for several days and plot the results in curves as in Fig. 215. Notice how very slowly water ascends in dry peat in spite of the latter's great capacity for water.

A similar experiment with dry cotton-grass peat will show that water may rise in the column only $5\frac{1}{2}$ inches in thirty-nine days. The diagram, Fig. 216, shows the amount of water absorbed by the different kinds of soil in this experiment.

Half-fill another set of tubes with similar soils, but in these cases add 50 c.c. of water from above. Note and mark at short intervals the rates of percolation (Fig. 217) of (1) fine sand, and (2) fine sand mixed with humus. Collect the water that escapes at the lower end of the tubes and notice in each case (1) rate of percolation, (2) time of appearance of the first drop of water from the bottom of the tube, (3) amount of water which escapes in a given time, (4) amount of water held by the soil, observing again the effect of humus on the water-content of a soil. Note that sand is made more coherent and less pervious when mixed with humus. If we consider these experiments and observations in connexion with our previous ones, we are able to understand why dark soil is better for plants than subsoil.

Make similar observations on cotton-grass peat. Note that it is greasy to the touch. Test samples with litmus paper and determine whether it is acid, alkaline, or neutral. How do peat and humus compare in these respects? Peat is acid, humus is neutral or alkaline.

Weigh out 10 grammes of fresh bracken peat, dry it, and determine the water-content; then burn it and weigh again. How much is lost by burning? Ten grammes of bracken peat contained 66·2 % water, 17·4 % organic matter, and 16·4 % mineral matter.

Peat, like clay, holds much water; it is therefore badly aerated, and root-respiration is difficult. Further, it contains little mineral food and a great excess of organic matter; and in addition it is acid or sour. No wonder, then, that it is a poor soil for plants, or that so few species grow on it.

Calcareous soils.—**Chalk and Limestone soils.** In a soil derived from chalk or limestone, carbonate of lime is always present. These soils support a vegetation differing in many respects from that growing on soils deficient in lime. The presence of lime in soils may be determined as follows:

Take 5 grammes of chalk soil, boil it in water, and pour off the liquid. Repeat this two or three times and so remove the soluble calcium salts (sulphate, chloride, and nitrate). Drain off the water or press between filter paper, and to the residue add dilute hydrochloric acid (1 in 10 parts of water). Note if there is any effervescence; if so, it denotes the presence of a carbonate, e. g. of magnesium, lime, or iron. If the gas is copious it may be led into lime-water and tested. What gas will it be?

Filter off the insoluble matter and add ammonium hydrate in slight excess to the filtrate. Filter off any precipitate which may result (e. g. iron or alumina) and add ammonium oxalate. Is a white precipitate formed? If so, it is calcium oxalate. Thus it is shown, not only that the soil contains a carbonate, but which specific one it is, namely, calcium carbonate.

Rain-water carries with it into the soil carbon dioxide, which has acid properties, and which, acting on the calcium carbonate, dissolves it, so that a soil over limestone may become relatively poor in lime, except for fragments of limestone rock in the soil. This denudation of mineral salts from a soil, and its consequent impoverishment, is called 'leaching'.

Effect of liming clay soils.—The liming of soils, especially in clayey districts, is common. Why is this done? Clayey soils hold much water, and are therefore badly aerated, and organic acids formed from the decay of plants tend to accumulate and make the soil 'sour'. The addition of lime lessens these defects. Two very simple experiments will show us the effect of lime on clay:

(1) Stir up a little clay in water; the very fine particles remain indefinitely in suspension. Add to the muddy liquid a little lime-water and note its effect. The particles run together in fluffy-looking masses and soon sink to the bottom, leaving the liquid clear.

(2) Bend two pieces of gauze each into the shape of a saucer or shallow cup and line them with wet clay. Pour lime-water into one and tap-water into the other, and place each on the top of a tumbler. Does the clay hold both liquids equally well? The tap-water does not percolate, but the lime-water permeates the clay readily and changes its properties. It becomes less sticky; it has clotted and become more porous, and it is no longer able to hold water. Other salts and mineral acids have a similar effect, and as lime also neutralizes the organic acids in a sour soil, we see that its effects are threefold: it renders the stiff, unworkable clay more workable, improves its drainage, and, as the farmer says, 'sweetens it' by neutralizing the organic acids. Clay is also rendered more pervious by mixing with humus.

State of the water in the soil: capillarity.—**How water ascends in the soil.** Previous experiments have shown that water ascends higher in loam than in sand, and higher still in clay. Microscopic examination shows that, of the three, sand consists of the coarsest particles, clay the finest, while loam is intermediate. The larger the grains, the greater will be the spaces between them. Will this in any way affect the power of a soil to absorb water, and if so, how?

Make a few fine tubes of different diameters by heating a piece of soft glass-tubing in a flame and drawing out portions to the thickness desired. Dip the ends of these into a coloured liquid and note how high it rises in each case. The liquid wets the inside of the tube and creeps up it, forming a concave surface film. The pull which this film exerts is proportional to the line of contact with the tube. As we reduce the diameter of the tube, we not only reduce its capacity, but at the same time reduce the pull, though we see that in such a narrow tube it will support a much higher column of water. In a wide tube the water rises very little. This property of the ascent of water in narrow tubes is called ' capillarity '.

Apply this principle to the different soils. Sand with its coarse grains and large spaces will correspond to the widest tube, loam to the intermediate one, and clay to the finest. The capillarity of humus is greater still.

From these observations we see that water is present in soils in different states :

(1) That which fills the free spaces, and, when in excess, displaces the air.

(2) That which is still retained in air-dried soil, and may be driven off at a high temperature.

(3) What may be called capillary water, which forms films over the particles in the soil and on the roots of plants, behaving similarly to that noticed in the capillary tubes.

The first readily drains away in a loose permeable soil ; the second, which is called the hygroscopic water, is held firmly by the particles and is of no use to plants ; the third is the most important, because it is able to move in the soil from wet to drier parts in any required direction, so that the effect of absorption by root-hairs is to provide a space into which more water is drawn by capillarity. By this means, water may be drawn upwards in the soil eight to ten feet.

In the case of trees with a wide-spreading root-system and very large leaf-surface, the amount of water drawn from the soil and given up to the atmosphere as vapour is enormous, and has an important effect in drying the soil and in increasing the humidity of the air. It is estimated that, during one season, a Beech wood gives off 354 tons of water per acre, and that suitable trees planted in marshy ground play an important part in draining it and bringing it ultimately into a state suitable for higher cultivation.

If the mineral salts needed by plants are in weak solutions, how is it that after heavy rains the soluble compounds and even added manures are not washed out of the soil? To some extent this does occur, as may be determined by noting the difference of residue after evaporating (*a*) rain-water, and (*b*) spring-water. We have seen (p. 326) how very soluble are the nitrates in the soil and how easily they are washed out. But chemical changes are constantly going on in the soil, which counteract this tendency to depletion of food-materials. By means of these changes soluble compounds like salts of potassium, magnesium, calcium, or ammonia displace the alkali in the silicates of the soil and insoluble compounds are formed, and for a time fixed in the soil. The next changes result in these insoluble compounds being slowly acted upon and re-converted into soluble substances which provide a steady supply of mineral food for plants.

In this manner the soil is constantly storing up plant-food in an insoluble form, preventing its escape, and then giving it up slowly to the plants in a soluble form. It is important, however, that the solutions should be very weak, for if they exceed a concentration of 3 % the plants are unable to absorb them.

Effect of hoeing.—Hoeing produces a loose, dry, well-aerated layer of soil which conducts heat badly and protects the soil below. At the same time it reduces evaporation, so that a hoed soil is cool and moist and therefore better adapted to plant-requirements. When the ground is covered with vegetation this receives the sun's heat, the soil is screened, and its temperature does not rise so much as that of bare soil.

Factors affecting the water-supply in the soil.—From our previous observations we have seen that any factor which tends to modify the water-supply of a plant will affect its growth. The more important factors are :

1. *Temperature.* A high temperature favours rapid transpiration, a low temperature renders root-absorption difficult. Much more heat is absorbed by a dark soil than a light soil. Water is a bad conductor of heat, and a wet soil is a cold soil.

2. *Air* is important to plants in several ways :
   (1) Its temperature determines the rate of absorption.
   (2) The amount of moisture present determines the rate of transpiration.
   (3) The oxygen of the air is necessary for respiration both by roots and shoots, and aeration of the soil is essential.
   (4) Carbon dioxide is necessary for photosynthesis by green plants.

3. *Wind.* Moving air, by increasing the volume affecting a given surface, increases evaporation ; hence the drying effect of winds. The rate of transpiration is great in plants having large thin leaves with stomata on both surfaces, while it is reduced to a minimum in plants with small rolled leaves with stomata only on the enclosed surface, and therefore in a 'still air' chamber.

4. *Precipitation.* As plants must obtain their mineral food in a weak solution, water is essential as a solvent

and diluent. It is obvious, therefore, that a suitable water-supply is the most important edaphic or soil factor. Water carries into the soil gases from the air, one of which, carbon dioxide, gives to water the power of dissolving part of the mineral matter of the soil, e.g. potash and lime. On the physical character of the soil depends the amount of water available for plants. The rainfall of a district will play an important part in determining the character of the vegetation.

We generally find that plants growing in a sour, wet, cold soil have peculiarly modified shoots. The leaves are reduced, and either up-rolled or back-rolled ; the cuticle is thick ; the stomata are sunk in grooves or pits ; and similar devices occur which serve to reduce transpiration. Plants growing in a wet soil may possess structures characteristic of those found in habitats liable to periods of drought. If the soil, though it contains much water, is too acid, too cold, or if the water is otherwise rendered difficult to absorb, it is said to be ' physiologically dry '.

CHAPTER XXVII

PLANTS OF HEDGEROWS AND WALLS

Uses and distribution of hedgerows.—Hedgerows provide endless material for the study of plants and offer numerous problems for solution. They are, however, fences introduced by man and not a natural feature of the vegetation of a country. Hedges are useful in many ways. They protect the crops and surface soil from the drying and tearing effects of the wind; they lessen the rate of evaporation, and in consequence help to maintain a higher temperature. To render them more effective, spiny shrubs like the Hawthorn are generally selected, and large trees such as Oak, Elm, Sycamore, Ash, and other forest trees are introduced at intervals to provide shelter for the cattle. So numerous are the trees that, when viewed from a distance, their appearance is that of an open wood. Under the shrubs and trees wild plants establish themselves, especially woodland species, which are well suited to the shade of the hedgerow.

Hedges are not of universal occurrence. They are absent from the uncultivated, sandy wastes and salt-marshes of our shores, and they are not found on our moorlands and mountains. They accompany man in his farming operations, but even here they have a peculiar distribution. In areas over sandstones and limestones, where the soil is shallow and the ground often stony, they are replaced by stone walls. On the other hand, in wet, low-lying fen-districts the 'fences' commonly take the form of ditches and drains. It is in better-drained areas and often over deeper soils that hedgerows predominate. Hence their geographical distribution is significant. Hedgerows are also essentially English, and give a character to the landscape without parallel in any other country.

Habitats of the hedgerow.—In a common type of hedgerow the shrubs and trees are planted on a bank, below which is either a ditch or a moist hollow, and beyond this is a grassy sward. Thus in a very limited area are the following distinct habitats: (1) Under the shrubs, shade, protection, and a soil containing much humus. (2) On the hedge bank, a well-drained, more exposed slope. (3) At the bottom either a ditch filled with water or a wet hollow. (4) A drier, flat, miniature meadow. In each situation characteristic plants occur, showing many interesting biological features. Note how numerous are the plants with

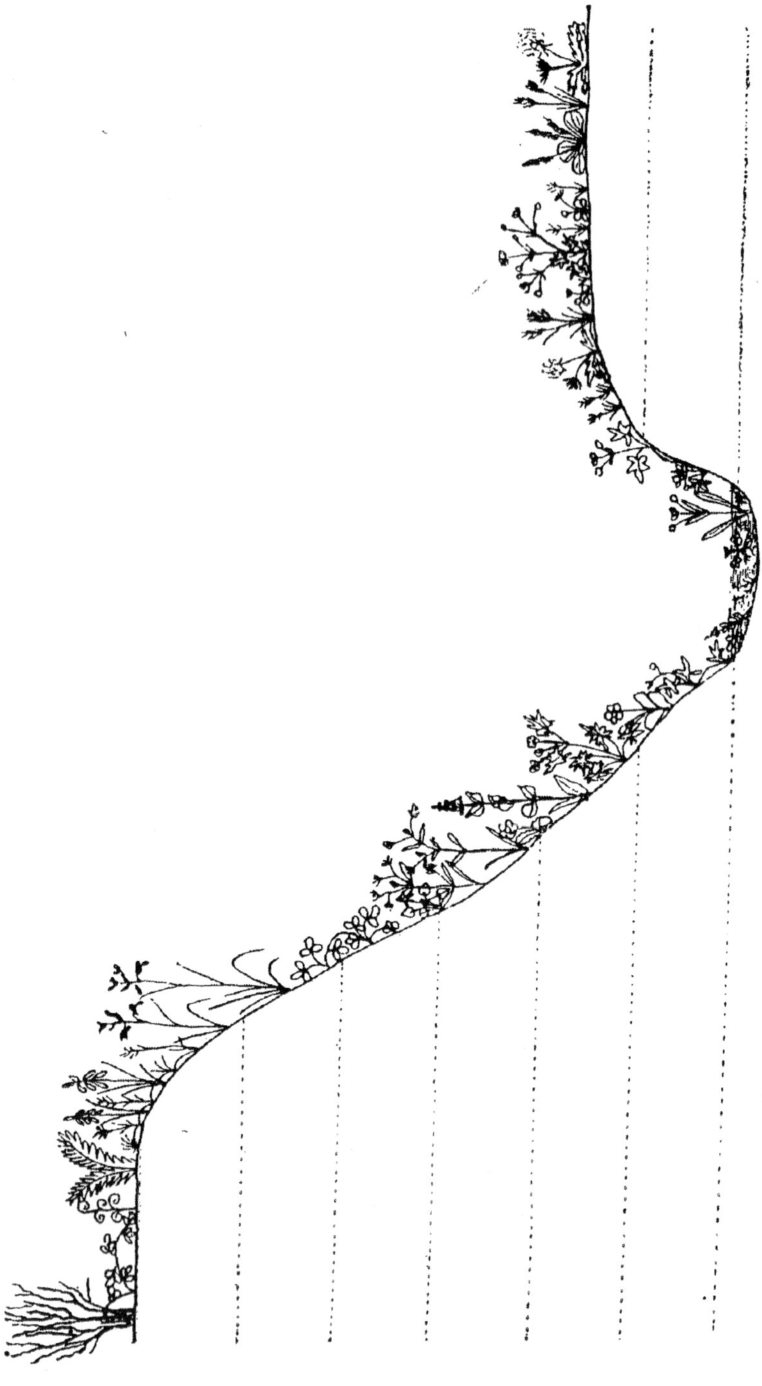

FIG. 218. TRANSECT THROUGH HEDGE, DITCH, AND SWARD.

climbing organs; also those with fruits dispersed by animals, especially birds.

Examine such a hedgerow and draw to scale a transect passing through the above-mentioned zones and indicate on it the species met with, as in Fig. 218.

Make lists of the species found in each zone and carefully compare the different forms.

Study the changes that occur throughout the year, e.g.:

Winter.—The deciduous and evergreen habits; branch-systems; bark of trees; winter-buds; protection; leaf-scars.

Spring.—Nature of bud-scales; opening buds; folding of leaves and their modes of growth; spring tints; characteristics of spring flowers; wind-pollination; plants with storage organs, tuberous roots, rhizomes, bulbs.

Summer.—Leaf-mosaics; comparisons of stems and leaves; flowers and their insect-visitors.

Autumn.—Fruits and fruit-dispersal; autumnal tints; leaf-fall; roads as highways for fruit-dispersal.

Many examples of organs adapted for special purposes will be found, and the following should be studied:

(1) Shrubs with thorns and prickles:
- (*a*) branch-spines—Hawthorn (Fig. 219, 1), Blackthorn, Gorse.
- (*b*) prickles—Roses, Brambles, Gooseberry (prickles on the leaf-base, Fig. 95).
- (*c*) leaf-spines—Holly and Barberry[1] (Fig. 219, 2, 3, 4).

[1] The Barberry has two kinds of shoots: (*a*) long shoots bearing leaves reduced to branched spines (Fig. 219, 2 and 4, *l.s*); in the axils of these arise (*b*) dwarf shoots (Fig. 219, 2, *d.s*) bearing several simple foliage-leaves, each having a joint near the base (Fig. 219, 3, *j*).

(2) Plants with nectaries on their leaves (extra-floral nectaries): Guelder Rose (Fig. 219, 5), Bird Cherry, Wild Cherry (Fig. 219, 6), and Bracken.

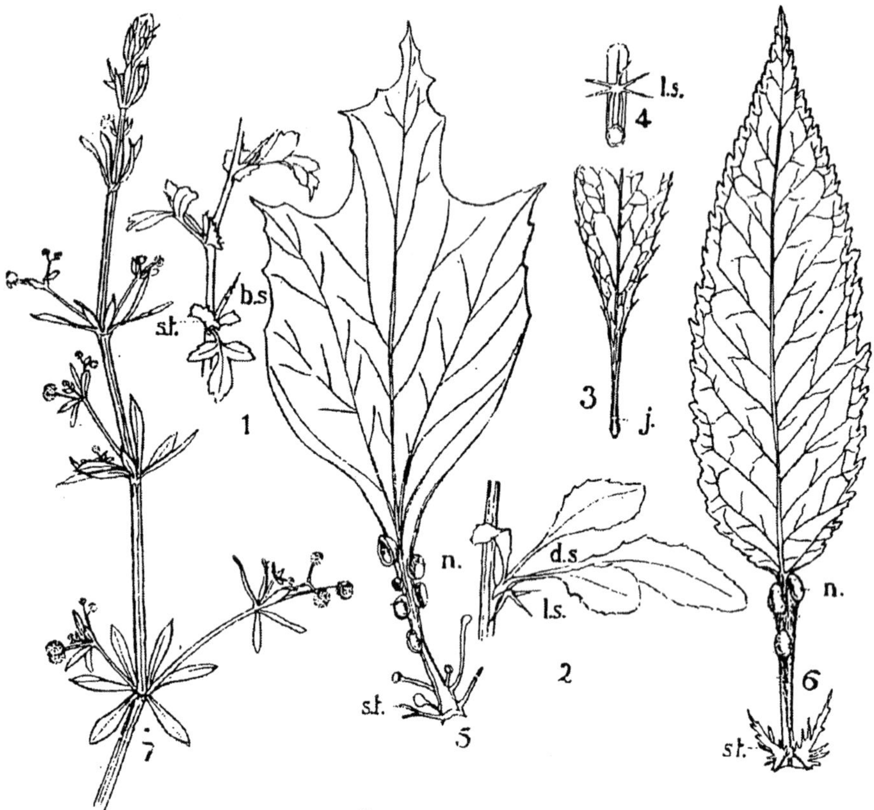

FIG. 219. MODIFIED SHOOTS AND LEAVES.—1, Hawthorn; 2, Barberry; 3, base of Barberry leaf; 4, spiny leaf on long shoot of Barberry; 5, leaf of Guelder Rose; 6, leaf of Cherry; 7, shoot of Cleavers with hooked fruits; *b.s*, branch-spine; *d.s*, dwarf shoot; *j*, joint; *l.s*, leaf-spines; *n*, extra-floral nectaries; *st*, stipule.

(3) Climbing plants:

These are very numerous, and all the main forms may be found and their modes of growth observed.

(A) Hook climbers or scramblers. Brambles, Cleavers (Fig. 219, 7).

(B) Root climber. Ivy (Fig. 222, 1).

FIG. 220. WOOD VETCH CLIMBING BY LEAFLET TENDRILS.

FIG. 221. WALL PENNYWORT.

(C) Twining stems, twining either
- (*a*) to right or left indifferently—Woody Nightshade. (Also by means of its leaf-stalks.)
- (*b*) clockwise (i.e. left to right or with the sun) —Hop, Black Bryony (Fig. 90), and Honeysuckle.
- (*c*) Contra-clockwise (i.e. right to left or against the sun)—Bindweeds (Fig. 89).

(D) Sensitive organs (tendrils):
- (*a*) Branch tendrils—White Bryony (*Bryonia dioica*) (Fig. 91).
- (*b*) Leaf-stalk tendrils—Clematis (Fig. 92).
- (*c*) Leaflet tendrils—Wood Vetch (Fig. 220) and Climbing Fumitory.

The Ivy is well adapted to the conditions of life in a hedgerow or wood, and possesses many points of interest. If its long slender branches spread out on the ground the leaf-blades face upwards. Commonly the branches climb the trunks of trees by pressing their groups of adventitious roots into irregularities of the bark, where they adhere firmly and serve as holdfasts (Fig. 222, 1). The leaves arise spirally on the axis, and in their broad bases the buds are partly embedded. The leaf-bases are sensitive motile organs, and a very slight movement in this region carries the blades at the ends of their long stalks through a wide arc away from the shade towards the light; by further movement at the top of the leaf-stalk the blades are so placed in relation to each other as to form good leaf-mosaics, and the tips all point downwards. The upper surface of the blade is glossy and concave, and the drainage-channels are directed towards the apex, which serves as a 'drip-tip'; thus rain and snow quickly drain away. Observe the behaviour of the shoots which overtop the tree-trunk or a wall and thus lose support and shade. On such shoots adventitious roots are not formed; the leaves arise in five rows as before, but being illuminated on all sides the bases do not bend, and the blades are not all directed to one side (Fig. 222, 3); they are also smaller and simpler in outline. In October or November these shoots produce

FIG. 222. IVY.—1, climbing shoot; 2, section through leaf-base, showing embedded bud; 3, flowering shoot; 4, flower-bud; 5, flower seen from above; 6, flower in side view; 7, vertical section of flower; 8, fruit; *a.r*, groups of adventitious roots; *b*, bud; *d*, disk; *l.b*, leaf-base; *p*, petals; *s*, sepals; *st*, stem.

umbels of flowers (3), which possess several peculiar features. Often one or two flowers are formed before growth of the umbel ceases, and these are left behind on the axis. The flowers (4, 5, 6, and 7) have five sepals, which are so small that the five green petals which protect the inner organs may be mistaken for a calyx; five stamens alternate with the petals and are proterandrous. The pistil is inferior, and consists of five carpels enclosed in a disk which becomes fleshy; the fruit is a berry (8).

Woodland species are so numerous in the hedges that hedgerows may be regarded as linear extensions of the woodland flora. The more typical herbaceous species are Bracken (*Pteris aquilina*) and other ferns, Soft-grass and other woodland grasses, Bluebell, Garlic, Purple Orchis, Dog's Mercury, Moschatel, Anemone, Wood Violet, Wood Sorrel, Primrose, Cowslip, Stitchwort, Herb Robert, Dead-nettle, Hedge Woundwort, Foxglove, Crosswort, Jack-by-the-hedge, Chervil, and Nettle.

Invaders from the meadows are represented by the Daisy, Yarrow, Clover, and cultivated grasses; while wind-dispersed composites like the Dandelion, Groundsel, Goat's-beard, and Thistles are also common.

Parasites are frequent, and we may find the strange-looking Toothwort, which lives parasitically on the roots of trees and shrubs; also Broomrapes on the roots of herbaceous plants (p. 359), or the Dodder (Figs. 230 and 231), twining round and absorbing food from the stems of Gorse or Clover.

The vegetation of the ditch varies with the water-supply. If much water is present, aquatic and marsh species occur, similar to those found in a pond and on its margin. Common species are Mud Crowfoot (*Ranunculus Lenormandi*), Lesser Spearwort (*R. Flammula*), Marsh Marigold (*Caltha palustris*), Water Cress (*Radicula Nasturtium*), Bog Stitchwort (*Stellaria uliginosa*), Water Blinks (*Montia fontana*), Water Starwort (*Callitriche stagnalis*), Square-stemmed Willow Herb (*Epilobium tetragonum*), Water Dropwort (*Oenanthe crocata*), Brooklime (*Veronica Beccabunga*), and Floating Mead-grass (*Glyceria fluitans*).

In addition to the above, many other problems may be studied in a hedgerow, e.g. how patches of bare soil or disturbed ground become covered by plants. The first to invade and colonize the ground are annuals with good seed-dispersal mechanisms; these are in turn succeeded by perennials, many of which have effective means of vegetative propagation. The effect of shade and of exposure on plant-distribution is obvious when we note how numerous are the species on the sunny side, and how few on the shady side, of a hedgerow. Similar differences may be seen in hedgerows on opposite sides of a road running east and west.

Walls

In some districts, walls become so overgrown that, when seen from a distance, they resemble hedgerows. This is often the case in areas covered with glacial débris and where stones of various shapes and sizes are abundant in the soil. When the land is brought under cultivation, the stones are removed and used for building walls, often of great thickness. The interstices are filled with small stones and earth, forming a soil which is very porous, and from it the water soon drains away. Seeds which are deposited in the crannies by the wind or animals may germinate, but as the plants grow they are liable to suffer periodically from drought. It is usual to find, therefore, that plants which succeed in such a habitat are provided with devices to reduce transpiration; the leaves are either very small or arranged in rosettes, and often modified as water-storage organs. Many of the species are familiar as rockery plants, e.g. fleshy-leaved forms like the Stonecrops, Saxifrages, and Wall Pennywort (Fig. 221). Heaths, Hair-grass, Gorse, and Broom have reduced and wiry leaves. The Ivy-leaved Toadflax has slender trailing stems and fleshy leaves; the stalks of its ripening fruits bend away from the light, lengthen, and carry the capsules into the crannies, where the seeds are shed. The Ferns found growing on the wall are all hardy kinds, e.g. Wall-rue, Black Spleenwort, and Polypody.

CHAPTER XXVIII

WOODLAND PLANTS

Features to observe in the study of a wood.—The vegetation of woods varies considerably in different localities, according to elevation, slope, and aspect, nature of the soil, and water-supply, but in all cases the striking species are the trees. In the study of woodland plants an attempt should be made to answer the following questions and to make the following observations:

(1) Which is the dominant species of tree: i.e. which species is it that on the whole makes its influence most apparent?

(2) What are the subordinate trees?

(3) Does their arrangement or the character of the species suggest artificial planting, or have the trees grown spontaneously and formed a natural wood?

(4) Look for seedlings, and see which trees are reproducing themselves from seeds.

(5) What are the dominant species of the ground flora?

(6) Is their distribution influenced in any way by the overshadowing trees?

(7) Compare the shade produced by the different species of trees, especially that of Beech, Elm, Pine, Oak, Common Ash, and Birch.

(8) Is the vegetation the same under trees with a close canopy, like the Beech and Elm (shade-endurers), as under trees with an open canopy, like the Oak, Ash, or Birch (light-demanders)?

(9) Compare the parts closely planted with the more open parts.

(10) What is the nature of the soil, (*a*) siliceous or calcareous, (*b*) coarse, stony, shallow, and dry, or fine-grained,

deep, and moist ? (*c*) Is the soil covered with humus, and if so, to what depth ?

(11) What is the aspect ? and how is drainage affected by slope, dip of the rocks, or other causes ?

(12) If differences are met with in these respects, do you find accompanying changes of species ?

(13) If the wood contains a wet hollow or if a stream runs through it, compare the plants (both trees and undergrowth) of the wet parts with the drier parts of the wood.

(14) Compare the undergrowth of the wood with that of adjacent fields. To what extent are the species similar ?

(15) How will such a comparison help to decide between (*a*) a recent plantation and (*b*) an old wood ?

Select a typical piece of woodland and mark out a narrow strip across it, as indicated in Fig. 223, 1, A to B. This may be subdivided into convenient squares and the parts studied along the lines indicated. Afterwards a more extended study may be made, and the results recorded on comparative maps, as in Fig. 223, 1, 2, and 3, of which 1 is a tree map, 2 a map of the common plants of the undergrowth, and 3 a soil map.

From studies of this kind, we learn that, by a combination of factors such as increased shade, a moister atmosphere, more humus in the soil, a more regular water-supply to the plants, and protection by the overshadowing trees, an assemblage of plants is met with in a wood which differs in many respects from the adjoining vegetation. We shall also find that if trees are planted on pasture-land, the undergrowth for a time will consist of pasture-species, but these will eventually give place to species which can better endure the shade, and the ultimate flora will be determined by soils, water-supply, and the other above-named factors.

Dry and moist Oak woods.—Fig. 223, 1 and 2, are maps showing the vegetation of a wooded escarpment with siliceous soils. Fig. 223, 3, is a soil map of the same wood. Note the form of the escarpment as indicated by the

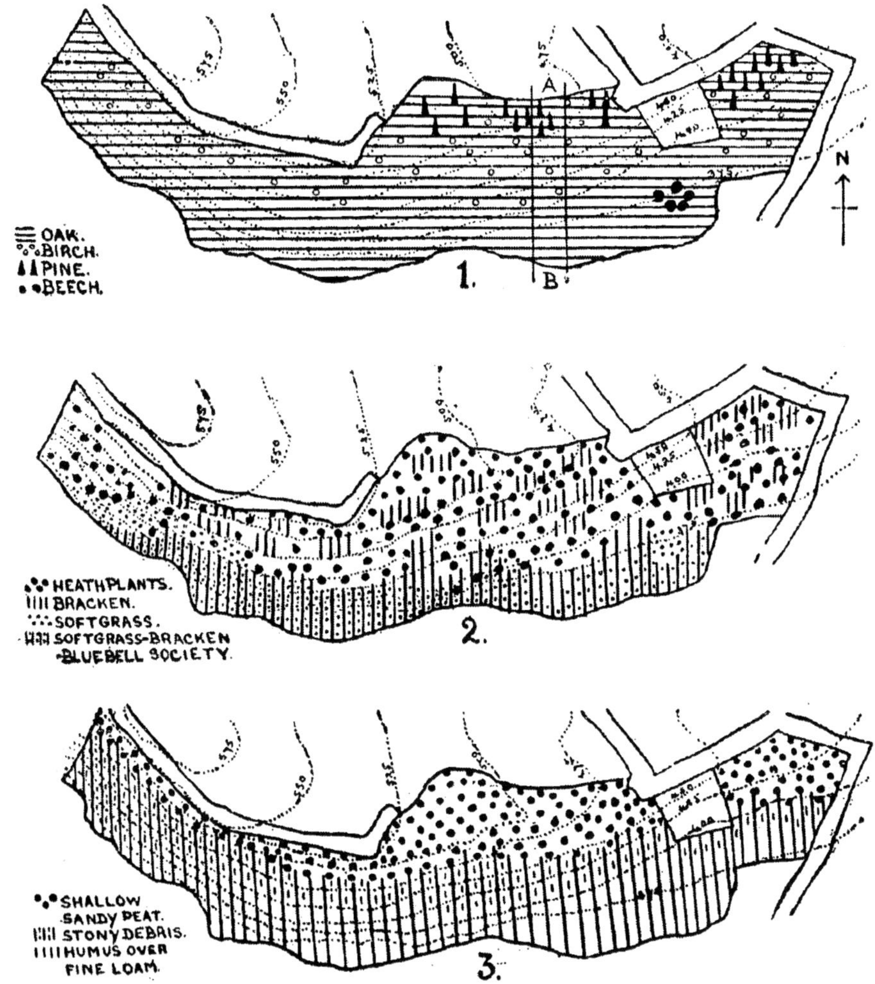

FIG. 223. COMPARATIVE MAPS OF A WOOD.—1, showing distribution of trees; 2, undergrowth; 3, soils.

contour lines. The highest part of the wood has a shallow soil of sandy peat resting on millstone-grit sandstone; the latter is coarse-grained and jointed, and thus permits rapid drainage (see Fig. 212). The steep slope is covered with stony débris and washed-down material from the weathered sandstone edge. This débris rests upon a bed of fine-grained impervious shales, but the steepness of the slope secures good drainage. The lowest part, beyond the influence of the weathered sandstone, has a deeper and moister soil formed from the shales, and is covered by humus several inches in thickness.

The dominant tree is the Sessile Oak, the subdominant ones Birch and Scots Pine, with occasional trees of Mountain Ash, Holly, and Hawthorn. There is also a small group of Beeches. In the upper part the trees are small, produce little shade, and the undergrowth consists of Ling, Bilberry, and Hair-grass. Bracken, which is also abundant, has thick hard leaves. A little lower, on the stony slope, Ling is less abundant, Hair-grass and Bracken are dominant, and there is also Bedstraw, Tormentil, Cow-wheat, &c. In the lowest part, where the soils are deeper and moister, the heath-plants give place to a society of Soft-grass, Bracken, and Bluebell, the Bracken here, however, having thinner and larger leaves than that of the upper part. In a wet hollow in an adjoining portion of the wood occur such trees as Alder, Willow, and Bird Cherry, as well as the characteristic trees of the wood, while the ground flora consists of Marsh Marigold, Bitter Cress, Lady's Smock or Cuckoo Flower, Greater Stitchwort, Marsh Stitchwort, Marsh Violet, Hogweed, Sanicle, Yellow Deadnettle, Narrow-leaved Thistle, Garlic, Rushes, Lady and Male Ferns, and twenty other species.

We thus see that in the drier, well-drained parts, the species are fewer, and often possess small, up-rolled or back-rolled leaves and other modifications, which enable them to withstand drought. But when the water-supply is constant, many more species are able to exist; and they possess no such extreme forms as do the species characteristic of the drier and more exposed parts. We observe,

FIG. 224. OAK WOOD IN SUMMER.

FIG. 225. OAK WOOD IN SPRING.—Ground flora of Soft-grass, Bracken, and Bluebell.

FIG. 226. THE SULPHUR TUFT TOADSTOOL (*Hypholoma fascicularis*) GROWING ON DECAYING WOOD.

also, that while some species are restricted in their distribution, others (e.g. Bracken) are much more plastic, and persist in very varied habitats.

In the typical moist Oak woods on siliceous soils the above-mentioned species, along with many others, occur, and the ground flora is often a bright flowery carpet.

Ash woods on calcareous soils.—In woods on calcareous soils the ground flora is still more varied. Oaks are rare, the Common Ash tends to occupy the first place, accompanied by shrubs like the Hazel, Wayfaring Tree, Spindle Tree, White Beam, Buckthorn, Dogwood, and Privet. Among the herbaceous plants the Dog's Mercury is a very abundant social species covering large areas. Others common or frequent are Primrose, Cowslip, Wood Crane's-bill, Blood Crane's-bill, Avens, Strawberry, Great Burnet, Stone Bramble, Spurge Laurel, Hellebore, Small Scabious, Lily-of-the-Valley, Solomon's Seal, several Orchids (e.g. Tway Blade, Helleborines, and Purple Orchis), False Brome-grass, and Brittle Bladder Fern. The Bracken is usually inconspicuous and often absent. Thus Oak woods on siliceous soils and Ash woods on calcareous soils form two well-recognized types of woodland.

As we have seen, woodland species extend along the hedgerows, and reference should be made to the groups of plants mentioned in Chapter XXVII for further examples, especially of climbing and scrambling plants, whose natural habitat is the woodland.

If we take a general view of the plants in a wood we see at once that they tend to occupy successive layers. The highest is the tree layer, below which is a layer of shrubs, and lower still are layers of tall, intermediate, and low-growing herbaceous plants (Fig. 224).

With such a succession of overshadowing layers it is obvious there must be considerable accommodation among the different species. To watch these layers throughout the seasons forms an interesting study. In the winter, tree and shrub layers stand out in sharp contrast, as in the case of leafless Oak and evergreen Holly. Still more striking is the spring aspect of Ash and Hazel, the early bright-green foliage of the latter being very conspicuous against the grey branches and black unopened buds of the Ash. The green winter carpet of grasses and mosses is followed in the spring by low-growing plants (Fig. 225) like the Bluebell, Anemone, Primrose, Cowslip, and Celandine, to be succeeded in the summer by the taller-growing Bracken and other ferns. Meanwhile, the leaves of the trees are expanding, and by their ever-deepening shade, protect the tender plants of the ground flora.

Complementary societies.—The diagram (Fig. 227) will help us to appreciate the significance of such adaptations in the ground flora of an Oak wood where the three plants illustrated commonly grow in close association. In the loose leaf-mould on the surface run the rhizomes of the Soft-grass. Beneath this, in the dark soil containing much humus, are the rhizomes of the Bracken, while in the fine yellow loam below are the bulbs of the Bluebell, though young bulbs on their way downwards may be found in the other two layers. In November the young green blades of the Soft-grass appear, and through the winter and early spring form a bright green carpet. Meanwhile, the leaves of the Bluebell come above ground, to be followed in the early spring by a wealth of flowers (Fig. 225). Towards the end of the flowering period, and as the fruits are ripening, the Bracken unfolds its fronds, and raised on tall stalks above the tops of the young Bracken are the flowers of the Soft-grass. In the late summer and autumn the mature Bracken-fronds form a continuous cover. Eventually they die down to form a warm winter carpet. Thus their soil-requirements, their modes of life, their periods of active vegetative growth, their times of flowering and fruiting, are for the

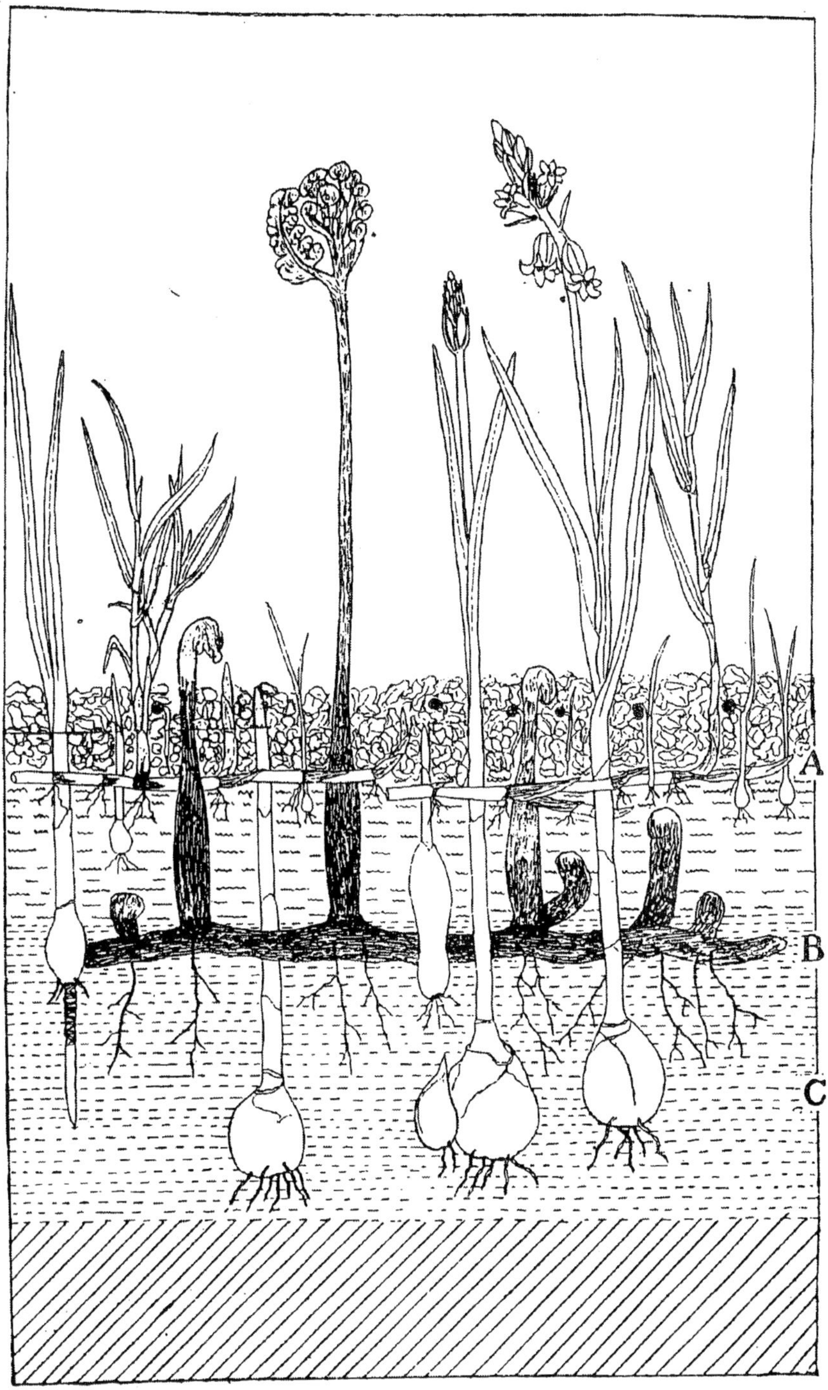

FIG. 227. COMPLEMENTARY SOCIETY IN AN OAK WOOD.
A, Soft-grass; B, Bracken; and C, Bluebell.

most part different. Species growing mutually together in this way form a **complementary society.** Reference to the maps (Fig. 223, 1 and 2) will show that these three species do not always grow together. In places you will find that one may grow to the exclusion of the others. Under the deep shade of the Beech, the Bracken seldom occurs, but the Soft-grass is frequent, and sometimes the only species is a weak form of Bluebell. The latter also occurs on the stony shallow soil among the Hair-grass.

Types of British woodlands.—There is very little natural woodland in the British Islands. Plantations are numerous, and some of these, being on the sites of native woods, preserve many of the features of primitive forest. The woodlands of Britain may be divided into two main groups, namely:

(1) *Woodlands on siliceous soils*: clay, loam, and sand.

- (*a*) Alder-Willow wood: a lowland type with a ground flora of marsh-plants;
- (*b*) Pedunculate Oak wood, with a flowery carpet of moisture-demanding species;
- (*c*) Sessile Oak wood, on drier, often sandy, soil: a type of which has been given above (Fig. 223);
- (*d*) Oak-Birch-Heath wood, with a ground flora of heath-plants;
- (*e*) Birch wood: characteristic of the northern uplands, in which the Birch is the dominant tree;
- (*f*) Pine wood: this type differs from the above deciduous woodlands in being composed mainly of evergreen coniferous trees.

Native Pine woods occur in Scotland, and were formerly extensive in England. The ground flora is usually of the heath type; and seedlings of the native Scots Pine often develop freely on heather moors, both lowland and upland, and form a Pine-Heath wood. Many of the present Pine woods are plantations, and exotic conifers are commonly planted, e.g. Spruce, Douglas Fir, and Larch; the latter, however, has deciduous leaves (see p. 275). In addition to the heath-plants the following interesting species occur in old Pine woods: Winter-greens (*Pyrola minor, P. media,* and *P. secunda*) and Cow-wheat (*Melampyrum*), which are semi-parasites; the Coral-root Orchid, a saprophyte (see p. 357), and Chickweed Winter-green (*Trientalis europaea*).

(2) *Woodlands on calcareous soils*: marls, limestones, chalk.

(*a*) Alder wood or Carr, with a ground flora of fen-plants. The Alder is the dominant tree. Other trees and shrubs are Sweet Gale, Creeping Willow, Black and Red Currant, Berry-bearing Alder, Buckthorn, Guelder Rose, Common Ash, and Birch (*B. tomentosa*). Common herbaceous species are: Marsh Fern, Tussock Sedge, Yellow Iris, Nettle, Meadow-sweet, and Marsh Marigold.

(*b*) Oak-Ash wood on marls. These woods are similar in many respects to the peduncled Oak wood, but have a more varied ground flora. Oak and Ash are the dominant trees. The shrub flora is abundant, and consists of Hazel, Wayfaring Tree, Spindle Tree (*Euonymus*), Traveller's Joy, Dogwood, Privet, Field Maple. The characteristic herbaceous species not found in the peduncled Oak woods are Herb Paris, Meadow Saffron (*Colchicum*), Gladdon (*Iris foetidissima*), Dog Violet (*V. sylvestris*), Nettle-leaved Bellflower (*Campanula Trachelium*), and the Orchids *Helleborine media* and *H. purpurata*.

(*c*) Ash wood on limestone, the characteristic plants of which have been given above, p. 349.

(*d*) Beech wood on chalk. Native Beech woods are confined to the Chalk Downs of the south of England, where they form 'hangers' on the steep slopes. The deep shade cast by the Beech tends to exclude other trees and shrubs, and the ground flora is very scanty (Fig. 190). The more common trees are Gean (*Prunus Avium*) and Yew; the latter sometimes forms a shrub layer, as do the Hazel and Holly in Ash and Oak woods. The characteristic species of the ground flora, especially in the lighter parts, are Dog's Mercury, Sanicle, Violets (*V. sylvestris, V. Riviniana*, and *V. hirta*), Strawberry, Enchanter's Nightshade, Helleborines, Large Butterfly Orchis, and the saprophytes, Bird's-nest Orchis and the Yellow Bird's-nest (*Monotropa*); also the Green and Stinking Hellebores, Deadly Nightshade, Spurge Laurel, and Butcher's Broom (*Ruscus aculeatus*).

CHAPTER XXIX

PLANT-LIFE IN HUMUS

Abnormal Modes of Nutrition

The work of Fungi in humus.—The surface layers of the soil contain much organic matter, and are commonly covered by the remains of plants which are the accumulations of successive years of growth. In meadows and pastures, this tangle of vegetable matter, living and dead, forms the turf. In woods, we know it as leaf-mould or humus; and it is often many inches in thickness. On the moors, it forms deep beds of peat. Examine the leaf-mould in a wood, and, on lifting it from the ground, note that the decaying leaves are often held together by a white, felt-like mass of mould-threads. This felt is the vegetative part of various species of Fungi and is known as the **mycelium**. From this mycelium arise the fruit-bodies of the Fungi, some of which we are familiar with as Mushrooms and Toadstools (Fig. 226). The threads (or **hyphae**) of the mycelium obtain their food from the dead leaves and other organic matter; and, along with organisms like bacteria, are responsible for their decay. The Toadstools, therefore, live upon the complex organic substances of dead leaves and twigs, or on animal remains and excrements; whereas the food of green plants consists of solutions of certain mineral salts and carbon dioxide. As there is much available energy in the organic compounds in humus, Fungi, by making use of this, do not need to absorb light-energy, and can thrive without developing either leaves or chlorophyll.

Saprophytes, Mycorrhiza, and Symbiosis.—Plants growing upon dead organic matter (animal or vegetable) are called **saprophytes** (Gr. *sapros* = rotten). Some fungal saprophytes are very restricted in their distribution, and occur only on the fallen leaves of particular species of trees.

You will find mycelia abundant in the damp leaf-mould of woods, and the roots of plants growing in the mould often become intimately surrounded by hyphal threads. Sometimes the hyphae enter the tissues of the roots and coil up within their cells (Fig. 229 A). They gain an entrance, not merely by mechanical pressure, but also by means of a ferment which they secrete and which digests the cell-walls of the roots with which they come into contact. Usually they do no harm to the roots, and possibly they convey to them useful materials from the humus.

Fungi thus do an important work; they decompose the cast-off leaves, prevent their accumulation, and convert their constituents into useful food for green plants. They may even convey this food into the tissues of the plants. In return they receive shelter and possibly some food from the roots.

Such a combination of fungal hyphae and the roots of plants is called a **mycorrhiza** (Gr. *mykēs* = a fungus, *rhiza* = a root), and a very large number of plants which grow in humus, trees and shrubs as well as herbaceous species, have a mycorrhiza on their roots. These plants have normal green leaves, and their mode of nutrition is, on the whole, the same as that of typical flowering plants; but by association with fungal hyphae they directly or indirectly utilize the humus. The union of two organisms whereby they mutually benefit is called **symbiosis** (Gr. *syn* = together, *bios* = life).

The best example of symbiosis is found in Lichens, so commonly seen as leafy incrustations on walls and tree-trunks, or as grey branching threads on the ground, like the Reindeer Moss. A Lichen, though usually regarded as a distinct plant, is really a colony of plants of two kinds: the predominant one is a fungus, and this entangles within the meshes of its mycelium innumerable green, algal cells. The fungus protects or imprisons the algae and supplies them with mineral food, out of which the latter, by virtue of their chlorophyll, build up organic materials, which in turn are absorbed as food by the fungus. We have already noticed (p. 326) the case of symbiosis in leguminous plants, where root-nodules are formed as the result of the action of bacterioids which enter the root-hairs. The Alder and Sea Buckthorn are further examples.

Plants that live entirely on humus or other dead organic matter are generally lowly forms like Fungi, and, as we have seen, have no need for and do not contain chlorophyll in their tissues. Some flowering plants grow in the humus of woods and, like the Fungi, live entirely upon it. Very few British plants are able to subsist in this way; some of them are Orchids, e.g. the **Bird's-nest Orchid** (*Neottia Nidus-avis*) and the **Coral-root Orchid** (*Corallorhiza*). Another, belonging to the heath family, is the Yellow Bird's-nest (*Monotropa Hypopitys*), and all have mycorrhiza on their roots or rhizomes.

These plants have become strangely modified in consequence of their saprophytic method of obtaining food, and they possess several features in common with the Toadstools: (1) their vegetative parts are embedded in the humus; (2) they obtain their food from the complex organic compounds of the humus; (3) they contain very little or no chlorophyll, and so are white, yellowish, or brownish in colour; (4) only their flowering and fruiting shoots come above ground. Under such circumstances green leaves are not needed, and they are reduced to mere scales.

The rhizome of the Bird's-nest Orchid gives off into the humus a tangled mass of underground stems and roots which resemble a bird's nest, hence its name. They are invested with a mycorrhiza, the hyphae of which are attracted to and enter the absorbing cells. The Coral-root (Fig. 228) does not develop roots at all; its rhizome (*rh*) is short and much branched, and from it many absorbing 'root'-hairs are given off.

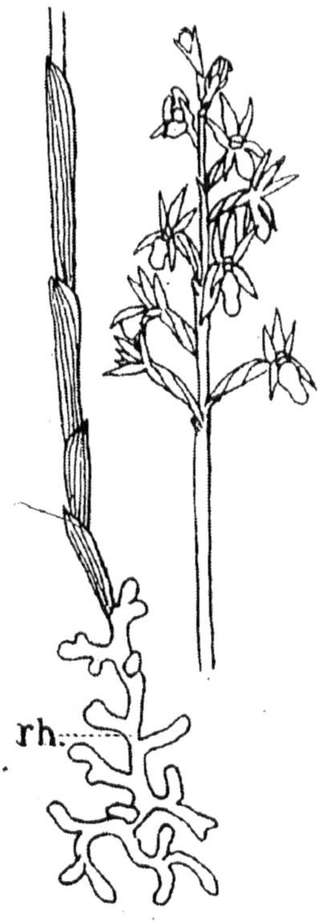

FIG. 228.
CORAL-ROOT ORCHID.
rh, rhizome.

The Yellow Bird's-nest (*Monotropa Hypopitys*) grows in the humus of shady woods, and it was on this plant that mycorrhiza was first discovered. From its underground stem is given off a dense mat of roots which are covered with fungal hyphae, and these provide food for the Yellow Bird's-nest from the humus they are decomposing.

It is common for Fungi to live at the expense of green plants, but it rarely happens that flowering plants are able to live upon the labour of the fungus.

Parasites.—Not only do dead and decaying leaves provide a habitat for many plants, but commonly living plants (and even animals) provide habitats for many species, especially Fungi. Familiar examples of this are the rusts, mildews, and blights, which often destroy valuable crops. A plant which lives at the expense of another organism is called a **parasite**; and the organism upon which the parasite preys is known as the '**host**'. The guest, however, is uninvited and takes advantage of the 'hospitality' to the injury of the 'host', and sometimes causes its death.

Living habitats are often selected by plants other than Fungi, e. g. many flowering plants prey to a greater or less extent on their neighbours. The most familiar example is the Mistletoe, which grows perched on the branches of the Apple, Poplar, and other trees. Birds eating the berries are unable to swallow the seeds because of the sticky material around them, so they scrape them off on to the branch, where they germinate. Suckers enter the branch and form a union with the wood of the 'host', from which they draw the mineral food for the mistletoe. The influence of this mode of nutrition is seen in its yellow-green leaves, which contain chlorophyll, and the products of photosynthesis enable the plant partly to maintain itself. The leaves, too, are evergreen, and are able to form organic substances at favourable periods throughout the year. These may be passed downwards and contribute somewhat to the nourishment of the host when the latter is not in leaf. The Mistletoe, therefore, is not entirely dependent on its 'host' for food, and may be regarded as a partial parasite.

Many flowering plants are partially parasitic, and attach themselves by means of suckers to the roots of neighbouring plants, especially grasses. They produce green leaves and closely resemble plants which obtain their food in the normal manner. Examples are Cow-wheat, Eyebright, Yellow Rattle, and Louseworts. The leaves of the Yellow

Rattle are, as in the Mistletoe, often pale-green, and those of the Louseworts are red or reddish-green.

A few flowering plants are entirely dependent on 'host'-plants for food. Some of these, like the partial parasites above mentioned, attach themselves by suckers to the

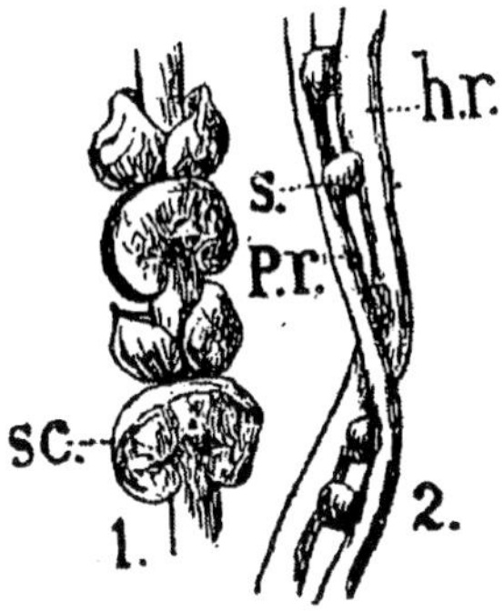

FIG. 229. TOOTHWORT.—1, part of underground stem; 2, roots; *h.r*, root of 'host'-plant; *P.r*, root of parasite; *s*, suckers attached to 'host'; *sc*, scale-leaves.

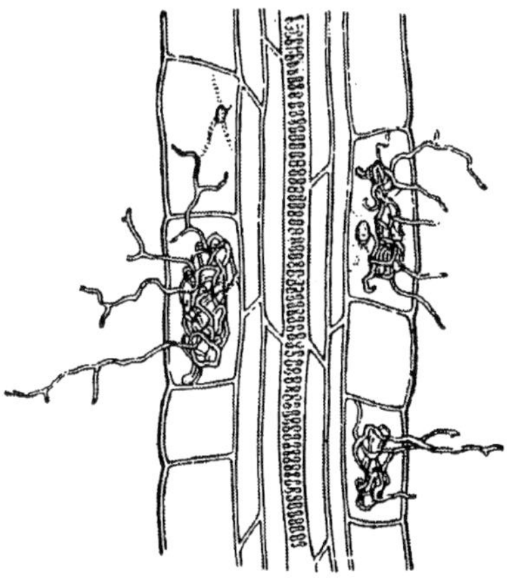

FIG. 229 A. SECTION OF ROOT OF LING SHOWING MYCORRHIZA.

FIG. 230. TWINING STEM OF DODDER ATTACHED BY SUCKERS TO THE STEM OF HOP.

roots of other plants, e.g. the Toothwort (*Lathraea squamaria*) and the Broomrapes (*Orobanche*).

The Toothwort, frequently found in hedgerows and woods, possesses many features in common with saprophytes like the Bird's-nest Orchid and Coral-root, e.g. its vegetative organs are underground, its leaves are reduced to scales, only the flowering shoot comes above ground, and the plant is a sickly yellow or purplish colour with little or no chlorophyll in its tissues. The rhizome is thick and bears four rows of curious scale-leaves (Fig. 229, 1), which are thick and fleshy, and back-rolled in such a way as to give rise to a branched cavity opening to the exterior by a narrow slit at the base. Very small animals often enter the cavities and die there, and the products of their decay may be absorbed; while special cells in the walls of the cavities may serve for the excretion of water. The roots form disk-like attachments (Fig. 229, 2 *s*) on the roots of trees such as Hazel, Elm, and Beech, and suckers from the disks enter the tissues of the host and absorb nutriment; it is also probable that some nutriment is absorbed from the humus after the manner of saprophytes.

Several species of Broomrapes occur in Britain and are parasitic on the roots of such plants as Broom, Gorse, Clovers, Hemp, and Ivy. They are dirty white or yellowish in colour, or tinged with pink and, like the Toothwort, send only their flowering shoots above ground.

An extreme example of parasitism is found in the Dodders. These, however, are stem-parasites, and they possess many remarkable features. The seed, on germination, sends its radicle a very short distance into the ground, while the slender stem nutates, and is sensitive to contact like a tendril (Fig. 230). If it comes in contact with a 'host', it twines round the stem and sends a sucker or haustorium (Fig. 231) into it, and both the wood and the bast of the haustorium form a union with the corresponding tissues of the host. By this means the Dodder can obtain the whole of the food it requires. The root now dies away, and the slender stem, which is without chlorophyll and does not bear any green leaves, subsists entirely at the expense of the plant on which it preys. As new branches arise they

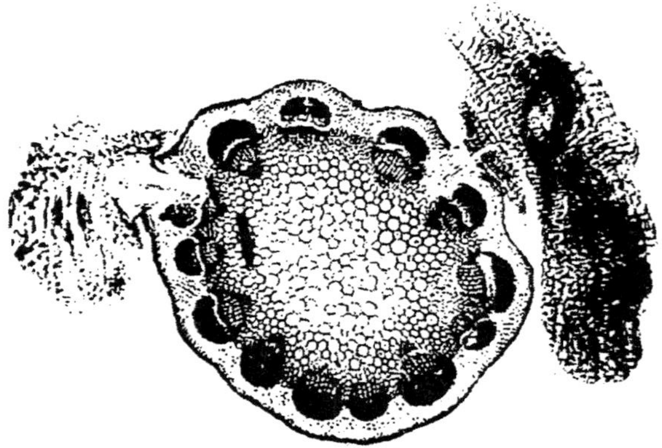

FIG. 231. TRANSVERSE SECTION OF STEM OF CLOVER PENETRATED BY TWO SUCKERS OF THE DODDER.

FIG. 232. PLANT OF ROUND-LEAVED SUNDEW.

spread over and twine around other stems; and it is an interesting fact that usually only a few turns are made around a single 'host'-stem, but from these coils many haustoria may be given off. In time, many bunches of small flowers are formed, each flower being like a tiny Convolvulus. Dodders (*Cuscuta spp.*) occur on a great variety of plants, such as Clovers, Vetches, Flax, Hop, Gorse, Ling, Bedstraw, Thyme, Thistles, and Nettles; and they may be so abundant as to do great damage to crops.

Total parasites, like total saprophytes, show how degenerate and modified the vegetative organs become in plants which have thus changed their mode of nutrition. The flowers show little or no modification except in colour, and much of the energy they derive from the 'host' is expended in the production of an abundance of seeds. Partial parasites and partial saprophytes are more common, and show intermediate stages of modification according to the degree of parasitism or saprophytism they have reached.

Insectivorous plants.—In a variety of ways plants may entrap small animals. In the Toothwort, as we have seen, there are cavities in the back-rolled leaves into which small animals may enter. In the Teasel, the bases of the opposite leaves are united in such a way as to form a cup which contains water, and insects, finding their way into the cup, may become drowned. Often flower-stalks are covered by glandular, sticky hairs to which small insects adhere and die in large numbers. Some species of *Silene* have received in consequence the popular name of 'Catchfly'. In other plants more specialized traps are found, and the insects or other small animals caught in them may contribute by their decay to the nutrition of the plant. Species which possess such peculiarly modified leaves or shoots and supplement their nitrogenous food in this way are known as **insectivorous** or **carnivorous plants.**

Many of them occur in boggy or marshy habitats, usually in soil poor in mineral food; while some of them are water-plants. In all cases, however, they contain much chlorophyll in their tissues, and their general mode of nutrition is like that of an ordinary green plant. The animal food obtained by means of their traps is, therefore, only supplementary. Still, their capturing devices are ingenious and curious. They may take the form of: (1) sticky hairs or sticky and sensitive tentacles; (2) trap-like bladders and pitchers; and (3) sensitive leaves which form rapidly-closing traps.

Two examples, the **Sundew** (Fig. 232) and **Butterwort** (Fig. 234, 1, 2), are locally abundant in boggy places on the moors, both in hilly districts and on lowland moors. Three species of Sundew occur, the most common one being the Round-leaved Sundew (*Drosera rotundifolia*). It is a small plant, only a few inches high, whose slender roots anchor the plant in the wet soil, or among bog moss, and the leaves form a rosette pressed close to the ground. At the end of the leaf-stalk is a circular, reddish blade, fringed with long, clubbed tentacles, and similar, but shorter, tentacles cover the upper surface. The tips secrete a sticky, viscid fluid, forming shining dew-like drops, attractive to insects. The secretion is mistaken for honey, and if a small insect alights on some of the tentacles at the edge of the leaf it is held firmly by the secretion. The tentacles are very sensitive to pressure, and the stimulus given by the insect in its struggles to escape results in the tentacles bending over; and the stimulus may extend to other tentacles and they also bend over, and so carry the insect to the centre of the blade, where it is brought into contact with other drops and eventually it is smothered (Fig. 233). The secretion of the tentacles is now changed, and a ferment is poured out which digests the protein compounds of the insect's body, and these digested materials are absorbed by the leaf, leaving only the indigestible remnants, such as wings, hairs, and claws, on the blade. After the 'meal' the tentacles bend outwards once more and again secrete the sticky and attractive, though deceptive fluid.

In the plant photographed (Fig. 232) some leaves were digesting food; others had 'scraps' and 'leavings' on them; some were fresh and open, waiting for prey; while in the centre of the rosette the youngest leaves were still unfolded, like a tiny hand with innumerable fingers tightly closed.

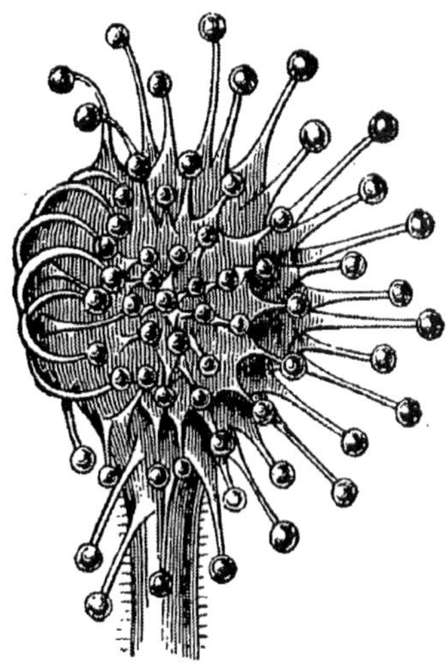

FIG. 233. LEAF OF SUNDEW.—Tentacles on the left curving as the result of stimulation (Pfeffer).

The Butterwort (Fig. 234, 1) has a rosette of radicle leaves pressing so firmly against the ground that when the plant is taken up they bend sharply backwards (Fig. 234, 2). Each leaf is yellowish-green and ovate, with its edges uprolled, and the flowers, though different in structure, resemble those of the Wood Violet. The surface of the leaf is greasy to the touch, due to a secretion from glandular hairs. Insects alighting on the leaf adhere to it, and under the stimulus the edges may roll farther inwards and more or less enclose the prey. A digestive ferment is now secreted, and the products are absorbed as in the Sundew.

In pools of the lowland moors, in stagnant ditches and slow-moving streams or drains, a curious insectivorous plant may be found with much-branched, thread-like, green shoots, bearing on them numerous small bladders (Fig. 234,

FIG. 234. INSECTIVOROUS PLANTS.—1, Butterwort; 2, the same showing leaves strongly reflexed when taken from the ground; 3, branch of Bladderwort; 4, bladder enlarged; 5, 6, leaves of Pitcher-plants; 7, leaf of Venus' Fly-trap, open; 8, the same closed; *bl*, bladder; *h*, sensitive hairs; *t*, tendril.

3 and 4). These have suggested both the common and scientific names for the plant—Bladderwort and *Utricularia*. Several species occur in Britain, and all are remarkable in that they never possess roots. The shoots float freely in the water, but when in flower, the inflorescence is raised into the air. The bladders are curious 'eel-traps', filled with water and provided with 'doors', which open only from the outside. Small aquatic animals may enter, but are unable to escape. After swimming about for a time, they die and are decomposed by the action of bacteria, and the products of their decay are then absorbed by the bladder. The Bladderwort probably does not secrete a digestive ferment and thus its mode of nutrition is that of a partial saprophyte.

In the Indo-Malayan region and elsewhere, a number of bog-plants occur which have large and remarkable water-pitchers, and many of these may be seen in hothouses and botanic gardens. Fig. 234, 5 and 6, shows the leaves of two of these **Pitcher-plants**, which are species of *Nepenthes*. The midrib of the large leaf-blade is continued as a long tendril (*t*), which serves as an organ of attachment to a neighbouring plant, and the tip develops into a large pitcher with water at the bottom and overhung by a lid. Round the mouth is a firm, smooth rim, projecting inwards and fringed with sharp teeth. The outer surface is blotched with various shades of red, brown, and green, and so is attractive to insects. At the entrance are honey-glands, and below them the surface is glazed and smooth, forming a 'slide-zone'. On reaching this, insects find it easy to descend into the water, where they are drowned. Digestion is brought about by ferments secreted by the glands of the pitcher.

Some pitchers are formed from whole leaves, as in the Side-saddle Flowers (*Sarracenia*) of North America; while in *Cephalotus*, an Australian Pitcher-plant, division of labour exists, some leaves having normal flat green blades, and others being transformed into pitchers.

A more complicated device is found in the **Venus' Fly-trap** (*Dionaea muscipula*) (Fig. 234, 7 and 8), a plant growing in mossy places in the woods of Carolina. Its leaves, like those of the Butterwort, form a rosette close to the ground; the leaf-stalk is winged (*phyllode*), and the blade, slightly bent upwards along the midrib, is fringed with long, comb-like teeth. Many glandular hairs cover the upper surface, and on either side of the midrib are three long jointed hairs (Fig. 234, 7 *h*). These are sensitive, but if slightly touched once no result is observable; if, however, a second stimulus is soon applied the blade suddenly closes, and if an insect supplies the stimulus it is at once entrapped. The teeth along the edges interlock, and the two halves of the blade draw close together. A digestive secretion is now poured out over the body of the insect and the digested materials are absorbed by the leaf. Later it expands in readiness for more food.

The advantage to insectivorous plants of this mode of nutrition is in the gain of nitrogen and nutritive salts supplied with the nitrogen compounds, and the plants thus supplied with animal food thrive better than those living solely on inorganic materials.

CHAPTER XXX

GRASS-LANDS: PASTURES AND MEADOWS

A LARGE part of the British Islands is devoted to pasturage, about one-half of England, three-quarters of Wales, one-half of Scotland, and three-quarters of Ireland being so utilized. Some of this is mountain and heath land, but the greater part is permanent pasture dominated by grasses.

FIG. 235. A GRASS MOOR WITH MAT-GRASS AND RUSHES.

Thus a greater area is covered by grasses than by any other type of plant. Favoured by good methods of vegetative increase, social habit, and great range of species suited to very varied conditions, they form one of the most important vegetation features of temperate regions, e.g. the extensive pastures and meadows of Europe, the steppes of Russia, and the prairies and savannas of America. They commonly form extensive carpets, as in our pastures and meadows, and their numerous leaves and the accumulated remains of rhizomes and felted roots form turf or sod.

Grass moors.—On the shaly siliceous Pennine Slopes large areas are dominated by grasses (Fig. 235), two species being especially conspicuous. The Mat-grass (*Nardus stricta*) (Fig. 236) in the summer forms large tussocks with grey-green, wiry, up-rolled leaves (Fig. 255, 3); in the autumn it turns a light yellowish-brown, colouring the mountain-sides and forming a conspicuous feature in the landscape; it is easily recognized in the winter by its one-sided empty spikelets.

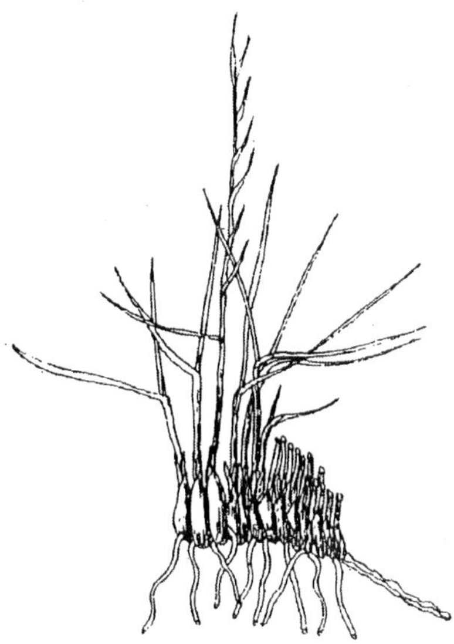

FIG. 236. MAT-GRASS.—Part of a tussock showing the closely-packed shoots on the rhizome.

Along with it, and often becoming dominant over considerable tracts, is another tussock-forming species, the Waved Hair-grass (*Deschampsia flexuosa*). Its leaves are still finer than that of the Mat-grass, and up-rolled so as to leave only a very narrow groove (Fig. 255, 4).

The Bent (*Agrostis vulgaris*) and Sheep's Fescue (*Festuca ovina*) (Fig. 255, 2), both with up-rolled leaves, are very abundant, especially at rather lower levels. Associated with these in the drier parts are many heath-plants, e.g. Ling, Cross-leaved Heath (*Erica Tetralix*), and Bilberry. In the wetter parts of these grass moors the ground is occupied by huge tussocks of the Purple Moor-grass (*Molinia caerulea*). In spring its pale yellow-green leaves, especially those following the burning of the moor, are very striking.

Calcareous grass-land.—Very different are the calcareous grass-lands of the hill-slopes of the mountain limestone and the great sheep-grazing grounds of England, the Chalk Downs. The Mat-grass, Waved Hair-grass, and Purple Moor-grass are absent, and in place of their tussocks we have a much shorter grassy turf dominated by the Sheep's Fescue-grass (*Festuca ovina*), and with it the Lesser Meadow Rue (*Thalictrum minus*), Lady's Fingers (*Anthyllis Vulneraria*), Horseshoe Vetch (*Hippocrepis comosa*), the Lesser Burnet (*Poterium Sanguisorba*), the Rock Rose (*Helianthemum Chamaecistus*), the Lesser Scabious (*Scabiosa Columbaria*), and the Hoary Plantain (*Plantago media*), all of which are absent from siliceous grass-land. Where much leaching has occurred, on calcareous slopes, heath-plants are found alongside typical calcareous species. They are small at first and hidden by the grass ; but in places they become dominant and form a limestone heath.

Neutral grass-land.—When grass-land is heavily grazed and manured, the number of species is, as a rule, reduced, and the typical plants of the two previous types of grass-land are rare or absent. This is called neutral grass-land, and is dominated by such species as the Rye-grass (*Lolium perenne*), Vernal-grass (*Anthoxanthum odoratum*), Cock's-foot-grass (*Dactylis glomerata*), Yorkshire Fog (*Holcus lanatus*). Many of these species, however, may occur within a limited area, and much may be learnt from a careful study of an old pasture or a meadow.

Meadows.—Because of their higher cultivation, meadows

possess fewer species than pastures; the grasses are commonly of species introduced by man, and since these are grown as a crop, relatively few other species occur. In a pasture, however, which in hilly districts may never have been under the plough, we have a greater variety of wild grasses together with a considerable number of flowering plants. In either case the species will tend to vary according to changes of soil, water-content, or other factors.

Survey of a pasture.—Examine a pasture, one with some variety of surface for preference, and select a line passing through typical parts of it. Study the plants along this line and note also any differences which occur as to changes of slope, soils, and their water-capacity, and drainage. Do you find any indication of change of species which corresponds to change of conditions? Compare the plants of the drier, well-drained parts with those of wetter parts, and notice any peculiarities in habit and form of leaf. Do the species persist throughout and under all these conditions, or are some more restricted in their distribution? Draw to scale a plan of the field; indicate changes of level by contour-lines, and mark on this plan by means of signs the more characteristic plants as they occur. A small field studied in this way has a steep slope marking the outcrop of a bed of sandstone; the lower part is flat, lies over a bed of shale, and is ill-drained and damp. This pasture is bounded above by a Hawthorn hedge, beneath the shade of which are woodland plants, e.g. Bracken, Male Fern, Bluebell, Anemone, and woodland grasses. The steep slope is covered by a sandy loam, and the fine, rolled-leaved Hair-grass is so abundant as to give a distinct aspect to the slope.

Careful examination shows that many other species grow along with it, such as Lady's Bedstraw, Tormentil, Purging Flax, Speedwell, Mouse-ear Hawkweed, Sheep's Sorrel, White Clover, Field Rush, Narrow-leaved Plantain, Mouse-ear Chickweed, and a few plants of Eyebright and

Yellow Rattle. The delicate roots of the last two species are often attached to the roots of grasses, from which they absorb part of their nutriment.

At the lower level, where the soil is finer-grained, wetter, and lies over a bed of shale, the plants are quite different. The rolled-leaved grasses are displaced by grasses with larger flat blades, such as Cock's-foot, Meadow Fescue, and Yorkshire Fog, while flowering plants like Lesser Spearwort, Ragwort, Knapweed, Lousewort, Yarrow, Red Campion and Wood Betony occur.

In a limestone pasture, however, species like Wild Thyme, Yellow Violet, Lady's Fingers, Burnet, Yellow Bedstraw, Small Scabious, Hoary Plantain, Blue Sesleria, Sheep's Fescue-grass, and others are found.

CHAPTER XXXI

WATER AND MARSH PLANTS

Vegetation of a pond.—Examine the vegetation of a pond or lake, and compare the plants of the banks with those growing along the water's edge, and also with those extending into the open water (Fig. 237). At the inlet, look for the stages in the development of a marsh, and notice how invasion of the pond takes place (Fig. 238). Draw a transect, e.g. Fig. 239, through the pond, including a portion of the bank, and show in a diagram the succession of plants met with. The upper part of the bank (A) is covered by meadow-species, but near the wetter soils below (B), these give place to Rushes, Purple Loosestrife, Water Dropwort, Iris, Marsh Marigold, Lady's Smock, Large Bitter-cress, Bog Stitchwort, Ragged Robin, Meadow-sweet, Lesser Spearwort, and Bog Starwort.

Nearer the water's edge is a belt of reed-like plants (C),

FIG. 237. WATER-PLANTS INVADING A LAKE.

FIG. 238. MARGINAL VEGETATION OF A POND; WATER BUTTERCUPS EXTENDING FAR INTO THE WATER.

many with erect strap-shaped leaves, such as Iris, Bur-reed, Flowering Rush, Arrowhead, the Reed Poa or Mead-grass (*Glyceria aquatica*), and the Common Reed (*Phragmites*). The last named is a good indicator of the direction of the prevailing winds; it has smooth leaf-sheaths, and all the upper exposed blades turn round in the direction in which the wind is blowing (Fig. 240).

Extending farther into the water (D) are the Smooth Horsetail, Water Plantain, and Mare's-tail. Usually these

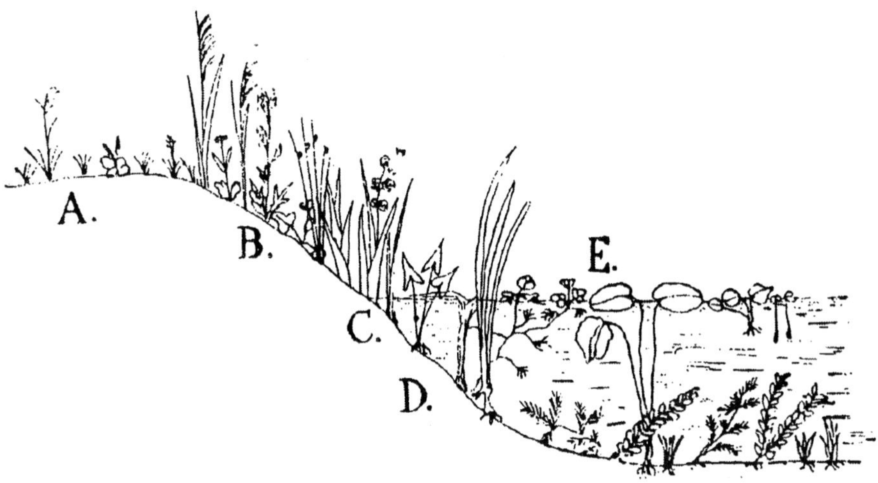

FIG. 239. TRANSECT OF A POND.

avoid complete submersion, their upper parts growing above the surface of the water. They are closely followed by species like Water-Lilies (E), Water Buttercups, Floating Mead-grass (*Glyceria fluitans*), and some Pond-weeds which, though rooted in the mud, have some leaves in the water, and others (usually of a different type) floating on the surface.

In the Water Buttercups the submerged leaves are dissected (Fig. 242); the floating leaves are entire. Submerged leaves of the Water-Lily are large and very thin, while the floating leaves are thick, and covered on the upper surface by a leathery cuticle (Fig. 241). Plants like these, which bear different kinds of leaves, are said to be *heterophyllous* (Gr. *heteros* = different). Heterophylly is common in water-plants and frequently occurs in land-plants.

The Frog-bit has floating leaves and much-branched roots hanging in the water, but often not rooted in the soil (Fig. 239, E). The Duckweeds, with their curious leaf-like stems (*phylloclades* or *cladodes*) and roots hanging in the water, are entirely floating. In the more open water many plants occur which are rooted and entirely submerged, e. g. Water Milfoil, Canadian Water-weed, Stoneworts, and many Pond-weeds; while plants like the Bladderworts have neither root nor floating leaves, but are suspended unattached in the water.

Growing among these is a rich flora of minute plants, the Algae, some in the form of a tangle of delicate green threads and others, though consisting each of a single microscopic cell, exhibiting, as in the case of desmids and diatoms, structural details which are both elaborate and beautiful.

Structural peculiarities of water-plants.—Specimens of the different types of water-plants should be obtained, and their varied forms studied with especial reference to their aquatic surroundings, on account of which water-plants form a distinct type of vegetation.

Cut sections of stems and leaves of the larger species and examine them with a lens. Note that the bulky cortex contains very large air-spaces (Fig. 243), and these occur not only in stems and leaves, but also in the roots. The roots growing amongst decaying organic matter, and in badly-aerated mud, are thus able to obtain a supply of air by diffusion from the shoots above. The vascular bundles of the stem, as in roots (Fig. 15), are often concentrated in the centre, which is a good position for resisting the longitudinal strain of running water. The

FIG. 240. WATER-PLANTS IN A DITCH. In the foreground Arrowhead, and beyond the Common Reed with leaves indicating the direction of the wind.

FIG. 241. FLOATING LEAVES OF THE YELLOW WATER LILY.

woody tissues are so poorly developed that many water-plants collapse when taken out of the water, showing how dependent they are on water for mechanical support.

The position and mode of growth of the underground stems and roots, even in the same species, vary considerably,

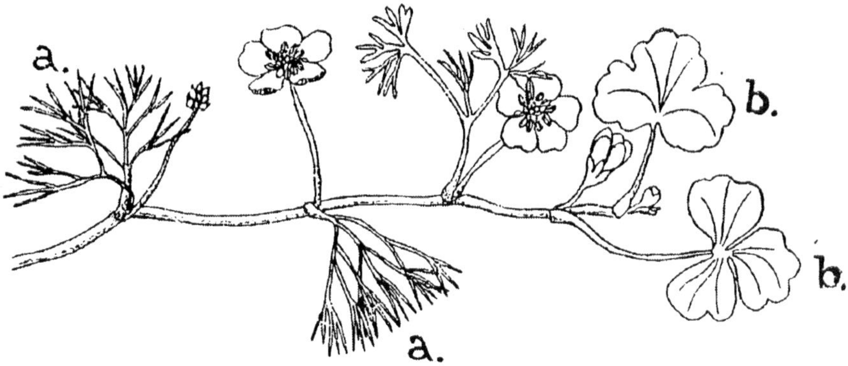

FIG. 242. WATER BUTTERCUP.
Submerged leaves; *b*, floating leaves.

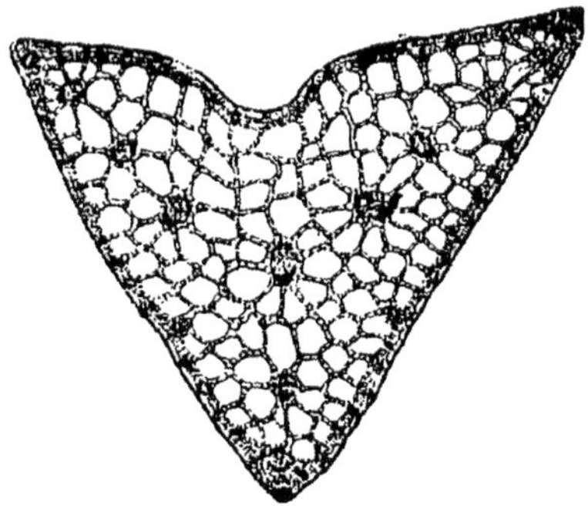

FIG. 243. TRANSVERSE SECTION OF LEAF OF FLOWERING RUSH, SHOWING LARGE AIR-SPACES.

and are determined largely by 'water-content' in the soil, and the supply of oxygen. The rhizomes and roots are often placed more or less horizontally, and in this position are near to the air-supply. As the soil becomes drier and better aerated, the roots take a more vertical course.

The stems and leaves of water-plants are often slimy,

and therefore not readily eaten by such aquatic animals as snails. The epidermis is usually very thin, and enables the plants to absorb through it much of their food in the form of mineral salts and carbon dioxide dissolved in the water in which they live.

Invasion. Water-plants as land-winners.—Plants growing in water are not subjected to such extremes as plants growing in the air, and consequently grow rapidly; the rhizomes plough their way through the mud and give off innumerable shoots into the water. This rapid vegetative growth enables the plants to spread quickly, and they are further aided by the ease with which detached shoots grow and form new plants. On the approach of winter, special winter buds are formed in many species, such as Pond-weeds and Frog-bit; these break off, fall to the bottom, and rest until the following spring. By such vegetative means, rather than by seeds (see p. 142), water-plants reproduce themselves extensively, and rapidly invade the water.

Interesting examples of this may be found in many ditches, ponds, and lakes. Reed-like plants may be seen to extend from the margin across the inlet. Their stems and leaves form a filter, keeping back the mud, which, together with the remains of successive seasons of plants, chokes up the channel and provides a soil over which the plants from the banks push their way farther and farther. Thus in time the pond or lake becomes converted into a marsh.

By such invasion, plants become important land-winners. The aquatic vegetation by its very luxuriance prepares the way for its own extinction. It contributes to the changes of conditions which favour a drier type of plant-life, and it is only a matter of time when it will be succeeded by the invaders. If further changes should result in improved drainage, the marsh type will itself be supplanted. In this way invasion and succession follow each other in natural sequence, and form a widespread phenomenon. A parallel example, as we have seen, is the invasion of sand-dunes by grasses and sedges, whose very long rhizomes bind and fix the sand and prepare a soil on which a grassy turf or a heath, and eventually a wood, may become established.

Flowers of water-plants.—The flowers of most water-plants are carried above the water, and open in the air. In a few cases they open and are pollinated in the water. This is seen in e.g. the Grasswrack, while some, like the Water Bistort, may be self-pollinated under water, but the flowers do not open, so that really they are pollinated in air and not in water. Usually, however, the flowers are at a disadvantage, owing to the scarcity of insect-pollinators on water, but this is compensated for, as we have seen, by the great powers which water-plants possess of vegetative reproduction.

Marsh-plants.—While there is every gradation between water-plants and marsh-plants, typical examples differ in several respects. The aerial parts of marsh-plants agree closely in form and structure with those of land-plants, and are exposed to similar conditions; the underground parts grow in wet, cold, badly-aerated soil. In consequence, as in aquatics, some marsh-plants have large air-spaces in their tissues; others have cylindrical stems and greatly-reduced leaves, e.g. Rushes and Horsetails. Interesting modifications, related to differences in transpiration, may be easily seen in the Meadow-sweet. The lower leaves, which are exposed to little evaporation, are without hairs; the intermediate leaves have patches of hairs on the lower surface; and the upper more-exposed leaves have a close covering of silky hairs, which materially reduce transpiration.

The following is a list of the common plants found growing in a marsh at the head of a small lake which has been invaded by the Pond-weed, Water Milfoil, Water

Horsetail, and Marsh Club-rush. Encroaching on these invaders were: Water Plantain, Common Reed, Bur-reed, Soft Rush, Marsh Horsetail, Marsh Bedstraw, Bog Stitchwort, Ragged Robin, Meadow-sweet, Hairy Willow-herb, Square-stalked Willow-herb, Lesser Spearwort, Great Valerian, Large-flowered Bitter-cress, Lady's Smock, Marsh Marigold, Brooklime, Water Forget-me-not, Marsh Thistle, Hemp Agrimony, Marsh St. John's-wort, Tufted Hair-grass. Numerous trees surround the marsh, but only two of the species have invaded it, the Alder and Willow. These thrive well in the wet soil, and the marsh is becoming converted into an Alder-Willow wood.

CHAPTER XXXII

WEEDS

A CAREFUL examination of a natural piece of vegetation produces the impression of a natural blend of colour and form, and both are in keeping with the habitat. Such an area contrasts sharply with a piece of cultivated land, which, whether garden or farm, shows an obvious selection and arrangement of plants suited to the needs or caprice of man. Even here nature cannot be ignored, and it impresses itself by topography, soil, and climate in a manner which compels even cultivation to keep along more or less definite lines. Nevertheless, man's aim is to substitute for the less useful native plants those needed by him. In this way much of the native vegetation is destroyed, but an interesting meshwork persists along the hedgerows (Fig. 244), roadside patches, and in ditches and streams; or some

Fig. 244. A Hedgebank with Beaked Parsley, a common meadow-weed.

peculiar topographical feature may render cultivation impracticable, e.g. a steep bank, a moor, copse, or wood, a mountain or fringe of the shore, and here the native plants maintain their footing.

(*a*) Cornfield Weeds

In order to render land suitable for crops, the farmer has not only to destroy the native plants, but to change the habitat as well, and this he does by ploughing, draining, manuring, and the like. On this freshly-made soil his seeds are sown, and in time germinate. Meanwhile the surface is exposed to invasion by the native plants which have escaped destruction. Those with a good dispersal-mechanism, will stand the best chance of spreading; others with runners, such as the Silverweed, or the quick-growing rhizomes of the Couch-grass, soon make headway. Unless care is taken to eradicate them they will occupy the soil intended for the crops, and, being stronger and sturdier, will gain the mastery. From man's point of view these are 'plants in the wrong place', and he calls them 'weeds'.

If, however, we make a collection of the weeds of arable land and examine them, we find that very few are of the same species as the native plants of the district.

The native plants are mainly perennials, while the weeds, like the plants man cultivates, are for the most part annuals. The ground is prepared for annual crops, and annual weeds find it a favourable soil, and thrive accordingly; and as long as man disturbs the ground, perennials have little chance of succeeding. Their opportunity comes when cultivation ceases; with a more stable soil, the sturdy natives soon invade the land, and the annuals, accustomed to rely on man for a suitable habitat, are succeeded by perennials which have migrated from the adjacent, more natural areas. Soon the land reverts practically to its original state.

Waste-heaps, quarry-tips, new embankments, or roadsides, furnish numerous examples of plant invasion and succession; and the plants of such habitats should be carefully studied as to their origin, means of dispersal, and increase both by seed and vegetative propagation.

Native annuals occur along the coast, on the rocks, and on shifting banks; and some weeds which have somewhat fleshy leaves may have come originally from such habitats, e.g. Fumitory (*Fumaria officinalis*), Hedge Mustard (*Sisymbrium officinale*), Charlocks (*Brassica Rapa* and *B. Sinapistrum*), White Charlock (*Raphanus Raphanistrum*), Goosefoot (*Chenopodium album*), Orache (*Atriplex patula*), Black Bindweed (*Polygonum Convolvulus*), Knot-grass (*P. aviculare*), Persicaria (*P. Persicaria*), all of which are annuals. The Mayweed (*Matricaria inodora*) is a biennial, and another frequent weed, the Bladder Campion (*Silene inflata*), is a perennial. Other annuals have doubtless been introduced with impure seed, and those with bright and showy flowers have probably come from sunnier climes, e.g. the Poppies (*Papaver Rhoeas* and *P. dubium*), Poor Man's Weather-glass (*Anagallis arvensis*), and the Corn Marigold (*Chrysanthemum segetum*). Many have been introduced in this way from North and Central Europe and Asia.

Other common annuals are:

Field Buttercup (*Ranunculus arvensis*), Red Poppy (*Papaver Rhoeas*), Shepherd's Purse (*Capsella Bursa-pastoris*), Corn Pansy (*Viola tricolor*), Corn Cockle (*Lychnis Githago*), Chickweed (*Stellaria media*), Spurrey (*Spergularia arvensis*), Soft-leaved Crane's-bill (*Geranium molle*), Cut-leaved Crane's-bill (*G. dissectum*), Herb Robert (*G. Robertianum*), Trefoil or Black Medick (*Medicago lupulina*), Hop Trefoil (*Trifolium procumbens*), Tare (*Vicia hirsuta*), Parsley Piert (*Alchemilla arvensis*), Fool's Parsley (*Aethusa Cynapium*), Cleavers (*Galium Aparine*), Field Madder (*Sherardia arvensis*), Cudweed (*Gnaphalium uliginosum*), Groundsel (*Senecio vulgaris*), Sow-thistle (*Sonchus oleraceus*), Corn Forget-me-not (*Myosotis arvensis*), Corn Scorpion Grass (*Myosotis versicolor*), Ivy-leaved Speedwell (*Veronica hederaefolia*),

Corn Speedwell (*V. agrestis*), Hemp Nettle (*Galeopsis Tetrahit*), Red Deadnettle (*Lamium purpureum*) and other Labiates, Sun Spurge (*Euphorbia helioscopia*), and Petty Spurge (*E. Peplis*).

A few are biennials, e.g.:

Spear Thistle (*Carduus lanceolatus*), Viper's Bugloss (*Echium vulgare*).

Perennial weeds are uncommon in cornfields; the following are examples:

Rest-harrow (*Ononis repens*), Bush Vetch (*Vicia Cracca*), Creeping Cinquefoil (*Potentilla reptans*), Silverweed (*P. Anserina*), Willow Herb (*Epilobium montanum*), Coltsfoot (*Tussilago Farfara*), Field Thistle (*Carduus arvensis*), Corn Sow-thistle (*Sonchus arvensis*), Bindweed (*Convolvulus arvensis*), Field Mint (*Mentha arvensis*), Broad-leaved Plantain (*Plantago major*), Dock (*Rumex obtusifolius*), Sheep's Sorrel (*Rumex Acetosella*), Stinging Nettle (*Urtica dioica*), Couch-grass or Wicks (*Agropyron repens*).

(*b*) Meadow and Pasture Weeds

Many plants, not usually regarded as weeds, but of little nutrient value, may occur so abundantly in meadows and pastures as to reduce considerably the value of the herbage, and so may be classed as weeds. While annuals are able to thrive in disturbed and prepared ground, they are ill-adapted for the struggle with turf-forming perennials, and so we find that the majority of meadow and pasture weeds are perennials with effective means of vegetative increase. Annuals are relatively few, and it is an interesting fact that among the latter the most persistent are semi-parasites, living to some extent on the roots of the grasses, e.g. Yellow Rattle (*Rhinanthus Crista-galli*), Eyebright (*Euphrasia officinalis*), Red Eyebright (*Bartsia Odontites*). Other common annuals in pastures are the Purging Flax (*Linum catharticum*) and Nipplewort (*Lapsana communis*).

Biennials are few, e.g. Goat's-beard (*Tragopogon pratense*) and the Soft Brome-grass (*Bromus mollis*).

The most abundant weeds of grass-lands are perennials, namely:

Upright Buttercup (*Ranunculus acris*), Bulbous Crowfoot (*R. bulbosus*), Jack-by-the-Hedge (*Sisymbrium Alliaria*), Mouse-ear Chickweed (*Cerastium triviale*), Rest-harrow (*Ononis arvensis*), Meadow Pea (*Lathyrus pratensis*), Tormentil (*Potentilla erecta*), Lady's Mantle (*Alchemilla vulgaris*), Burnets (*Poterium Sanguisorba* and *P. officinale*), Earth-nut (*Conopodium majus*), Beaked Parsley (*Anthriscus sylvestris*), Wild Carrot (*Daucus Carota*), Yellow Bed-straw (*Galium verum*), Field Scabious (*Scabiosa arvensis*), Daisy (*Bellis perennis*), Yarrow (*Achillea Millefolium*), Ox-eye Daisy (*Chrysanthemum Leucanthemum*), Ragwort (*Senecio Jacobaea*), Knapweed (*Centaurea nigra*), Great Knapweed (*Centaurea Scabiosa*), Cat's-ear (*Hypochaeris radicata*), Autumn Hawk-bit (*Leontodon autumnale*), Dandelion (*Taraxacum officinale*), Cowslip (*Primula veris*), Germander Speedwell (*Veronica Chamaedrys*), Self-Heal (*Prunella vulgaris*), Hoary Plantain (*Plantago media*), Ribwort Plantain (*P. lanceolata*), Sorrel or Green Sauce (*Rumex Acetosa*), Field Rush (*Luzula campestris*), Spring Sedge (*Carex caryophyllea*), Yorkshire Fog (*Holcus lanatus*).

We thus see that for a weed to succeed among the turf-forming plants of meadows and pastures it must possess a similar mode of vegetative growth and reproduction, and be a perennial. Very few annuals succeed here, and of these the most successful are root-parasites, such as the Eyebright and Yellow Rattle. But arable land, regularly disturbed and prepared for the cultivation of annual and biennial crops, is well adapted to the requirements of annual weeds.

CHAPTER XXXIII

VEGETATION OF THE SEA-COAST

A GLANCE at a geological map of Britain shows, not only how varied are the rocks inland, but also how varied they are along the coast. Further, just as large inland areas are covered by superficial deposits of ice-borne materials such as clay, sand, and boulders, obscuring the solid rocks beneath, so along the coast, thick beds of such materials may be found, sometimes forming high, easily-denuded cliffs. Steep rocks offer a jagged resistant line to the tearing action of the waves, and are covered with spray at every high tide. At the other extreme, low ground may pass gradually seawards and deaden the force of the incoming waves. The varied materials of the coast are exposed to the ceaseless efforts of the sea and atmospheric weathering, and the products of denudation are spread out in a characteristic manner along the coast, and form a somewhat unstable soil for plants. In some places it is finely pulverized and muddy, in others it is coarser and sandy, and heaped up into banks or dunes; or the still coarser pebbles and boulders may form banks of shingle. The coast-line thus offers a great variety of surface and soil, and we might expect the influence of these variations to be reflected on the coast-vegetation. But the factor which most powerfully influences plant-life along the coast is the presence of salt water, and the plants exposed to its influence show many peculiarities both in colour, form, and structure.

In consequence of these conditions, which of necessity are confined to a narrow belt around the coast, we find certain types of vegetation which present a strong contrast to the vegetation immediately beyond it. The best-marked plant-formations of the coast are those of the sand-dunes and salt-marshes, the vegetation of which impresses us at once by its peculiar, blue-grey colour.

Seaweeds.—In the sea, or along that part of the coast often covered by sea-water, seaweeds abound. If the coast is rocky, brown seaweeds, like the Bladder-wrack and other species of *Fucus* and *Pelvetia*, often form a long belt, the plants being anchored to the rocks or to stones by peculiar holdfasts. Farther seawards are the light, yellow-brown straps of *Laminaria*; and in the rock-pools and in parts always covered by sea-water are numerous red seaweeds. All these are Algae, a group of plants of much more lowly organization than the flowering plants, and differing widely from the latter in not having true roots, stems, or leaves, and producing no seeds.

Salt-marshes.—Between the tide-marks is a belt destitute of vegetation; here the ground is regularly under the influence of the waves. If we listen to the roar of sand and pebbles as the waves roll inwards and retreat, and watch the constant movement of the surface, we can realize how difficult it is for plants to secure a root-hold in such unstable ground. In areas where the ground is bare for longer intervals, as along the shores of sheltered bays, in estuaries, or the banks of tidal rivers, land-plants establish themselves. The muddy soil which accumulates in such localities provides a peculiar habitat for plants. It is badly aerated, liable to be covered by sea-water at very high tides, and there is often much salt in the ground-water. On the other hand, during exposed intervals, much evaporation may take place, leading to a concentration of salt in the soil. Or the reverse is possible; during heavy rains much salt may be washed out of the soil. Under such conditions it is not surprising to find the ground occupied by a peculiar type of vegetation.

A habitat of this kind is known as a salt-marsh (Fig. 245). The plants growing here have usually fleshy leaves covered with a waxy bloom or grey hairs; some develop short hairs which break off and form a mealy covering over the surface. Most salt-marsh plants have a reduced transpiring surface, and store water in their fleshy tissues. They thus possess many of the characteristics of xerophytes; some of them, however, e. g. the Glassworts (*Salicornia spp.*) and Sea Aster (*Aster Tripolium*), transpire freely, and are even able to absorb water by their green surface. Salt-marsh plants are known as **halophytes.**

FIG. 245. SALT-MARSH ON THE BANKS OF A TIDAL RIVER.

FIG. 246. SAND-DUNES, THEIR CRESTS COVERED WITH MARRAM-GRASS.

FIG. 247. OLDER SAND-DUNES.—In the foreground Dwarf Willows; in the centre is seen the invasion by Marram-grass.

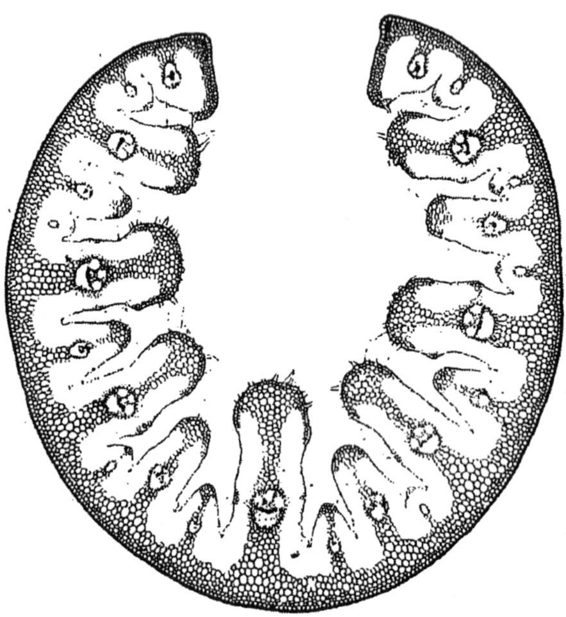

FIG. 248. TRANSVERSE SECTION OF LEAF OF MARRAM-GRASS.

The plant best able to withstand salt water is the Glass-wort (*Salicornia*), which acts as a pioneer, but near the sea the plants are few and occur only at wide intervals. The Grasswrack (*Zostera marina*) often occurs in quantity, and is interesting in that it flowers under water, and has pollen-grains which float in and are carried by water from the anthers to the stigmas. Other plants generally met with on the salt-marsh are the following: Sea Poa (*Glyceria maritima*), Sea Arrow-grass (*Triglochin maritimum*), Sea Aster (*Aster Tripolium*), Glasswort (*Salicornia radicans*), Sea Oraches (*Atriplex spp.*), Sea Purslane (*A. portulacoides*), Sea Spurrey (*Spergularia maritima*), Sea Lavender (*Limonium vulgare*), Sea Plantain (*Plantago maritima*), Sea Blite (*Suaeda maritima*), Scurvy-grass (*Cochlearia officinalis*); and other species, e. g. *Cochlearia anglica*, Sea Rush (*Juncus maritimus*), Sea Milkwort (*Glaux maritima*), Buck's-horn Plantain (*Plantago Coronopus*), Sea Couch-grass (*Agropyron pungens*).

These species vary much in distribution with the nature of the ground, as to whether it is muddy or sandy; low-lying and wet; or raised, exposed, and drier. The associations of plants occurring in these different parts of the salt-marsh, together constitute the salt-marsh formation.

Rock-plants.—Many of these plants occur also on the rocks of the coast, especially those washed by the spray. The more characteristic are: Samphire (*Crithmum maritimum*), Sea Lavender (*Limonium vulgare*), Buck's-horn Plantain (*Plantago Coronopus*), Sea Pink (*Statice maritima*), and Sea Campion (*Silene maritima*).

Sand-dunes.—If we examine the sand along the shore which is subjected to the action of the waves, characteristic ripples will be found on the surface; but, in addition to wave-ripples, other similar ripples may be seen, especially on sand over which wind has blown for a few hours. The wind rolls the sand-grains before it and heaps them up in little ridges ; the windward slope of each ripple is gradual, but the lee side is steep (Fig. 246). If we watch the ripples during high winds we can see them shift their position ; the sand is blown up the gentler slope to the crest of the ridge, then rolls down the leeward slope where it is protected against the wind and is likely to lodge. Thus, while the windward slope is wearing away, accumulation is going on on the leeward side. In this way the little ripples move steadily forwards in the direction in which the wind is travelling. If in its progress a plant (or even a pebble) is encountered, a little heap of sand is built up around it, and it may eventually be buried. If, however, the plant by continued growth is able to keep its tip above the surface, the mound may increase in size and form a miniature sand-dune. It is in this way that sand-dunes are built up, and they are thus the result of two causes: (1) the action of wind on mobile sand, and (2) the binding action of plants which establish themselves on the sand so transported.

The plant which is best able to overcome the difficulties of shifting sand is the Marram-grass (*Psamma arenaria*), which, by virtue of its long perennial rhizomes and deep-growing fibrous roots, helps to bind the sand together, while its shoots, by being able to grow and keep above the surface, not only maintain the existence of the plant, but aid considerably in building up the dune. By the decay of the older parts of the plants, humus is added to the soil, so that the Marram-grass becomes a valuable pioneer of coast-vegetation. As we have seen (p. 127), the vegetative mode of reproduction of this plant is put to practical use along miles of our coasts as a 'land-winner' (Figs. 247, 80, and 81).

The windward side of a dune is usually bare (Fig. 244), but on the crest are the Marram-grass and Sea Lyme-grass, which not only serve as sand-binders but provide protection and a suitable soil for other species. On the more sheltered leeward side, plants secure a better footing, and more numerous species are found, which have usually long rhizomes or deep-growing roots, e. g. Sea Purslane (*Arenaria peploides*), Sea Holly (*Eryngium maritimum*), Ragwort (*Senecio Jacobaea*), a variety of the Dandelion (*Taraxacum erythrospermum*), Hawkweed (*Hieraceum umbellatum*), Cat's-ear (*Hypochaeris radicata*), Sea Bindweed (*Calystegia Soldanella*), Sea Spurge (*Euphorbia Paralias*), Sea Buckthorn (*Hippophaë rhamnoides*), Sand-sedge (*Carex arenaria*), Fescue-grasses (*Festuca rubra* var. *arenaria* and *Festuca uniglumis*), Sea Couch-grass (*Agropyron junceum*), and Lichens.

The vegetation of the sand-dune differs in several important respects from that of the salt-marsh. The soil-water of sandy shores does not usually contain much salt, and the plants growing there, though subjected to the influence of salt spray from the sea, are not halophytes. The mobility of the sand, unsuited to most perennials, is the main factor which determines the character of the vegetation. Our previous experiments on soils have shown that the capillarity of sand is less than that of ordinary soil, that water percolates quickly, and that the water-capacity of the sand is slight. White sand reflects the heat of the sun, the surface layers become rapidly heated, water is quickly driven off, and the air around is hot and dry. At night rapid cooling occurs and the surface conditions resemble those of a desert. Although much organic matter may be strewn over the surface, rapid oxidation takes place, and the sand in consequence is poor in humus. Rapid percolation of water may also tend to deplete it, and thus the sand is poor in food-materials. Further, if a soil which dries rapidly contains 1 per cent. of salt it may act as a poison to most plants, though they may be able to withstand two or three times that amount in a soil which does not rapidly dry. If the surface layer of sand is removed, the lower layers are found to contain much moisture even in dry seasons. This moisture is probably derived from internal dew in the sand. The plants able to withstand salting and to endure the severe conditions along the shore are few in number and for the most part xerophytes, with tough, leathery, rolled or reduced leaves (Fig. 248). In some the leaves are fleshy and often coated with wax, others are spiny, and plants with branch spines are not uncommon. Grey-green is the prevailing colour of the vegetation.

The names 'shifting dunes', 'travelling dunes', 'grey dunes', and 'white dunes', by which sand-hills are commonly known, are suggestive of their most characteristic features. Beyond the shifting dunes, and farther from the influence of mobile sand, a grassy vegetation develops, which forms a sod, covering and protecting the sand. The dunes are low, the sand is more firmly fixed, and contains more humus; there is greater variety of soil and surface, and in consequence a much more varied flora is supported. On the one hand are the dry sand-banks with their grey-green xerophytes, and on the other, wet hollows supporting a marsh vegetation. To this part of the coast the name 'fixed dune' is given; nevertheless, the area is liable to be covered during high winds by blown sand, and this has an influence on the character of the vegetation. Cultivated crops farther inland often suffer materially from the effects of blown sand. The dominant plants of the fixed dune are: Sand-sedge (*Carex arenaria*), Fescue-grass (*Festuca rubra* var. *arenaria*), and sometimes Sea Couch-grass (*Agropyron junceum*). Associated with these and sometimes abundant are Sea Cat's-tail-grass (*Phleum arenarium*), Rest-harrow (*Ononis repens*), Stork's-bill (*Erodium cicutarium*), Bird's-foot Trefoil (*Lotus corniculatus*), together with species occurring on the shifting dunes, and numerous species which have migrated from adjacent pastures and meadows. In places, bushes are so abundant as to form thickets, e. g. Dwarf Willow (*Salix repens*), Sea Buckthorn (*Hippophaë*

Fig. 249. A Shingle Beach invaded by Orache and Saltwort.

Fig 250. Cotton-grass Moor.—Single-headed Cotton-grass in fruit.

rhamnoides), Burnet Rose (*Rosa spinosissima*), Brambles (*Rubus spp.*), Honeysuckle (*Lonicera Periclymenum*), Elder (*Sambucus nigra*). Sometimes moorland plants like Ling (*Calluna vulgaris*) and Heaths (*Erica*) occur, and may eventually invade the ground to such an extent as to give rise to a typical heath. The association of the shifting dune thus differs from that of the fixed dune, and within each association various societies occur. These societies and associations developed on sand-dunes form the sand-dune formation.

Strand-plants.—Along the strand between the line of shifting dunes and high-water mark a few plants occur very sparingly. They are easily overlooked, being almost buried beneath small dunes an inch or two high. These have all the characteristics of salt-marsh plants, e. g. Sea Rocket (*Cakile maritima*), Sea Kale (*Crambe maritima*), Sandwort or Sea Purslane (*Arenaria peploides*), several species of Orache (*Atriplex*), Goosefoot (*Chenopodium*), Saltwort (*Salsola Kali*), and Sea Knot-grass (*Polygonum Raii*). These form an association of strand-plants.

The strand and a considerable part of the dunes have a scanty vegetation ; much of the ground is bare, and there is no competition among the plants. Associations of this kind are called open associations. On the other hand, in areas completely covered by vegetation the association is said to be closed. Examples of closed associations are pasture, heath, and woodland.

Shingle Beaches

Along many miles of the English coast is a fringe of shingle, consisting of water-worn stones carried from the wasting shore, piled up into banks by the alongshore waves and currents, and driven landward by onshore gales during high tides (Fig. 249). If the bank is a low one and covered at high tides by the waves, it is devoid of vegetation, and, like the sand between tide-marks, the shingle is moved to and fro freely by the advancing and retreating tides. On such a shifting surface, plants cannot grow. Even in larger banks, and when the crest is beyond the reach of the highest tides, the shingle is more or less mobile by reason of the heavy impact of the waves and the readiness with which the sea-water percolates and buoys up the loose materials of which the bank is composed.

The shingle bank has a steep seaward slope, and a more gradual landward slope. On the seaward slope very few plants occur, and the general appearance is that of a bare bank of rounded stones. The bank is apparently dry, but if a few of the stones are examined they will be found to be wet even in very dry seasons, and often covered with lichens. The water, too, is fresh, not salt. Drifted materials, such as seaweed and animal-remains carried up by the waves, especially during storms, accumulate between the stones and form a black soil on which flowering plants from neighbouring ground can establish themselves and form an open association.

Shingle-binding plants.—Before a plant-covering is possible, the mobile beach has to be rendered more stable, and several species of plants do great work as pioneers and shingle-binders, e.g. the Shrubby Sea Blite (*Suaeda fruticosa*), Sea Campion (*Silene maritima*), and the Sea Purslane (*Arenaria peploides*). As with sand-binders, the essential character required is the power of the plant to grow through the shingle and regain the surface when buried, and the Shrubby Sea Blite possesses this power in a remarkable degree. The plant, when grown on stable ground, is much branched, three to four feet high, and has a stem an inch or more in thickness; the leaves are small and fleshy and covered with a waxy bloom. On mobile shingle, however, the young plants quickly anchor themselves by means of long roots, and produce stems, at first erect, but which are soon laid prostrate by shingle rolled over them by the waves. From the horizontal branches, new erect shoots arise and grow above the surface; these in turn are bent over and covered, and so the process is repeated. The plant, thus growing along the line of the moving shingle, travels obliquely to the crest of the bank, where it establishes itself. From the prostrate stems, a tangle of adventitious roots arises, which, together with the mat of shoots, serves to prevent the removal of shingle as the water runs down the bank. Further, the shoots arrest the landward flow of shingle, and the crest becomes raised beyond the reach of the highest tides; thus a sea-wall is formed which effectually checks the force of the waves. The Sea Blite, therefore, may, by its vegetative growth and power of rejuvenescence, become a valuable protector of the land against the incursions of the sea.

Some of the plants on the shingle beach are halophytes, e.g. the Oraches and Beet, Sea Blite, and Sea Campion, which are able to grow in parts influenced by the salt water. Others are characteristic of sand-dunes, especially if the beach has been formed on a bed of sand, e.g. Sea Purslane, Horned Poppy (*Glaucium luteum*), Biting Stonecrop (*Sedum acre*), Viper's Bugloss (*Echium vulgare*), and Sea Pea (*Lathyrus maritimus*); or strand-plants, e.g. the Saltwort (*Salsola Kali*). Frequently plants from the neighbouring fields and cultivated ground occur, their seeds having been carried to the beach by the wind or by birds, e.g. Elder (*Sambucus nigra*), Woody Nightshade (*Solanum Dulcamara* var. *marinum*), Curled Dock (*Rumex crispus*), Creeping Buttercup (*Ranunculus repens*), Herb Robert (*Geranium Robertianium* var. *purpureum*). On the more sheltered and stable landward slope of the beach, plants are more abundant, and their remains form a humus on which eventually a grassy carpet may form In this way the open association of the shingle beach may be replaced by a closed association of pasture species.

The conditions chiefly affecting plant-life on the coast are mobile soil and salt water. On the disturbed ground the plants are usually annuals; anchorage is secured by deep-growing tap-roots, and the long rhizomes and numerous adventitious roots of perennials serve as binders for the loose soil. The shoots are liable to be buried, but their great power of rejuvenescence enables them to keep above the surface and aid in building up banks of sand and shingle. The plants have usually a low habit, whereby the tearing effect of the wind is reduced. The shoots are modified in many ways: the stems and leaves may be spiny; the leaves are often small or reduced by rolling; fleshy plants are common, and the epidermis is covered by a grey waxy bloom or by hairs. In some species the hairs become detached and form a 'meal' on the surface. By such modifications the stomata are protected, transpiration and radiation reduced, and the water-supply conserved. Plants possessing these modifications are able to survive the conditions of the habitat, and they give a characteristic appearance to the vegetation of the coast.

CHAPTER XXXIV

MOORLAND AND ALPINE PLANTS

Many of our English moorlands occupy the sites of former woodland or scrub, and are extensively developed on the Pennines and on the Cleveland Hills. Two distinct types occur: (1) the Cotton-grass moor (Fig. 250) and (2) the Heather moor (Fig. 251), but these are connected by several intermediate phases or transition types.

Fig. 251. A Heather Moor, dominated by Ling.

Fig. 252. A Bilberry Moor.

Cotton-grass moors.—A large part of the Pennine plateau is covered with deep, very wet, acid peat, which contains much organic matter and is very poor in mineral salts. It is composed of the remains of previous generations of Cotton-grass and other moorland plants, and, in places, of Bog-moss and Hair-moss, while buried at the base of it are numerous remains of Birch and other trees. Growing on this is a monotonous stretch, many miles in extent, of closely packed 'tussocks' or 'hassocks' of the Cotton-grass; its growth being favoured by a high rainfall of 45 inches or more. Locally these Cotton-grass areas are called 'mosses', and very many place-names are derived from them.

Very few other plants are found here; the chief are the Cloudberry, Crowberry (Fig. 253, 1), and Bilberry (Fig. 254, 1). Here and there are patches of Bog-moss, growing over which are the slender branches of the Cranberry (Fig. 254, 3). Two species of Cotton-grass are met with. One, by far the more abundant (Fig. 250), has narrow leaves permeated by two rows of large air-channels (Fig. 255, 6), and when in fruit, bears a single cottony tassel. The other kind, often confined to wet channels or hollows, has broader leaves and large air-spaces (Fig. 255, 5), and bears several cottony tassels on its fruiting shoot (Fig. 158, 1). The stomata in both species are on the exposed surfaces; and this feature, together with the large air-channels, recalls structures we have met with in water-plants. Another point of comparison is that the Cotton-grasses grow in a wet, badly aerated soil; very rich in organic matter.

At the edges of the Cotton-grass moors, where drainage is better and the peat drier and shallower, the Cotton-grasses are replaced by Bilberry (*Vaccinium Myrtillus*), and with it, but growing less abundantly, are Cowberry (*V. Vitis-Idaea*), Crowberry (*Empetrum nigrum*), and other heath-plants. Often long Bilberry edges thus arise (Fig. 252). On the steeper, more sheltered slopes, extensive stretches of

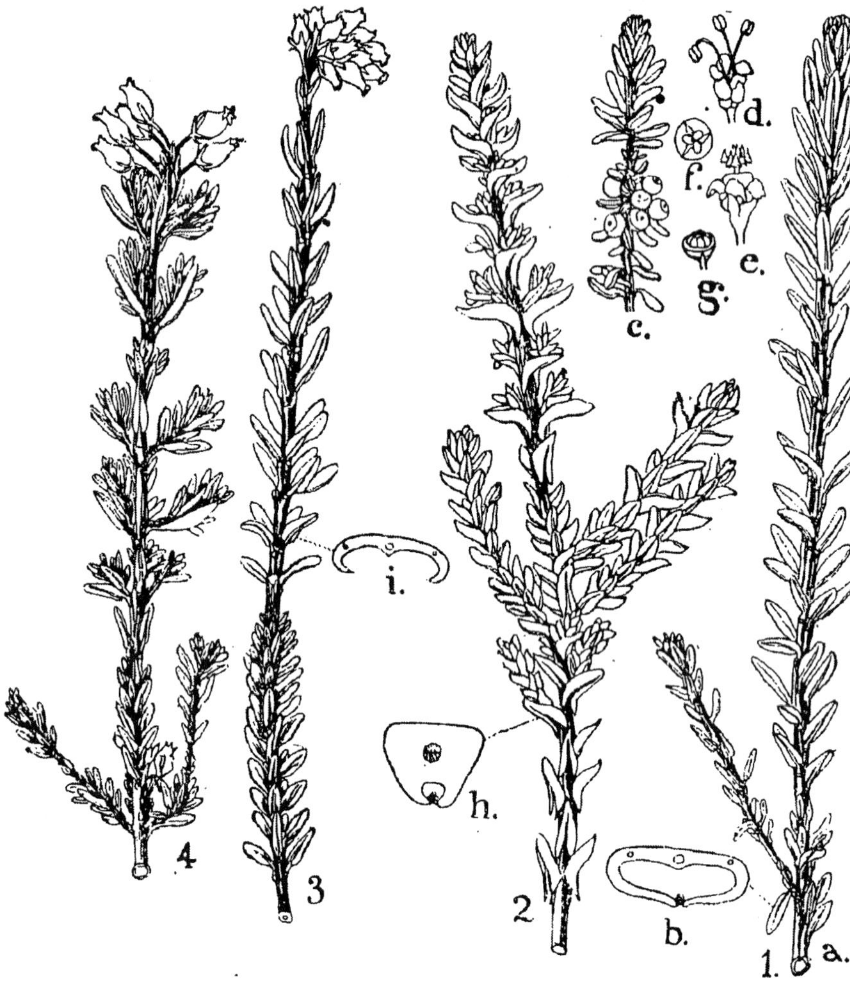

FIG. 253. HEATHS AND HEATH-LIKE MOORLAND PLANTS WITH BACK-ROLLED LEAVES.—1, Crowberry; 2, Ling; 3, Cross-leaved Heath; 4, Fine-leaved Heath; *a*, leafy shoot; *b*, transverse section of leaf; *c*, fruiting branch; *d*, male flower; *e*, female flower; *f*, fruit from below; *g*, fruit cut open to show the seeds; *h*, section of leaf of Ling; *i*, section of leaf of Cross-leaved Heath.

Bracken are found, and in places wiry rolled-leaved grasses like the Mat-grass (Figs. 235, 236, 255, 3) and Hair-grass (Fig. 255, 4) are abundant, and appear as conspicuous features in the landscape.

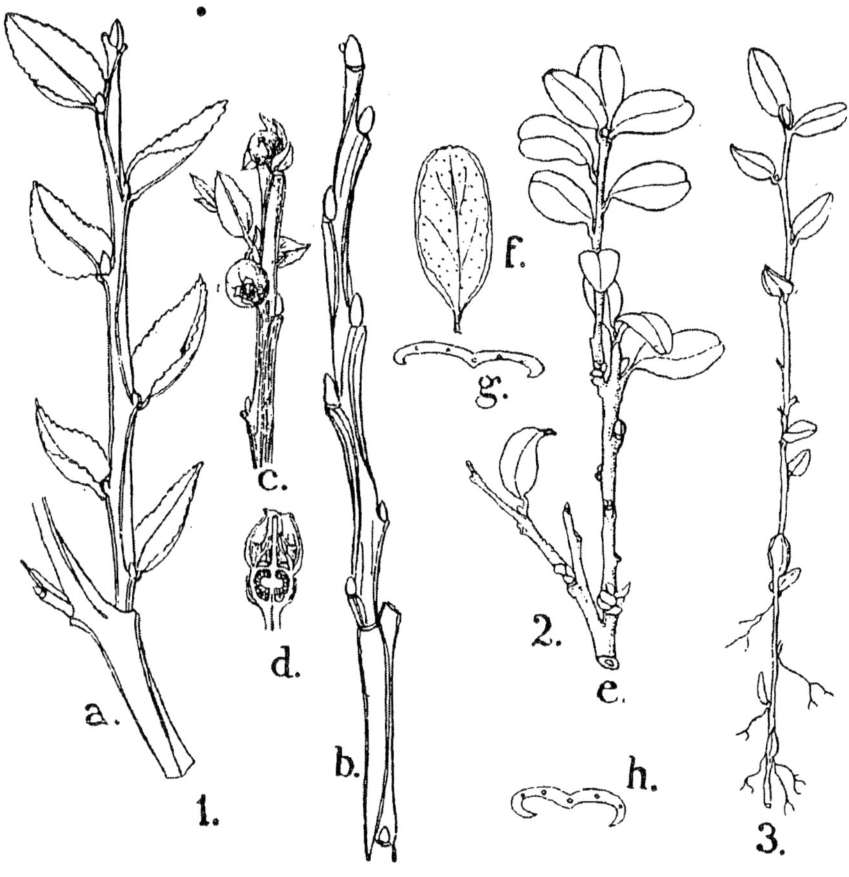

FIG. 254. MOORLAND PLANTS.—1, Bilberry; 2, Cowberry; 3, Cranberry; *a*, leafy shoot; *b*, winter shoot; *c*, flowering shoot; *d*, vertical section of flower; *e*, leafy shoot of Cowberry; *f*, under side of leaf showing pits; *g*, section of leaf; *h*, section of Cranberry leaf.

Heather moors.—The above form a transition region to the Heather moors (Fig. 251). They stand in strong contrast with the Cotton-grass moors, which are a dull green in early summer, forming, later on, snowy patches when in fruit, and becoming rusty brown in the autumn. The Heather moors are a lighter green in summer, assuming a rich purple towards autumn, when the plants are in flower. This is a region of dwarf evergreen shrubs, the dominant species being the Ling (Fig. 253, 2).

The peat of the Heather moor is shallow, often sandy, and acid, and is frequently developed over sandstones or other areas with a shallow, well-drained siliceous soil; but Heather moors also occur, as we have seen, over limestone.

At the base of the peat is often found a hard gritty layer called the 'moor pan', consisting of sand-grains bound

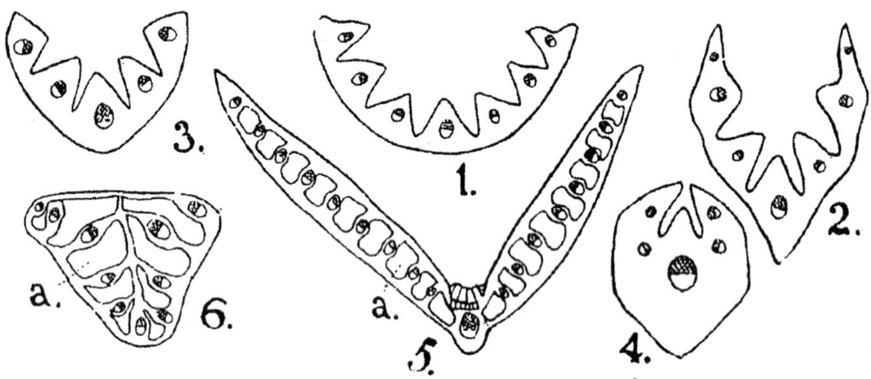

FIG. 255. SECTIONS OF UP-ROLLED LEAVES OF MOORLAND GRASSES AND SEDGES.—1, Tufted Hair-grass; 2, Sheep's Fescue-grass; 3, Mat-grass; 4, Waved Hair-grass; 5, Many-headed Cotton-grass; 6, Single-headed Cotton-grass; *a*, air-spaces.

together into a compact bed, a few inches in thickness, by either oxide of iron or humus-compounds. This pan is often so hard that roots of young trees cannot penetrate it.

The plants found on heaths have many features in common. The shrubs, with the exception of the Bilberry, are all evergreen, with greatly reduced, back-rolled leaves, as shown in Fig. 253. An interesting series comprises Cowberry (Fig. 254, 2) with edges slightly curled back, Cranberry (Fig. 254, 3), and Cross-leaved and Fine-leaved Heaths (*Erica Tetralix* and *E. cinerea*), farther back-rolled (Fig. 253,

3 and 4). In the Crowberry (Fig. 253, 1) the hairy edges meet, while the Ling has the smallest leaf, and its under surface is confined to a narrow groove. In all these cases the stomata occur only on the under surface.

The Bilberry (Fig. 254, 1) usually sheds its leaves in the autumn, but its bright-green angular stems (Fig. 254, *a*, *b*, *c*) render it functionally evergreen. Small stunted forms, however, of the Bilberry retain their leaves through the winter. The Bilberry is able to withstand great extremes; it ascends to a greater altitude than the other species, and is frequently the dominant plant on high, exposed, rocky summits.

Moorland grasses have also rolled leaves (Fig. 255). Along their upper surfaces are ridges, between which are lines of large, water-containing cells. When water is abundant and these cells are distended, the blade spreads out; but in times of drought, water is withdrawn from the cells, the ridges in consequence fall together, and so the blade rolls up. In the grasses, therefore, the blades are up-rolled, and not, as in the heaths, back-rolled. In the grasses, too, the stomata are on the sides of the ridges on the upper surface, and are absent from the lower more exposed surface, which is protected by a thick cuticle. In either case the rolled leaf encloses a chamber of still air, and as the stomata are in this, they give off very little water. In these ways plants are well adapted to withstand the severe conditions of life on the moors.

The water-logged, acid peat decomposes very slowly, and the mineral substances it contains are not readily available for plant-food. Peat-plants, however, agree in one respect with those growing in humus: they are able to subsist by the aid of mycorhiza, which is present on the roots of most moorland plants.

A plant often abundant on Heather moors and on mountain slopes is the **Gorse** or Whin (Figs. 132 and 256), and it exhibits many interesting modifications which should be carefully studied. What is the nature of the short spines? Where and how do they arise? Distinguish between leaves, and branches. Cut a section across the stem, and note how much the ridges on it increase the green, assimilating surface-tissues (Fig. 256, 1). Seedlings may easily be grown, and the history of the leaves and spines determined (2 and 3).

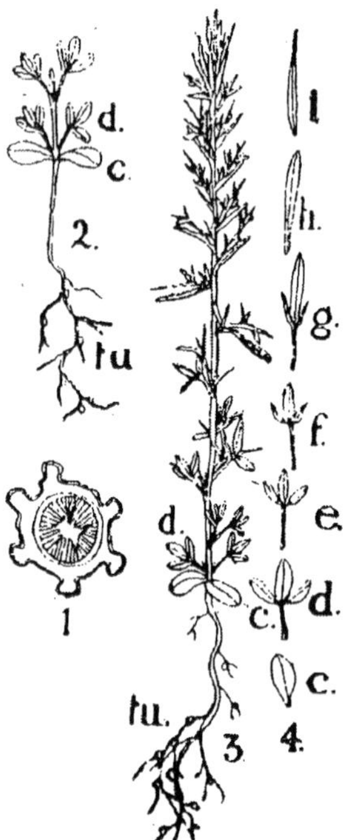

FIG. 256. GORSE SEEDLINGS.—1, transverse section of stems; 2, young seedling; 3, older seedling; 4, different kinds of leaves on a seedling; *c*, cotyledon; *d* to *i*, transition from trifoliate leaf to needle-leaf; *tu*, root-tubercles.

The cotyledons (2, 3, 4, *c*) are entire, but the first foliage-leaves (*d*) are trifoliate. As the plant grows, the newer leaves tend to produce smaller lateral leaflets, and eventually the centre leaflet only is produced, and this is very narrow and sharp-pointed, but flexible. Buds arising in the leaf-axils grow into shoots with narrow, undivided, and sharply-pointed leaves, and the branches also become sharply pointed, harden, and so form a branch-spine. The development of trifoliate leaves in the seedlings suggests that the Gorse has probably descended from an ancestor with compound leaves, which now only persist in the seedling stage. We have already noticed the interesting way in which seeds of the Gorse are dispersed by ants (p. 226, Fig. 161).

The Sphagnum bog.—The Sphagnum bog is dominated by species of Bog-moss (*Sphagnum*) and other peat-forming mosses such as the Hair-moss (*Polytrichum*). It is developed on an impervious soil in situations where the air is very moist, either at low or high levels; and an essential condition seems to be a soil poor in lime and other mineral salts; it thus differs from a marsh, which is richer in mineral salts. The Bog-moss grows rapidly; its closely-packed shoots, and the narrow channels in its leaves, form a series of capillaries which enable the plant to draw water up to a considerable height, and hold it firmly. As the upper branches continue their growth, the lower, older parts die, and, decaying very slowly, form beds of peat often of great thickness. If a piece of such peat is examined the remains of the Bog-moss are clearly seen. On bogs thus developed are found the Sundews (Fig. 232), Butterwort (Fig. 234, 1), Cotton-grasses, White Beak-rush (*Rhynchospora alba*), Purple Moor-grass (*Molinia*), Bog Asphodel (*Narthecium ossifragum*), Marsh Andromeda (*A. Polifolia*); and on the drier parts many shrubby heath-plants, e.g. Cross and Fine-leaved Heaths, Ling, Crowberry (Fig. 253), Bilberry, Cowberry, Cranberry (Fig. 254), Cloudberry (*Rubus Chamaemorus*), also the Sweet Gale or Bog Myrtle (*Myrica Gale*) and Creeping Willow (*Salix repens*).

Alpine Plants

Plants growing on the tops of mountains have many difficulties to contend with. The air is rarefied; the winds, often very drying, are at other times moisture-laden; cold driving mists alternate with bright sunshine; hot days are followed by cold nights; snow lasts long, especially in the hollows, and the soil is thin and well drained.

In spite of these varied and fluctuating conditions, many plants grow here, but they develop dwarfed, tufted, and other xerophytic growth-forms, well adapted to such a habitat. Many of the species occur only at great altitudes and in the Arctic regions. A few are found on the sea-coast, but they are absent from the intervening lowlands, e.g. Scurvy-grass (*Cochlearia spp.*), Sea Campion (*Silene maritima*), Sea Plantain (*Plantago maritima*), and Sea Pink (*Statice maritima*). The leaves are often up-rolled as in grasses and sedges (Fig. 255), or back-rolled as in heaths (Fig. 253), with stomata sunk in pits or grooves. The blades are usually small, leathery, hairy or fleshy, and often arranged in compact rosettes, forming the cushions so familiar in rockery plants like the Saxifrages (Fig. 257), Houseleeks, Cushion Pinks (*Silene acaulis*), or the woolly leaves and flowering shoots of Edelweiss (*L. alpinum*). Some plants form dense mats of interlacing trailing stems. Such are the procumbent Azalea, Crowberry (Fig. 253, 1), Alpine Club Moss (*Lycopodium alpinum*), the dwarf forms of Bilberry, and Alpine Willows. These form a flat carpet with their numerous rhizomes matted in the soil. Sections of the Willow-stem may show many (from fifteen to twenty) annual rings, but in spite of their age the plants only grow a few inches from the ground.

Insects are scarce, and usually of the lower types; yet many alpine plants produce large, showy flowers like the Pinks, Saxifrages, and Gentians. Vegetative reproduction by means of offsets, runners, and rhizomes is prevalent. Viviparous plants (see p. 141) are not uncommon on the mountains. Examples are Sheep's Fescue (*Festuca ovina*), Alpine Poa (*Poa alpina*), Alpine Bistort, and some sedges.

The study of plants and their distribution reveals the great power of adaptation which they possess. In relation to the factors of the environment, plant-organs are modified in a great variety of ways and present remarkable contrasts in form and structure, e. g. water-plants and plants of the sea-coast or the Heather moor; plants of the moist, shady woodland and the dry, sunny desert; trees of the lowland forest and the flowery cushions of the mountains. Our knowledge of the conditions under which these varied forms grow, leads us to conclude that the main features of the vegetation of a country are determined by the conditions of the habitat.

The plants of a given association, however, react on the habitat, and, by changing the conditions of the habitat, prepare the way for new forms and the extinction of the

Fig. 257. A Cushion of Saxifrage.

older ones. Such changes may easily be studied in the vegetation of waste-heaps, quarry-tips, or on the scree-covered slopes of our mountains. Bare ground becomes invaded by microscopic plants, Mosses, Ferns, and annual flowering plants (see p. 220). These are succeeded by open associations of perennials, which at first are all herbaceous species, but later, closed associations, with shrubby perennials, are developed, and finally trees appear, with a ground flora characteristic of the forest.

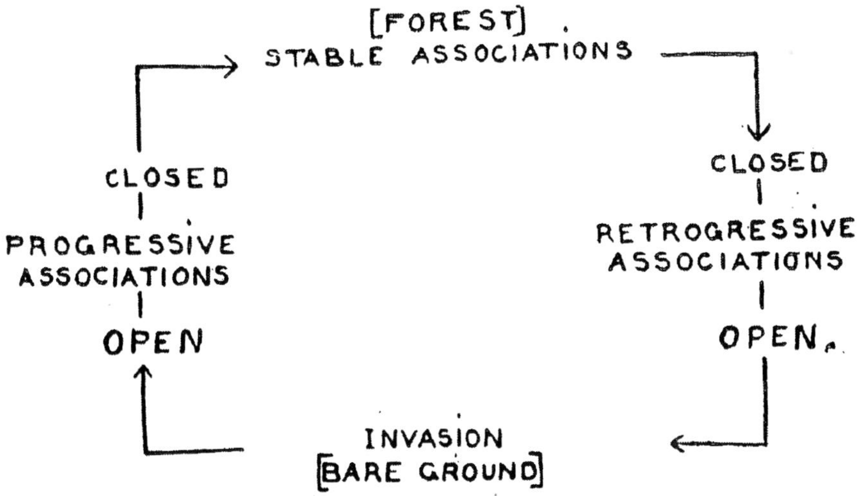

FIG. 258. LIFE-CYCLE OF VEGETATION.

The woodland or forest type is the highest and most stable phase in the development of vegetation, and persists for a long period. Eventually changes occur, due to natural or artificial causes, e. g. earth movements, fires, or the cutting of gaps in the forest for roads or railways. In consequence of the increased exposure, degenerative or retrogressive changes set in, and finally the ground becomes denuded of plants. Re-invasion then occurs, and the life-cycle of the vegetation is completed by the advent of progressive associations, which increase in complexity and stability until the forest is once more developed. In this way the vegetation of the earth is ever changing.

APPENDIX

EXAMINATION PAPERS

UNIVERSITY OF OXFORD

Local Examinations, July 1912

Senior

1. Describe concisely the specimen provided. [In describing an inflorescence or flower, candidates should *illustrate* the relative positions of the various parts both by a horizontal and by a vertical plan.]

2. State briefly how the presence of starch in green leaves and in underground stems is to be explained.

3. What special structures are found in the wood of a Dicotyledon? Explain how the form and nature of these structures fits them to carry on their special work.

4. Describe the mode of pollination and the relation of the parts of the flower in two of the following: Willow, Deadnettle, Evening Primrose, Sunflower.

5. Three leaves of an india-rubber plant are cut off, and vaseline is rubbed on the under surface of one, on the upper surface of another, and on both surfaces of a third. They are then left for a week, hanging in air. State the result in each case. Give a brief explanation and an account of the physiological process concerned.

6. Describe briefly the characteristic fruits of Cruciferae, Scrophulariaceae, and Labiatae. Describe in one case the mode of seed-dispersal.

7. Give an account of the vegetation to be found in a corn-field, *or* a shingle beach, *or* a sandy heath. Mention six characteristic plants from the situation you select.

8. Describe three methods of vegetative reproduction among flowering plants, giving examples of each case. What are the advantages and disadvantages of this method as compared with reproduction by seeds?

9. Give an account of the structure (omitting microscopical details) and situation on the tree of the male and female cones of a Pine. How is pollination brought about in the plant?

UNIVERSITY OF OXFORD

LOCAL EXAMINATIONS, JULY 1911

Junior

Section II

5. Describe in simple language the specimen provided, sketch and name its parts.

Section III

6. Describe the germination of a seed with endosperm (albuminous) and of a seed without endosperm (exalbuminous).

7. Describe the uses of the various parts of the flower, and point out how the nature of the various parts fits them for their special work.

8. What is transpiration? How would you prove its occurrence in a green leaf? What circumstances cause an increase in the rate of transpiration?

9. Describe the inflorescence, flower, and fruit, of any plant belonging to the Labiatae.

10. Explain why it is that leaves are usually *green, thin,* and *flat.*

UNIVERSITY OF OXFORD

LOCAL EXAMINATIONS, JULY 1911

Preliminary

1. Describe the flower of *either* the Buttercup, *or* the Pea, *or* the Wild Rose. State, as far as you can, the uses of the various parts.

2. What is the use of the fruit to a plant? Make drawings of two fruits which you have examined and name the different parts.

3. Why does a plant spread its leaves to the light and air? Describe two experiments you have seen which prove what you say.

4. Describe fully how Daisy plants spread over a lawn.

5. Describe a seed, and say what changes you would see as it grows into a young plant.

6. Describe any winter-bud which you have examined. What becomes of the different parts in spring?

UNIVERSITY OF CAMBRIDGE

LOCAL EXAMINATIONS, DEC. 1911

Senior. A

A 1. Write a concise description of the specimen M [=*Bouvardia*]. Show by means of sketches the relative positions of the parts of the flower; and draw attention to any characters which have special reference to its method of pollination.

A 2. Describe some simple experiments to illustrate the effect of light on the growth and development of a green plant.

A 3. What do you understand by vegetative propagation? Give three examples of structures specially adapted for this purpose.

A 4. With what habitats would you associate any four of the following: Gorse, Stone-crop, Dandelion, Whortleberry (or Bilberry), Sundew, Cleavers (or Goose-grass)? Mention any characters of the plants selected which you might regard as developed in relation to their respective habitats. (Candidates not in the United Kingdom may, if they wish, substitute the following question for question A 4:

Mention three climbing plants, and give some account of the means by which they gain the required support.)

B

B 1. Write a botanical account of the specimen N [=Acacia seedling]. Explain, so far as you are able, any notable peculiarities.

B 2. Describe with the aid of sketches the form and arrangement of the parts of the flower of (*a*) a Buttercup, (*b*) a Wild Rose, and point out the chief differences between the two. (Candidates at centres not in the United Kingdom may, if they wish, substitute the following question for question B 2:

Describe with the aid of sketches the form and arrangement of parts of some papilionaceous flower, and suggest the method by which pollination is effected.)

B 3. Give some account of the phenomenon of leaf-fall and the changes in the leaf which precede its occurrence.

B 4. Describe the structure of the seed of some monocotyledonous plant (e. g. Wheat), and give some account of the changes which take place in it during germination.

UNIVERSITY OF CAMBRIDGE

LOCAL EXAMINATIONS, DEC. 1911

Junior

1. Make labelled diagrams to illustrate the structure of specimen K [= bud of Brussels Sprouts] and its parts.

2. Dissect specimen L [= fruit of Acer] and describe its structure by means of labelled drawings. Mention in a few sentences any points of biological interest which strike you.

3. Why do grasses flourish, although they are closely cropped by grazing animals? Explain how it is that grasses make a firm soft turf and why the grass on a lawn should be kept short.

4. Give some account of the ways in which plants are adapted to take full advantage of the light which falls upon them.

5. Some trees are evergreen, others for part of the year are leafless. Mention one tree of each kind, and explain as far as you can the meaning of these different habits.

6. What is a stoma? In what important functions do stomata play a part?

7. Name four dehiscent fruits, and describe in each case how the seeds are enabled to escape.

8. How do water-plants obtain their food?

UNIVERSITY OF CAMBRIDGE

LOCAL EXAMINATIONS, DEC. 1911

Preliminary

1. Describe an experiment which shows that water is given off from the leaves of a plant during a warm day. From what source is this water obtained, and how is it carried to the leaves?

2. Give two examples of plants which accumulate large reserves of food at some period of their lives. What is the advantage of these stores of food to the plants concerned?

3. What are the chief differences in external features between roots and stems? Name a plant which has an underground stem.

4. Describe briefly the parts of some brightly coloured flower. Give the name of the flower, and state precisely the functions of the parts you describe.

5. Describe as fully as you can the appearance presented by the surface of the stump of a tree after the trunk has been cut down.

6. The root of a plant grows downwards and the shoot grows upwards. Explain the importance of these facts in the life of the plant.

UNIVERSITY OF LONDON

MATRICULATION EXAMINATION, SEPT. 1911

1. Draw a floral diagram and a median, longitudinal section (on a large scale and showing only the parts cut) of the flower A provided, naming the different organs. In both cases the structure of the ovary should be included. Refer the plant to its natural order, giving your reasons.

2. Make a series of annotated sketches illustrating the structure of B. Draw from memory a longitudinal section of the flower from which it has been derived, indicating clearly the parts which have withered and those which have undergone further development. All the drawings should be as nearly as possible on the same scale.

3. Select any five of the following plants and in each case (*a*) make descriptive notes of the habitat, (*b*) state the month or months of flowering, (*c*) mention briefly any peculiarities of habit: Plantain or Waybread (*Plantago major*), Common Reed (*Phragmites communis*), Ling (*Calluna vulgaris*), Sundew (*Drosera*), Marram-grass (*Psamma*), Thistle (*Carduus*), Sloe or Blackthorn (*Prunus spinosa*), Honeysuckle or Woodbine (*Lonicera Periclymenum*).

4. Make a careful drawing of a portion of a branch of any plant which climbs by means of tendrils. Indicate the morphological nature of the tendrils of the plant you select (giving full reasons for your conclusions) and describe exactly the way in which they perform their function.

5. Draw three successive stages in the germination of any seed, showing precisely how the different organs of the embryo behave. How would you show experimentally what conditions are necessary for successful germination?

6. Growing plants are continually absorbing fresh supplies of inorganic substances from their environment. Describe in detail how you would find out which chemical elements are necessary to the life of a green plant. State how and in what forms the plant obtains each of these elements.

7. Mention *four* plants that are pollinated by butterflies or moths and state how their flowers are adapted to these special insect-visitors. Draw a longitudinal section of the flower of *one* of these plants, indicating the positions of the nectaries, stamens, and pistil, and the position taken up by the pollinating insect.

8. Enumerate (illustrating the points with the aid of diagrams) the resemblances and differences of structure between a Tulip bulb and a Horse-Chestnut bud. What is the morphological nature of each, and from what sources are the principal food-supplies obtained when growth recommences?

BOARD OF EDUCATION TEACHERS' PRELIMINARY CERTIFICATE

EXAMINATION, APRIL 1911

Special Section C. Botany

7. Describe carefully the structure of the flower of the Orchis, Violet, or Dandelion. Make a drawing of a longitudinal section of the flower to show clearly the position of the various parts, and draw the floral diagram.

8. Compare the structure of the corm of a Crocus with that of the bulb of a Hyacinth, and show, with the aid of sketches, how new corms and bulbs are found on the old ones.

9. Draw and describe a leaf of Ivy and compare it with a leaf of Primrose. Explain any obvious points of structural difference with reference to the habit of the plant and its environment.

10. A stone lying in a field is removed and the grass beneath is found to be discoloured. Account for this, and describe experiments to show what effect the conditions which produce discoloration may have upon the growth of the plant.

11. Describe, with the aid of diagrams, the structures likely to be found upon a branch of Horse-Chestnut in May and June.

12. What conditions affect the rate of transpiration in plants? Describe any experiment which may be made to determine the rate of transpiration under various conditions, and sketch the apparatus used.

BOARD OF EDUCATION TEACHERS' PRELIMINARY CERTIFICATE

EXAMINATION, MARCH 1912

Special Section C. Botany

7. Describe, with the aid of sketches, the two kinds of flowers which are to be found on Violet plants. What explanation can be given of the presence of these two kinds of flowers?

8. How do plants obtain their food and in what form? State clearly the experimental evidence upon which your statements are based.

9. Give an account of the life-history of the Potato, and show clearly, with the aid of sketches, on what part of the plant the tubers are formed. Describe the structure of the Potato tuber.

10. What experiments could you devise to show how the rate of transpiration of water from a leafy plant compares with the rate of absorption of water by its roots? Upon what external conditions does the activity of transpiration depend?

11. Describe some of the common methods by which weeds are spread. Illustrate your answer by description of three specific instances of different methods.

12. Describe the structure of the flowers in two plants which are cross-pollinated by the wind, and in two plants which are cross-pollinated by insects. State clearly what experimental evidence there is for the statement that, in many cases, cross-pollination is more effective than self-pollination in the production of seeds.

BOARD OF EDUCATION TEACHERS' CERTIFICATE

Examination, Nov. 1911

Special Section C. Botany

7. On cutting across the stem of a plant, such as the Sunflower, Begonia, or Pelargonium, it is found, after a short time, that water exudes from the cut end of that part of the stem which remains attached to the root. Account so far as you can for this phenomenon, and describe any experiment which will throw light upon it.

8. What is meant by 'vegetative reproduction in plants'? Give examples of four different methods of vegetative reproduction. Illustrate your answer by sketches.

9. How would you demonstrate the conditions which determine the formation of starch in leaves?

10. Give sketches of the leaves of any three of the following: Gorse, Dog-Rose, Wood Sorrel, Marsh Marigold, Primrose. Account for any peculiarities of stiucture observed.

11. Compare the structure of the seeds of Maize, Wheat, Oak, and Bean. Show in what respects they resemble or differ from one another. How would you classify these seeds in respect of fundamental differences of structure? Give the essential characteristics of each class.

12. Mention the names of British plants which possess the following characteristics, giving an example in each class:

(*a*) Foliage-leaves in pairs on short special branches.

(*b*) Reproduction by means of small buds which fall off and develop into new plants.

(*c*) Seeds (not fruits) which are dispersed by the agency of wind.

(*d*) Nodules on the roots.

(*e*) Flowers with stamens and pistil but no calyx or corolla.

(*f*) Special adaptations of the leaves to withstand drought.

CENTRAL WELSH BOARD

Annual Examination, July 1911

Botany

1. Describe a typical flower. Explain the uses of the various parts and show how they are adapted to fulfil their functions.

2. What part of a flower is a fruit derived from? Name three kinds of dry and three kinds of succulent fruits, and show in each case how the dispersal of the seeds is brought about.

3. Describe the vegetation of *one* of the following localities: (*a*) a hedge, (*b*) a river-side, (*c*) a wood.

4. What are the distinguishing characters of the order Rosaceae? Mention three wild and three cultivated plants belonging to the order.

5. Give two examples of underground stems. By what external marks would you distinguish them from roots?

6. What is meant by respiration? What parts of plants respire? Why should plants not be kept in bedrooms?

7. Mention four wind-pollinated plants. How would you recognize a wind-pollinated flower ?

8. Point out the difference between the shoots (i.e. the stems with their leaves) of a Palm, an Oak, and a Carrot. Of what advantage is a tall stem to a plant ?

UNIVERSITIES OF MANCHESTER, LIVERPOOL, LEEDS, AND SHEFFIELD

Matriculation Examination, July 1911

1. Describe the structure of a mature sporangium of a Fern, and explain the functions of its various parts. Why is the Fern-plant termed the sporophyte ?

2. The embryos of phanerogams have laid up for them, within the seed-coat, certain stores of organic material to start them in life ; what is the chemical nature of these reserves, and in what precise situations are they found ?

3. Give an account of the morphology of any carnivorous plant with which you are acquainted. Describe the secretory glands and the nature of the secretion in the type you select.

4. Write a short essay on the dispersal of fruits and seeds by water, giving illustrative examples.

5. How would you demonstrate practically that plants respire ?

6. Indicate the distinctive characters of any *three* of the following trees : Oak, Elm, Ash, Pine, Willow. Arrange your answers under the following headings : (*a*) Mode of growth, (*b*) Flowers, (*c*) Bark.

OXFORD AND CAMBRIDGE SCHOOLS EXAMINATION BOARD

Higher Certificate

(*d*) *Botany*

1. How would you demonstrate by experiment the effect of the stimulus of gravity on the direction of growth ?

2. Give some account of the process of respiration. What is its meaning in the life of a plant ?

3. What differences would you expect to find in the seedlings of Beans which have been grown respectively under ordinary conditions and in the dark ?

4. Give some account of the part which is played in plant-nutrition by (1) sugar, (2) starch.

5. Describe how you would demonstrate experimentally the growth in length of a root.

6. Give some account of the tissues which are concerned in the transport of water in a plant.

7. Write botanical descriptions of the specimens A and B; carefully pointing out in what ways they (1) resemble and (2) differ from each other.

OXFORD AND CAMBRIDGE SCHOOLS EXAMINATION BOARD

Higher Certificate

(e) *Special Botany*

1. Give an account of the tissues which are concerned in the transport of water through the stem of an ordinary land-plant.

2. What do you understand by photosynthesis ? Show how the form and structure of a typical green leaf are specially adapted to this function.

3. Give three examples of modifications of the shoot for food-storage. Point out in each case the purpose of the store.

4. Describe with the aid of sketches the spikelet of a species of grass, and state the function of the various parts which you describe.

5. Give some account of the methods of seed-dispersal in the family Rosaceae. State precisely what is the morphological nature of the various structures to which you refer.

6. *Either* Give some account of the method of growth and the life-history (excluding details of fertilization and embryo-development) of Coltsfoot ; *or* State what you understand by the term plant-association, illustrating your answer by one or more examples.

7. Write a botanical account of the specimen A, and draw special attention to any structures which bear on its method of pollination.

OXFORD AND CAMBRIDGE SCHOOLS EXAMINATION BOARD

Elementary for Lower Certificates

1. Describe briefly the functions of a foliage-leaf.
2. Draw a diagram of a Potato, naming all the structures observable. Give your reasons for deciding whether it is a stem-structure or a root-structure.
3. Explain clearly the difference in origin between the edible portions of a Cherry and an Apple.
4. Describe, with diagrams, the flower of the Buttercup.
5. Describe, with examples, the various methods of seed-dispersal with which you are acquainted.
6. Explain why the growth of a plant is checked by transplanting it carelessly.
7. Make a careful drawing of the specimen provided, naming the parts.

COLLEGE OF PRECEPTORS

Certificate Examination, Christmas 1911

Second Class

1. Describe the specimens before you as fully as possible. Make diagrams of the floral arrangements (not drawings of the parts of the flowers). Refer the plants to their classes and divisions (not orders), giving your reasons.
2. Write down the characteristic features of the orders Malvaceae and Primulaceae, and show how they differ in the structure of the pistil.
3. How does the Common differ from the Round-leaved Mallow, and the Primrose from the Cowslip ? What is the importance of these differences ?
4. How are the different species of Buttercups adapted to living in (*a*) the open meadow (*Ranunculus bulbosus*), (*b*) on

mud (*R. hederaceus*); and (*c*) submerged (*R. trichophyllus*)? State the meaning of the Latin names.

5. What are the functions of ordinary green leaves, and why have some plants, as the Dodder and Broomrape (*Scrophularineae*), no green leaves at all?

COLLEGE OF PRECEPTORS

Certificate Examination, Christmas 1911

First Class. Advanced Section

1. Name and describe the flowers of the British genera of Corylaceae.

2. Explain, with a diagram, the origin of the vascular bundles of the stipules of the Geranium and Bedstraw.

3. When wild plants are cultivated, what are the most prominent effects of the change of conditions? Give examples.

4. Give the life-history of a Moss.

COLLEGE OF PRECEPTORS

Certificate Examination, Christmas 1911

First Class. Elementary Section

1. Describe the specimens before you as fully as possible. Make diagrams (i. e. cross-sections) only of the floral arrangements, and refer the plants to their classes and divisions, giving your reasons for so doing.

2. What genera of Rosaceae are cultivated? Explain the changes which cultivation has caused in them.

3. Give examples of roots which have adapted themselves to other purposes than that of absorbing water from the soil.

4. Describe any experiments in germination. What do they teach us?

5. How do plants get rid of superfluous moisture, and why?

6. Describe the anatomy, as seen in cross-section, of any ordinary root and the uses of its several tissues.

UNIVERSITY OF BIRMINGHAM

Matriculation Examination, Sept. 1911

1. Describe the specimen A provided. Refer it to its natural order, giving your reasons. Illustrate its floral structure by horizontal and vertical diagrams.

2. Describe the morphology of the two specimens B and C provided, and explain any peculiarities of structure they show.

3. Say what you know concerning the formation of starch by a plant.

4. Give an account of the development of a leguminous flower from the bud stage to the formation of a seed-pod.

5. Write an account of the nectaries found in the Ranunculaceae. What is nectar? Of what use is it to plants?

6. Describe the movements of organs exhibited by plants belonging to the Leguminosae and suggest any advantages the plants may derive from them.

7. Write a description of the various kinds of fruit seen in the Rosaceae.

8. Describe the seed-leaves of the Castor Oil, Pea, and Maize, and say how these differ from the leaves subsequently developed. Why are foliage-leaves and seed-leaves so different?

UNIVERSITY OF WALES

Matriculation Examination, Sept. 1911

1. Make a large-scale drawing of the flower of A as seen in median, longitudinal section and name the parts; draw also a floral diagram.

2. Describe, with a sketch, the external form of some Fern with which you are familiar, and compare it in this respect with such a plant as a Wallflower. Describe the place of growth of the Fern you select.

3. Describe a simple apparatus to measure the rate of absorption of water in a cut, transpiring, leafy branch. How would you (*a*) increase, (*b*) decrease the rate of absorption?

4. Describe the various stages in the opening of the bud of some tree.

5. What is respiration and what purpose do you suppose it serves in the life of the plant? How would you prove that germinating peas respire? What do you know of the respiration of green leaves in the light?

6. What is an insectivorous plant? Mention three native insectivorous plants and sketch and describe one of them.

7. Describe and compare, with careful sketches, the pollination of two flowers, one of which is pollinated by the wind and the other by insects.

8. What are the advantages and disadvantages of climbing as compared with the manner of growth of ordinary plants? Describe three different ways in which plants may climb.

UNIVERSITY OF DURHAM

MATRICULATION EXAMINATION, SEPT. 1911

1. Write a short account of some one flower which you have studied, and explain briefly the functions of its various members.

2. Describe carefully the structure of syncarpous and apocarpous fruits. Give examples of some common fruits which come under these heads.

3. Explain clearly how starch is formed in the Potato tuber, and the source from which this substance is derived.

4. By what experimental evidence can it be shown that a healthy shoot is continually absorbing water? How is the rate of absorption influenced by exposure to light or darkness?

5. Contrast the external forms of plants grown in darkness, in shade, and in full illumination.

6. Enumerate and briefly explain the principal means whereby plants are able to climb, and explain the advantages which accrue from a climbing habit.

7. Explain clearly the meaning of the knots and lines which are present in a deal board.

8. Describe as fully as you can the external appearances of the roots of annual and biennial plants.

UNIVERSITY OF BRISTOL

Matriculation Examination, July 1915

Only six questions to be answered, of which one must be No. 1. Answers should be illustrated by drawings wherever possible.

1. Examine and report anything of special interest in the specimen provided. In what locality was the plant growing?
2. Describe minutely the shoot of some tree in its winter condition.
3. Describe the vegetation of a hawthorn hedge in summer.
4. Of what importance is water to the green plant?
5. Mention the various parts of a green plant, briefly referring to their chief functions.
6. Describe the process of pollination in some plant with which you are familiar.
7. Give an account of as many British parasites as you know.
8. Can you give some account of seaweeds *or* bacteria *or* toad-stools?

INTERMEDIATE EDUCATION BOARD FOR IRELAND

Examination 1911 (June)

Honours Paper (Third Year)

Section A

1. (*Obligatory*) Describe the structure of the bulb of some bulbous plant. What substances does it contain? Where do they come from? What becomes of them?

2. When do 'evergreens' shed their leaves? Compare the leaves of 'evergreens' with those of deciduous trees. Can you suggest any reasons for the difference?

3. From what sources do the opening buds of spring draw their supplies of material? Mention the materials in each case.

4. Describe how the different regions of the root are formed. What are their uses to the plant? Give reasons for your answer.

Section B

5. (*Obligatory*) Of what uses are green eaves to plants? Describe experiments and observations which support your answer.

6. Describe the conditions under which seeds germinate best. Point out why these conditions are necessary.

7. Describe the appearance of any Fern. Point out some of the chief differences between a Fern and a flowering plant.

8. What is pollen? Where is it formed? What function does it perform?

INTERMEDIATE EDUCATION BOARD FOR IRELAND

Examination 1911 (June)

Special Paper (Third Year)

Section A

1. (*Obligatory*) How do the following insects affect the growth of plants: Humble-Bee, Aphis (or green fly), grub of Daddy-long-legs, Wireworm? State in each case what part of the plant is acted on by the insect, and whether the result is beneficial or injurious to the plant.

2. Describe the special characteristics of grasses that inhabit sand-dunes, and show how these are suitable to the conditions under which the plants live.

3. How can you tell the age of the branch of a hardwood tree (e. g. Beech, Lime, Elm) by external and internal features? Explain how these features are produced.

4. Give a general account of the autumn tints and the relative times of leaf-fall of the Oak, Ash, Larch, Beech, and Horse-Chestnut. Give the order also in which these trees come into leaf in spring.

Section B

5. (*Obligatory*) Give an account of the upward and downward flow of sap in a plant. State the source of each and the general nature of its composition. What is the function of each kind of sap ?

6. What is meant by the 'sleep' of plants ? When and why do they sleep ? Compare the 'waking' and 'sleeping' positions of the leaves of the 'Shamrock' with those of Wood Sorrel.

7. Describe the inflorescence of the Dandelion, and also one of its flowers. How is it pollinated, and how is self-pollination prevented ?

8. Describe the mode of growth and appearance of the common mushroom or any other fungus of a similar habit. How does it get a living ? What do you know of 'fairy-rings' in the grass ?

INTERMEDIATE EDUCATION BOARD FOR IRELAND

Examination 1911 (June)

Honours Paper (Fourth Year)

Section A

1. (*Obligatory*) What are the successive steps you would take in examining the following plants in flower so as to place them in their orders : Buttercup, Wallflower, Pink, Rose, Pea, Wild Parsley, Dandelion, Potato, and Stinging Nettle ?

2. Suppose you were to take a class of pupils to a meadow for a first lesson in the general classification of herbaceous flowering plants into two great classes, to what features in the plants would you draw attention ?

3. Describe how tree trunks increase in thickness. As trees increase in age what changes do the wood and bark undergo?

4. Give an account of the occurrence of root-nodules on the roots of some plants. How are these nodules caused, and how do they affect the life of the plant?

Section B

5. (*Obligatory*) Describe an experiment to show that plants exhibit root-pressure. How is this pressure produced, and of what advantage is it to the plant? What natural phenomena in herbaceous plants are due to root-pressure?

6. Describe the method of water-culture to prove the essential inorganic food-materials required by green plants.

7. Draw the vertical section and floral diagram of the Wild Rose, and name the parts of the section. Draw and describe the leaf of the Rose.

8. Give examples of varieties, hybrids, and double flowers, and describe how each may be developed.

INTERMEDIATE EDUCATION BOARD FOR IRELAND

EXAMINATION 1911 (JUNE)

Special Paper (*Fourth Year*)

Section A

1. (*Obligatory*) What is respiration? How may it be shown that plants respire? What is the use of respiration?

2. How would you proceed to find where the most rapid growth in length of the root and stem takes place?

3. Describe the changes which take place in the endosperm of a wheat-grain during germination.

4. Mention some causes of bending and curvature in plants. How do these causes affect the different parts? Describe observations illustrating your answer.

Section B

5. (*Obligatory*) A square is punched from a Sunflower leaf before sunrise and carefully weighed. A similar square is punched from a leaf after sunset, and its weight also recorded.

What difference do you expect to find? Suggest the causes of the difference. How would you test your suggestions?

6. How would you compare the amount of water given off from the leaves of a branch still attached to a tree, and that from the leaves of a cut branch supplied with water? Which would you expect to transpire the larger amount? State fully the reasons on which you base your answer.

7. How is it that young tender seedlings are able to stand up rigidly? Why is it that this stiffness disappears when the seedlings are killed?

8. What surmises would you form as to the method of pollination of a flower which has a large white corolla, narrowed below to a tube, with a honey-gland at its base, and which emits a perfume towards evening?

INDEX

The illustrations are indicated in brackets ; the scientific names of plants are printed in italics.